SCIENCE

IN THE

FEDERAL

GOVERNMENT

SCIENCE
IN THE
FEDERAL
GOVERNMENT

A. Hunter Dupree

ARNO PRESS

A New York Times Company
New York • 1980

Editorial Supervision: Steve Bedney

———————

Reprint Edition 1980 by Arno Press Inc.

© Copyright 1957 by the President and Fellows of Harvard
  College.

Reprinted by permission of Harvard University Press.

Reprinted from a copy in the University of Illinois Library

THREE CENTURIES OF SCIENCE IN AMERICA

ISBN for complete set: 0-405-12525-9

See last pages of this volume for titles.

Manufactured in the United States of America

———————

Library of Congress Cataloging in Publication Data

Dupree, A   Hunter.
    Science in the Federal Government.

    (Three centuries of science in America)
    Reprint of the ed. published by Belknap Press
of Harvard University Press, Cambridge, Mass.
    Includes bibliographical references and index.
    1. Science and state--United States.
2. Science--United States--History. I. Title.
II. Series.
[Q127.U6D78 1980]   353.008'55   79-7959
ISBN 0-405-12540-2

SCIENCE
IN THE
FEDERAL
GOVERNMENT

SCIENCE
IN THE
FEDERAL
GOVERNMENT

A HISTORY OF POLICIES AND ACTIVITIES TO

1940

A. Hunter Dupree

1957

THE BELKNAP PRESS OF

HARVARD UNIVERSITY PRESS

Cambridge, Massachusetts

Sponsored by
The American Academy of Arts and Sciences
under a grant from
The National Science Foundation

Advisory Committee

I. Bernard Cohen, *Chairman*

Edward C. Kirkland   Arthur M. Schlesinger, Sr.
William F. Ogburn   Richard H. Shryock

PREFACE

THE aim of this study is to trace the development of the policies and activities of the United States government in science from the establishment of the federal Constitution to the year 1940. To produce a rounded synthesis, I have been forced to enter many fields of science that I could not possibly know thoroughly, and to describe many government agencies whose internal histories I could not examine completely. Although this invasion may dismay some devoted scientists and administrators, they should remember that mastery of their specialty could have been purchased only at the expense of other sections. If this study serves as a guide and a stimulus to those best qualified to preserve the history of their own activities, it will have fulfilled a part of my hopes for it.

This subject is only one of a number which must be explored before the relations of science to society in America are placed in proper perspective. The history of science in the United States is still largely unwritten. A real comparison between the experiences of the United States government with those of other countries would require studies in Europe which are yet to appear. Until such comparisons can be made, generalizations must be fragmentary. The story of the changing relation between science and technology is by no means completely reconstructed. Although the temptation has been great to attack these pressing questions, I have felt that this study could not answer them all at once. I hope that a faint trail through the wilderness will encourage others to attack it with the confidence that much remains to be learned.

Even more than is usual, this book is the product of coöperation among many institutions and individuals. I am aware of my obligation to more people than I can possibly list in this acknowledgment. I am happy, however, to mention a few of them.

Dr. Leslie H. Fishel, Jr., as research associate and consultant, contributed greatly to the project by his professional services. He did the research on a number of specific subjects and also discussed with me every major problem and interpretation. Since he was familiar with the research notes, his reading of the manuscript resulted in innumerable constructive comments.

The National Science Foundation provided the initial impulse for the project and paid its expenses by means of a grant. During my periods of research in Washington, the Foundation allowed me the use of many facilities and gave me every attention that might conceivably help me. At the same time, I have not been an employee of the Foundation and have been left free to exercise my independent judgment at all times. I am personally grateful especially to Mr. Charles G. Gant and to his successor, Dr. John C. Honey, who have had immediate cognizance of the study.

The American Academy of Arts and Sciences of Boston and Cambridge, Massachusetts, has sponsored the project, taking the responsibility under the grant from the National Science Foundation. The many details of administration and local arrangements have been cared for through the American Academy. I am especially indebted to Mr. Ralph W. Burhoe, the executive officer.

The advisory committee, consisting of Professors I. Bernard Cohen, chairman, Edward C. Kirkland, William F. Ogburn, Arthur M. Schlesinger, Sr., and Richard H. Shryock, has been throughout a source of aid and inspiration. The members of the committee have all given the manuscript a critical reading, bringing to it their accumulated experience.

At Harvard University, I am particularly indebted to Dr. Keyes D. Metcalf, Director of Libraries, for office space as well as for the use of the great library resources of the University, without which this study would have been difficult if not impossible. I am grateful to Professor Reed C. Rollins for many courtesies at the Harvard University Herbarium.

My staff has been a source both of invaluable aid and of great satisfaction throughout the life of the project. Miss Elizabeth Cook has performed routine secretarial drudgery and intricate research with the same care and intelligence. Her service has been an essential factor in completing this book, especially within the narrow limits of time allowed.

At the National Archives I enjoyed guidance by Dr. Nathan Reingold, who generously placed at my service his detailed knowledge of relevant documents.

A number of busy people have freely given their time to me in interviews on specific subjects. I wish especially to mention Dr. Allen V. Astin, Dr. Vannevar Bush, Mr. Charles Campbell, Mr. Watson

Davis, Mr. P. J. Federico, Dr. Arno C. Fieldner, Dr. A. Remington
Kellogg, Dr. Waldo G. Leland, Dr. Thomas G. Manning, Mr. Ernest
G. Moore, Dr. Edwin B. Wilson, and Dr. Raymund L. Zwemer.

Despite assistance from so many sources, sole responsibility for
the work is mine, including the conclusions and interpretations and
any errors which may have escaped my vigilance.

For permission to quote copyrighted material I am indebted to
Harcourt, Brace and Company, Inc., for quotations from Gifford
Pinchot, *Breaking New Ground*; to J. B. Lippincott Company for
quotations from W. H. Dall, *Spencer Fullerton Baird*; to Prentice-
Hall, Inc., for quotations from Robert A. Millikan, *Autobiography*;
to Charles C Thomas, Publisher, and to the author for quotations
from John F. Fulton, *Harvey Cushing: A Biography*. Dr. Clifford
K. Shipton, Custodian of the Harvard University Archives, granted
permission for the use of letters in the Benjamin Peirce Papers.

I wish also to thank my wife, who has lived with this project as
long as I have.

A. HUNTER DUPREE

Cambridge, Massachusetts
June 21, 1956

CONTENTS

I   First Attempts to Form a Policy, 1787–1800   1

II   Theory and Action in the Jeffersonian Era, 1800–1829   20

III   Practical Achievements in the Age of the Common Man,
      1829–1842   44

IV   The Fulfillment of Smithson's Will, 1829–1861   66

V   The Great Explorations and Surveys, 1842–1861   91

VI   Bache and the Quest for a Central Scientific Organization,
     1851–1861   115

VII   The Civil War, 1861–1865   120

VIII   The Evolution of Research in Agriculture, 1862–1916   149

IX   The Decline of Science in the Military Services, 1865–1890   184

X   The Geological Survey, 1867–1885   195

XI   The Allison Commission and the Department of Science,
     1884–1886   215

XII   Conservation, 1865–1916   232

XIII   Medicine and Public Health, 1865–1916   256

XIV   The Completion of the Federal Scientific Establishment,
      1900–1916   271

 XV   Patterns of Government Research in Modern America,
      1865–1916   289

XVI   The Impact of World War I, 1914–1918   302

XVII Transition to a Business Era, 1919–1929   326

XVIII The Depression and the New Deal, 1929–1939   344

XIX   Prospect and Retrospect at the Beginning of a New Era,
      1940   369

      Chronology   383

      Bibliographic Note   387

      References   395

      Index   443

SCIENCE
IN THE
FEDERAL
GOVERNMENT

I

FIRST ATTEMPTS TO FORM A POLICY

1787 – 1800

SINCE World War II the relation of science to the United States government has received ample recognition as a problem of profound national importance. The atomic bomb and the hydrogen bomb are only the most dramatic of the many symptoms of this dynamic factor in national life. The government has emerged as a great user of science and a great support to many lines of research. On the other hand, all the institutions of the country in which science exists have found that the actions of the government in conducting research and in contracting for it are factors of first importance. As an integral part of military power, science is recognized as a necessity for the nation's survival.

Since 1940, when most government administrators and scientists first became aware of the magnitude and complexity of their mutual relations, a vast literature has reflected new interest in the subject and widespread emphasis on its momentous implications. Running through this outpouring of contemporary writing is a theme — sometimes explicit but usually implied, as if the authors sensed rather than analyzed its presence — that science is not a new thing in the federal government. Nearly everyone can recall a few examples of activity in some federal bureau long before 1940. A few even have felt that these had more importance than mere isolated anticipations of current problems. But the great record of science in the government, as a living body of experience on which the present could build, has lain so scattered that few have appreciated either the extent or the meaning of the relation.

From the beginning the federal government has rendered honor to science and profited from it. Almost as early, the support given by the government was a significant source of strength to science in America. The institutions that grew up in and near the federal structure have been prominent in all periods of the nation's history. Indeed, before the rise of the universities, private foundations, and industrial laboratories, the fate of science rested more exclusively with the government than it did later.

All this historical experience with science was an asset in the crises after 1940. But to the development of American civilization it has a more profound significance. Science has been a formative factor in making both the federal government and the American mind what they are today. The relation of the government to science has been a meeting point of American political practice and the nation's intellectual life. This conjunction has been continuous from 1787 onward and has interacted with both contributors. On the one hand, American democracy's very essence has been influenced by the presence of science. On the other, the institutions that harbored and fostered science in America would have been different, and much poorer, without the efforts of the government spread over many decades. The resulting picture gives an additional dimension to American history.

Across the century and a half beginning with the Constitutional Convention, science has meant different things to each successive age, and at any time it has meant different things to a wide variety of groups. It is the interplay among these various concepts, not a logical choice of only one of them, which provides the full historical definition of science. In practical terms, this story is concerned with the natural sciences — physics, chemistry, biology — in all their specialties and variations. Since the emphasis is on the institutional setting, science here means education, communication, and organization as well as the creation of new knowledge by means of research. It also includes the growing and changing profession of the scientist. The social sciences have generally followed so different a path and chronology for entering the government that they deserve a fuller treatment on their own merits than they could get under the shadow of the older disciplines. Sometimes, however, the natural and the social sciences have grown up closely intertwined. Here the institution, not an arbitrary line, determines the boundary.

The distinction between basic or fundamental science and applied science is much too useful to avoid, but it must be made only with the understanding that the terms represent two ends of a continuous spectrum, with innumerable shadings in between. The term "development," so common in the mid-twentieth century, has little usefulness through the several generations that did not employ it, although the concept behind it put in an early appearance.

Since World War I, science has become dominant in generating and directing changes in technology. In earlier years, however, the relations between science and a trial-and-error technology were sporadic, with practice influencing theory more often than the reverse. This story includes a consideration of technology to the extent that it is directly related to science.

As the history of the republic unrolled in the years down to 1940, scientific institutions within the government gradually solidified into a permanent establishment, the unsuccessful and abortive making their mark along with the rest. And the ideas behind them were quite as important as the actual organizations. Hence this is a story both of institutions and of ideas. Because the federal government provided the setting, ideas tended to become policies. Through all the twists and turns of the political history of the United States, and through the immense changes wrought by 150 years of rapidly expanding scientific knowledge, the policies and activities of the government in science make a single strand which connects the Constitutional Convention with the National Science Foundation.

The idea that the federal government should become the patron of science was easily within the grasp of the framers of the Constitution. As educated men of the eighteenth century they knew that European governments had often supported science, and their set of fundamental values led them to hold all branches of philosophy in high regard. Hence, as they went about their political task of reconciling the great interests of the new nation, they gave some consideration to the constitutional position of science in the government they envisaged. The problem was even more important than the setting up of some specific research activity, for it involved the whole later constitutional mandate for the government in science.

The Constitutional Convention

The framers had among them one of the great scientific men of the time in Benjamin Franklin, and, if he said nothing on the subject, others came forward with proposals that would have placed beyond argument the new government's duty and ability to encourage learning in both the arts and the sciences. Charles Pinckney's plan included power to "establish seminaries for the promotion of literature and the arts and sciences," to "grant charters of incorporation," to "grant patents for useful inventions," and to "establish public institutions, rewards and immunities for the promotion of agriculture, commerce, trades, and manufactures."[1] Had these provisions reached the final document, a national university devoted to advanced scientific training, societies chartered by the government, technical schools, and prizes and direct subsidies for creative effort could all have become realities — symbols that enlightenment was the first glory of a free republic. James Madison's proposals more succinctly called for the power to establish a university and to "encourage by premiums and provisions, the advancement of useful knowledge and discoveries."[2]

From these suggestions the committee of detail extracted most of the pith, reporting out only a clause for patents. Since science itself was not challenged by these cultivated men, the difficulties of a national university or of federal charters for societies stemmed from larger issues of more immediate concern than pure learning. These became clear in the convention's closing debates, which centered around a concept of great importance in the early history of the republic. "Internal improvements" were public works of all sorts. They might equally be roads and canals or universities and scientific societies. Whatever their object, internal improvements financed by the federal government were major political issues, and their fortune determined the degree to which science was supported.

The states naturally retained full freedom of action in this field, but the question of whether the federal government should participate had a special urgency because the very success of so large a union might well depend on the efficiency of the transportation system. When Dr. Franklin moved to add a power "for cutting canals where deemed necessary," he was speaking as a representative of the large state of Pennsylvania for strong action by the central government in all sorts of internal improvements. Objections immediately came from Roger Sherman of Connecticut, representing the small states and a restricted view of the powers of the central government. Madison then tried to return to his earlier idea of granting charters of incorporation "to secure an easy communication between the States," which immediately aroused the apprehension of those who feared powerful central organizations. But they suggested banks, not observatories, as the possible abuses of this power. When the motion on canals reached a vote, only large states favored it. Local forces were too strong, states were too jealous, and agrarian interests too much distrusted monopolies and special privileges.

Since science was the relatively innocent victim of this failure, Madison and Pinckney moved to empower Congress to "establish an University." Gouverneur Morris tried to smooth this over by saying it "is not necessary. The exclusive power at the Seat of Government, will reach the object." This statement Madison carefully recorded for future use. Pinckney's influence probably brought over North and South Carolina to join Pennsylvania and Virginia, but Connecticut split, leaving the vote six against, four in favor, and one tied.[3] This debate, although secret for many years, left some clear impressions on the members of the convention and also some murky misunderstandings. All agreed that a national university had failed to gain a specific place in the document, but some claimed with Morris that the government already had that power through its control of the federal district. Even when these proponents conceived an institution with professorships of science, museums, laboratories, explorers, and agents to plunder the latest knowledge of Europe,[4] they saw it with headquarters at the national capital. Others, such as Roger Sherman, publicly proclaimed that, since the national university had been rejected, Congress had no power to create scientific institutions.[5] Yet both sides could agree that universities and learned societies were in fact internal improvements. Hence the question of constitutionality would for a long time tend to rise or fall with fortunes of federal sponsorship of canals, roads, and banks.

Thus, although the Constitution as known to the public and put into use in April 1789 did not grant specific power for a university, it nevertheless contained several implied benedictions on national scientific institutions. The charmed word "science" actually appeared once, in the power of Congress to "promote the Progress of Science and useful Arts, by securing for limited Times to Authors and Inventors the exclusive Right to their respective Writings and Discoveries." A stop at the first comma would have given the new government plenary power, but the qualifying phrases not only suggest merely the English practice of protecting new inventions for a limited time, but carefully avoid the word "patent" as suggestive of the royal prerogative to create monopolies. There is no hint of Pinckney's and

Madison's additional suggestion of premiums for discoveries nor any indication whether such payments were implicitly prohibited.

Other parts of the document seemed to take for granted that science would be a handmaiden of the government. That the power over coinage, weights, and measures would necessarily entail highly technical expert advice and scientific experimentation was axiomatic to educated men. A census was provided not so much because of curiosity as because the political compromises made it necessary; nevertheless the men of that time could visualize scientific uses for it.[6] The powers over territories and over the seat of government involved both Congress and the executive in matters such as surveying, which within their own borders were left to states.[7]

That the more basic powers of the new government might involve science was readily evident, but their exact potential meaning had to wait for future decisions about the fundamental characteristics of the organization described in the Constitution. Whether the power to tax, to regulate commerce with foreign nations and among the several states, to establish post offices and post roads, to raise and support armies, to provide and maintain a navy, would either affect scientific institutions or cause the government to call learned men to its assistance depended on the collective judgment concerning issues far transcending science. What "general welfare" or "necessary and proper" meant in the Constitution varied with the balance of political forces at work in a federal union. Equally changeable was the state of science itself. For the ability of the government both to aid science and to use it rested not solely on political power but on the horizons of knowledge that men could see.

Science in the Early Republic

The American Revolution by 1790 had modified little the attitude of the former colonies toward science. As in Europe, the new United States found a knowledge of the natural world residing in an organized form largely within its upper classes. As in Europe, ideas stemming from science, in particular the laws of Isaac Newton, were tremendously influential in shaping the mental outlook of cultivated men. The natural law to which the colonists appealed in 1776 and the faith in reason which they trusted for deliverance from both political and clerical despotism sprang in part from science and established a climate congenial to its growth. The great experiment, the republic itself, was a product of rationalism and attuned to the supposed laws of nature. Many others besides Franklin and Thomas Jefferson were true Europeans of the Age of Enlightenment who responded to new scientific discoveries and took real delight in dabbling in natural history and natural philosophy in the same spirit in which they worked for a new political order.

If educated Americans were still Europeans, they were also colonists, depending on the mother country for both equipment and ideas. Benjamin Rush, an ardent patriot, thanked an English correspondent who in 1783 sent him a load of books for "an act of charity to pour upon a benighted American the discoveries of the last eight years in Europe." [8] Even the natural resources of the North American continent were studied more authoritatively by European travelers such as André Michaux than by residents. University education tended to be synonymous with study in Europe.

The new country did not entirely lack institutions of its own. The American Philosophical Society in Philadelphia had much to offer besides the accomplishments of Benjamin Franklin, its leading spirit. Boston, under the influence of John Adams and the wartime alliance with France, had established its American Academy of Arts and Sciences. The membership of these societies embraced most of the nation's scientific figures, including also leaders in the humanities. Science was not separate from philosophy, the arts, or literature in either organization or personnel. Within the framework of natural philosophy and natural history, the particular fields of physics and chemistry, botany, zoology, and mineralogy were clear, but nobody imagined that a man should devote his whole time to one of them. Indeed, almost none of the members were even professional scientists. Many were doctors, lawyers, or clergymen, making their living and spending much of their time in ways unconnected with science. Medicine provided perhaps the nearest approach to a scientific profession. But the physician of that day had no scientific basis for much of his work, and the research he did was usually collecting objects of natural history. While this diffusion and amateurishness severely limited the amount of work any individual might do, it also meant that men of affairs, often in their own persons, brought science into high councils. Thus a leading scientist could be president of a country that would be hard put to find a professional to hire as a chemist or a metallurgist. Lack of specialization also blurred any distinction between pure and applied research. Franklin could move easily from philosophical inquiries concerning the nature of electricity to the lightning rod. David Rittenhouse could make clocks and orreries, survey the heavens or canals, all with indifferent excellence. Science, thus unspecialized in intent as well as in fields of knowledge, was at the same time useful and ornamental, specific and universal.

The utilitarian aspect of science enjoyed high repute at the end of the eighteenth century. Physics and astronomy had become sufficiently advanced to convince ship captains and army officers that science could do something for them. Natural history had a close if not always fruitful alliance with medicine. Agriculture was undergoing its first great period of improvement in several hundred years. But even in these branches science as an organized body of knowledge only fitfully produced tangible results, and the practical activities of man moved along as best they could, supported by the tradition of craft and modified by untutored ingenuity. The textile machinery which had changed the face of England and was already finding its way surreptitiously to America, the cotton gin, and even the steam engine owed relatively little to the science of the day. The praise of the utility of science which rose so easily to eighteenth-century lips had in it a defensive note, a tacit admission that much invention, change, and innovation was taking place by those adept at trial and error but untutored in science.

In spite of their national names, the American Philosophical Society and the American Academy of Arts and Sciences hardly qualified as national institutions. If they aspired to emulate the great European societies, they nevertheless related more to local Philadelphia and Boston. The conjunction of the American Philosophical Society, William Bartram's botanic garden, Rittenhouse's observatory, and Charles Willson Peale's museum gave Philadelphia a good claim to being the cultural center of the new republic as it had been in the later colonial period, but neither in membership nor influence did the society reach the whole country evenly. Boston, with a more meager tradition and without a central geographic position, tended to influence only New England.

Both private endowments and government aid were familiar but minor supports to research. Although the colleges that were inherited from colonial times helped somewhat in providing centers of learning and equipment and employment for scientists, these benefits were largely casual and unplanned. No advanced training in science was offered anywhere, and those faculty members who undertook research often did so in the same spirit as did doctors and lawyers. Throughout the length of the new republic no body of professional scientists existed. The citizens of the United States in 1789 had been familiar with a few types of government assistance to science in their past experience. The assembly of Pennsylvania had helped to finance the observations of the transit of Venus in 1769. Patents as an inducement to discovery had been issued by the governments of various colonies, and in the period after 1776 several states as well as the Congress had taken an interest in them.[9] But most of these precedents might serve equally well for state action as for the federal government.

One factor remains to account for the principal resources of science in the United States in 1789 — the hope and expectation that the new political order would usher in a new and glorious era for the pursuit of knowledge. Joseph Priestley, the great chemist who fled from Britain because of political and religious passions aroused by the French Revolution, stressed this when he first addressed the American Philosophical Society. "I am confident . . . from what I have already seen of the spirit of this country, that it will soon appear that Republican governments, in which every obstruction is removed to the exertion of all kinds of talent, will be far more favourable to science, and the arts, than any monarchial government has ever been." [10] This great theme, worthy of the best in the traditions of both freedom and science, would recur in every following generation of Americans and of those immigrant and exiled scientists who reached the New World, but it tells only part of the story. For science is not often the sudden blossoming of the flower of genius, even in the soil of freedom. It is a group activity carried on by limited and fallible men, and much of their effectiveness stems from their organization and the continuity and flexibility of their institutional arrangements.

Precedents During the First Decade

When the Constitution went into effect, everything had to be done at once by men keenly aware that all their actions made precedents. What, for instance, did the clause mean in granting power "to promote the Progress of Science and useful Arts" by issuing patents? Individual citizens did not hesitate to try to force an answer. John Churchman, an ingenious surveyor from the Eastern Shore of Maryland, claimed to have solved the age-old puzzle of finding longitude by the magnetic variation of the compass. This hope, as old as Columbus, had baffled such men as Edmund Halley and has since proved impracticable; but the First Congress was hardly to blame if it did not recognize this, for it had no way of passing on questions of scientific fact. Churchman sought a patent on his spheres and charts, and also requested "the patronage of Congress to enable him to perform a voyage to Baffin's Bay, for the purpose of making magnetical experiments." [11] Had such a request gained acceptance, Churchman, who was sufficiently reputable to impress Sir Joseph Banks in England, might have launched an expedition in the great tradition of Cook, Vancouver, and Bougainville, who had under government auspices added so much to eighteenth-century knowledge. The young republic, by granting a subsidy under the patent clause, might have won renown as the friend and patron willing to spend money on science. The committee, with good reason in the period before Alexander Hamilton had effected any financial measures of importance, said that the government could not afford the Baffin's Bay undertaking. But they left the way open for Churchman to apply later, thus giving some encouragement to the idea that the patent clause made the government a patron of science in the positive sense, and that the premiums which Pinckney and Madison suggested were sanctioned by the Constitution after all.

Churchman found in the House of Representatives an ally of science in John Page of Virginia, early schoolmate of Jefferson. When the application came up again in 1791, Page cited as precedents the British support of Edmund Halley's work on magnetism and the prize of £20,000 offered by Parliament to anyone who would go to the North Pole. He felt that ingenuity deserved the backing of the government whether or not the theories advanced were correct, because "whatever can contribute to the discovery of longitude must be worthy of encouragement." [12] Other members also felt that "We ought to be cautious how we hastily decide on the views and experiments of philosophical applicants, and ought to take warning from the disgrace of other nations whom history has held up for their premature rejection of enterprises and schemes of science." [13] The opponents, not nearly so articulate, suggested that sea discoveries in the Arctic were "hopeless, since Europeans have failed in their attempts." Page's own statement indicated some misgivings about Congress's power to send the expedition, for he argued that a favorable committee report would recommend the theory "to the legislatures of the sister States whose authority and ability . . . cannot be doubted." [14] A committee under Page's chairmanship reported that "as the United States are peculiarly interested in whatever can adjust or prevent disputes between their citizens, and can improve geography and navigation, the Congress of the United States may, with great propriety, patronize such a person as Mr. Churchman, and grant such aid as may be necessary to enable him to prosecute his laudable inquiries to good effect." [15] The report left to the House the tougher immediate decision whether a grant should actually be made. Despite favorable reception of the memorial and the favorable report of the committee, action abruptly ceased when it came time to appropriate money. The scheme fell through without benefit of debate or any real vote. Churchman left for England soon after at the invitation of Sir Joseph Banks and spent most of his few remaining years seeking support in Europe.[16] Instead of support from the government of the United States, he got only the personal subscription of President George Washington to his *Magnetic Atlas*.[17] Despite this rebuff, the possibility remained that the government might become a patron of science on the model of European governments by means of the patent clause.

Meanwhile, the new government had many decisions to make concerning patents themselves. To what extent did they apply to scientific discoveries? And to what extent did the government take the responsibility of testing inventions to make sure the claims were valid and the devices really novel? Congress, soon having more petitions than it could handle individually, passed the first patent act in 1790 at the request of President Washington. The secretaries of state and of war and the attorney general constituted a board to pass on inventions. Since the records fell to the State Department as one of its home functions, its secretary was the principal responsible officer. Some in Congress had hoped to have the validity of patents determined by a jury or perhaps even by a board of experts,[18] but fear that such devices would deprive inventors of their rights placed this duty in the hands of high officers of the government. The board had full authority to refuse patents because of lack of novelty, utility, or importance, which placed the heavy responsibility of making technical decisions on three of the four leading men of Washington's official family.

By chance the administrator of the patent law of 1790 was himself an inventor and an enthusiast for science, even though chosen for the post of secretary of state because of his diplomatic experience. Thomas Jefferson had a double attitude toward patents for inventions. His dislike of monopolies of all kinds had at first included even patents of limited duration for inventions, but he came to see their usefulness in this one exceptional case, not as a natural right but as a grant from society to encourage inventors by giving them some chance of financial return for their work. Nevertheless, he was always on guard to restrict patents to real novelties and to protect the public from having familiar devices long the common property of all men subject to a levy by a patentee. Thus a principle abstracted from a machine was not patentable, but only the device itself.[19] These assumptions had two very important influences on Jefferson's conduct of the patent law. First, science itself was rigidly excluded from patents. Second, and paradoxically, all the techniques of science should be applied by the government to a patent application in an active effort to protect the public from unwarranted exclusiveness.

There is something sublime and pathetic in the spectacle of the secretary of state and a battery of professors from the University of Pennsylvania gathered around a distilling apparatus in the secretary's office to test the efficiency of a mixture supposed to help make salt water fresh. By a series of well-turned experiments, Jefferson proved that the fresh water came from the distilling process, long known and even used at sea, and that the mixture added did not enhance its efficiency. Nevertheless, Jefferson suggested to Congress that instructions for building an evaporator be printed at government expense and distributed to all shipmasters.[20]

Several aspects of this incident appear at first glance to foreshadow later developments. The use of the faculty of the University of Pennsylvania, made possible because the capital was then Philadelphia, was a very early example of the use of scientists for their expert knowledge. The protection of the public interest by preventing an ancient principle from being tied up in a monopoly because of the supposed action of a secret mixture had in it a certain element of regulation. That Jefferson should propose the dissemination of the knowledge thus incidentally called to his attention suggests that the federal government had a duty to promote the general welfare by broadcasting this useful bit of information. But a closer look indicates that the whole machinery set up by the patent law of 1790 was administratively fantastic and impossible. Even then the men who were rapidly becoming cabinet officers as well as heads of departments had other things to do than get involved in complex patent proceedings. The salt water affair shows that they tried. That only three patents were granted in the first year shows that they were sufficiently critical. The rules that they developed have since proved in general workable. But the demands on principal officers were impossible.[21] Jefferson complained that the "subjects are such as would require a great deal of time to understand and do justice by them, and not having that time to bestow on them," he was "oppressed beyond measure by the circumstance under which he has been obliged to give undue and uninformed opinions on rights often valuable, and always deemed so by the authors." [22]

The alternative was either to set up an efficient separate office under its own responsible official, or for the executive to abdicate the function of judgment over patents and allow the courts to decide whether they were any good or not after they had been issued. In the debates on the revision of the law in 1792, John Page and others tried to create an office of the director of patents. But some felt that an officer already conveniently on the payroll — the secretary of state or the director of the mint — could do it, or they feared centering the office in the seat of the government, preferring the district courts. These opponents of bureaucracy prevailed, and in 1793 a new act left the secretary of state in charge of patents, but with no discretion to reject them if the fees were paid and the forms properly filled in. A single clerk in the State Department cared for patents in the 1790's, and after 1800 Dr. William Thornton, a man of accomplishments, ran the office without much concern for orderly records or for businesslike methods, remitting fees whenever he felt sorry for an inventor.[23] The practical men such as Oliver Evans and Eli Whitney who tried to turn patents into commercial gain found themselves immersed in endless litigation, while the courts struggled with an increasing number of cases, a burden not lightened by patents issued to people who but sat down in the model room and, copying machines already deposited there, then turned in their own applications.[24] Only inventors who had machines of great potential sale and charlatans could profit greatly from this system. One of the few general rules which remained intact and grew stronger in this period of neglect was that "no patent can issue to guaranty *a mere principle*, it 'must be for the vendible matter.' " [25] Science that could not be reduced to a machine of potential cash value must look elsewhere for government aid. Even the practical inventor got little enough comfort.

If patents could not help, some still clung to the hope that had driven John Churchman on. Petitions seeking financial aid for research — on the "fundamental law that rules our solar system," on erecting a chemical laboratory to prepare sal ammoniac and Glauber's salt, on finding longitude by lunar observations — continued to engage the attention of Congress on occasion.[26] In 1796, when one Frederick Guyer made his try to get some help for his lunar observations, the committee of the House of Representatives reported not only that his improvements were without merit, but that "it is their opinion that application to Congress for pecuniary encouragement of important discoveries, or of useful arts, cannot be complied with, as the Constitution . . . appears to have limited the powers of Congress to granting patents only." [27] One member, who was not prepared to fight for the hapless Guyer's lunars, said he was "sorry to have it established as a principle, that this Government cannot Constitutionally extend its fostering aid to the useful arts and discoveries," but he did not press for a discussion.[28] Such an exchange does not make constitutional law in the grand manner. But the tide ran clear and strong against the establishment of any power of the government to subsidize science directly. The failure of the patent clause to become what it might have been — a basis for scientific activity of all sorts — well illustrates the reluctance of the Congress in this period to become active in science, even where, as in the case of patents themselves, they had to make no appropriation. The people's representatives settled into a silent groove of strict construction of the Constitution concerning a subject where popular enthusiasm did not measure up to the cost of action both in controversy and money.

A National University and Other Efforts

A variation on the same theme marks the vicissitudes of the national university, which, whatever else it might undertake, always included in the minds of its proponents advanced scientific training. Washington early advocated the idea before Congress and proposed to give the institution nineteen acres of land and fifty shares of canal company stock. Jefferson, ever the opportunist at picking up scientific bargains, found out that the University of Geneva had lost its government support because of the revolution and proposed to Washington that the whole faculty be brought over for the national university, although he seems to have envisioned some state aid and the possible location of it in Virginia rather than the District of Columbia.[29] In retirement and out of sympathy with the centralizing tendencies of the Federalists, Jefferson in this case made a rather considerable concession, much more than he had when opposing a military academy as unconstitutional.[30] The ideal of science and learning was one on which patricians could agree even while party bitterness was driving them apart.

When Washington wished to put his gift for this purpose into the hands of the Government in 1796, near the close of his presidency, James Madison showed the same spirit. Although already a leader of the opposition in the House of Representatives, he headed the committee that reported favorably on authorizing someone to receive the donations. He assented that "all men seem to agree on the utility of the measure," [31] but he found in debate that this general sentiment did not allay hostility and suspicion. One congressman, who preferred small academies to a big university, expected little from private gifts and felt that this was the first step to the United States' patronizing an institution with tax money. John Nicholas of Virginia approved the principle, but felt the time not ripe, questioning whether distant parts of the country which could not send children so far as the capital without injuring them morally should support the institution. Another claimed that land in the District of Columbia was for public use only, and "such institutions are not public, but private concerns." In vain Madison and the friends of the proposal pointed out that Congress had power over the District of Columbia, that no tax money was called for, and that the resolution did not mention a national university by name. A vote to postpone the resolution finally carried against Madison, and indirectly against Washington, by thirty-seven to thirty-six.[32] Had anyone been so foolhardy as to propose a direct appropriation, this debate indicates its overwhelming doom. The great aspirations for a national institution of learning, which few succeeding generations have voiced as eloquently and as cogently as did Washington, Jefferson, and Madison, were perhaps even further from accomplishment than they realized.

Washington's request that the government support agriculture —

the occupation of most of the population — brought forth in Congress a ringing declaration on the government's place in spreading information. "The only method which a Government can with propriety adopt, to promote agricultural improvement, is to furnish the cultivators of the soil with the easiest means of acquiring the best information respecting the culture and management of their farms, and to excite a general spirit of inquiry, industry, and experiment." [33] Since local societies were on too limited a scale, the "patronage of the General Government" was necessary to effect "national improvement." An American Society for Agriculture would animate state societies, provide exchange of information, give honorary and money premiums for discourses, and make a "complete statistical survey of the United States." Here again was a noble dream which illustrates how clearly the men of the 1790's could conceive of scientific institutions. And here again was the impossibility of effective action. The report asked for a "salary of a secretary and for stationery," but if that was too hard on the treasury, "it might be carried into effect without pecuniary aid." [34] Thus Congress could play the benefactor of agriculture simply by setting up the society in the District of Columbia. But even that painless gesture never came to anything.

Since the government spent most of the decade of the 1790's in Philadelphia, it lived close to disease. The great yellow fever outbreaks of 1793 and 1798 especially interrupted the government's business, and the arguments that raged over whether the pestilence had a local or a foreign origin had certain implications for the power which controlled commerce with foreign countries and among the states. President John Adams, in his annual message to Congress in 1798, asked Congress to establish regulations to reinforce state health laws, "for these being formed on the idea that contagious sickness may be communicated through channels of commerce, there seems to be a necessity that Congress, who alone can regulate trade, should frame a system which, while it may tend to preserve the general health, may be compatible with the interests of commerce and the safety of the revenue." [35] Although the health of the nation remained essentially under the police power of the states, some very tentative action did take place. In 1796 collectors of revenue were required to assist in the enforcement of state quarantine laws.[36] In 1798 the Congress passed a law financing the medical care of merchant seamen by deducting twenty cents a month from their wages. This system was a kind of crude compulsory government health insurance for a limited segment of the population. In the House of Representatives support for the measure came from the seaboard areas, where the presence of sick sailors worked a hardship on local agencies. While British practice and the fact that the system would not burden the national treasury were cited, little active desire to get into the hospital business and no intention of taking responsibility for the health of the nation seem to have arisen.[37] The United States Public Health Service traces its beginnings to these early hospitals, which had for many years after their founding few attributes of scientific institutions. The collectors of customs at various ports, without any central direction, either set up a hospital or farmed out the business by contract. Even though the facilities were always inadequate, some feared that the government would launch costly building programs.[38] Perhaps equally chilling to action was the uncertain state of medical science itself, which did not have the key to the prevention of most diseases, including yellow fever. A little later this embarrassment became clear to President Jefferson when Dr. Benjamin Rush called for investigations into the supposed domestic origin of yellow fever.[39]

No constitutional difficulties interfered with the setting up of a mint or the establishment of a standard of weights and measures. Coinage was an urgent necessity, and in the absence of anyone who knew anything about that rare but highly skilled craft a general assumption prevailed that a man of science could best handle it. Perhaps for this reason the mint found its way into the State Department under Jefferson, who, following the same assumption, appointed as its first director David Rittenhouse, the most distinguished astronomer America had yet produced and the successor of Franklin as president of the American Philosophical Society. The problems were indeed great and required considerable knowledge of metal, both for the manufacture of new coins and for assaying the foreign coins that continued to circulate.[40] Rittenhouse fondly pointed to the parallel between himself and Newton, who had spent his declining years as a master of the mint. This allusion was perhaps the only way in which the appointment added to Rittenhouse's reputation. Beset by administrative troubles, he had to stave off criticisms that his surveying activities and ill health prevented him from spending much time at his job. Without any experience in coining or in industrial operations, the aging savant never succeeded in producing enough coins.[41] The first experiment in enlisting an outstanding expert in government service has had a well-deserved lack of publicity. In addition to Rittenhouse's personal troubles, the question arises whether eighteenth-century science represented by a natural philosopher of general attainments could deliver results in a highly specialized field. This unfortunate beginning did not shake the idea that a man of science should supervise the mint. Elias Boudinot pleaded "want of Chemical Knowledge" but took the job anyway.[42] When he retired in 1805 Jefferson refused to appoint Benjamin Rush specifically because of his lack of mathematics.[43] From that time until 1853 all directors had some sort of scientific background.[44]

Like virtually all other problems of a scientific nature, the establishing of a standard of weights and measures gravitated into Jefferson's hands. He prepared a comprehensive report on the whole subject which recommended the length of a rod oscillating at latitude 45° in 1 second of mean time as a fixed standard of length and a decimal system of relations for all coins, weights, and measures.[45] David Rittenhouse, on more familiar ground here, contributed a good many details to Jefferson's report, which embodied much of the best French theorizing on this subject and also the passion then current in France to break with the past completely and set up a rational system. But in the absence of great political and social upheaval in America, Jefferson's ideas had little chance against the general indifference and the resistance to change. In 1796 the House of Representatives, while declaring that the standard of measure should not depart appreciably from the foot then in use, wished to give the President authority to employ "such person, of sufficient mathematical and philosophical skill," to make experiments.[46] Everyone seemed to feel that Rittenhouse was the one man who could do the work, that it could be done for $1000, and "the enlightened world would say $1,000 were never better expended." Although some members ridiculed the idea, no constitutional issue could arise, and the resolution passed easily, only to have the Senate postpone action.[47] Although the power of Congress was ample, although the utility to commerce of a uniform system was apparent, although the scientific knowledge was available, the record on weights and measures before 1800 is negative.

Throughout the 1790's the presence of the capital in Philadelphia made it possible for the American Philosophical Society to fill a semi-public role. Jefferson, vice-president of the Society from 1791, became its president at about the same time he took up his few duties as Vice-President of the United States in 1797.[48] This made it possible for him to introduce scientific schemes of general importance in the Society without facing the constitutional difficulties or the suspicion of a Congress now in the hands of a political party unfriendly to all his acts. He proposed a study of the Hessian fly, a plan for a national weather service, and an exploration of the Missouri River. The Society over his name petitioned the government to make the census into a survey "to determine the effect of the soil and climate of the United States on the inhabitants."[49] All these suggestions either involved government action or were plans which implied that the Society had a semiofficial place something like its English counterpart, the Royal Society. That none amounted to any more than the schemes proposed within the government indicates a general inability of American science to organize and support large-scale activities. Had the capital remained in Philadelphia, a thing that Jefferson strongly opposed, the local institutions such as the Society might have put themselves in the service of the government as a first step toward a gradual change into national scientific advisory bodies.

As the eighteenth century and the first decade of the Constitution drew to a close, the new government had few tangible accomplishments in science and had made little headway in developing permanent institutions either to use science in its own operations or to disseminate it among the people.[50] The unhappy fate of most proposals in Congress indicated doubt as to the constitutional position of science, lack of understanding of its possibilities and limitations, and indifference at best concerning its purpose. But this negative record is less than surprising in the light of the lack of administrative experience in public life generally and the very modest accomplishments of science in America outside the government. More important than the negative factors are the startlingly comprehensive ideas concerning the role of science, the clarity with which the institutions were conceived, and the energy which leading statesmen expended on fostering these ideas. Although only a minority saw the advantages of an alliance between science and the federal government, that small group included some of the most influential men in public life. Science has had a place in the government continuously since 1789.

II

THEORY AND ACTION IN THE
JEFFERSONIAN ERA

1800–1829

THE year 1800 brought a physical change of political origin to the government which had profound cultural implications. When the capital shifted from Philadelphia — with its societies, college, gardens, and museums — to Washington,

> This embryo capital, where Fancy sees
> Squares in morasses, obelisks in trees,[1]

the whole apparatus of civilization, and hence all facilities for science, remained behind. With this one move, ardently desired by many such as Jefferson, who, although friends of science, wanted insurance against urban hostility, the government lost the possibility of handy and informal appeal to the most promising body of consultants in existence in the United States. After this time, the problem of creating scientific institutions connected with the government was compound. National institutions that could serve the whole nation were still the aim, but the urgent necessity of furnishing local scientific services often took practical precedence. A city as well as a capital had to be made from an ill-assorted collection of boardinghouses set in a sea of mud, reducing every discussion of large institutions to miserable local realities. Until Washington had museums, observatories, libraries, and a group of scientists in residence, talk of national universities sounded Utopian.

President Jefferson and Science

The acrimonious election of 1800 elevated to the presidency one of the greatest savants of the day. Thomas Jefferson was "a pure republican, enlightened at the same time in chemistry, natural history, and medicine" as well as "a Citizen of the World and the friend of universal peace and happiness." [2] He embodied all the strong features of the science of his day, viewing knowledge as a single whole, moving gracefully across special subjects as if they had no barriers between. He had great faith in the usefulness of science, not only of practical inventions but of researches as esoteric as paleontology. He believed that science had no national bounds and that its followers "form a great fraternity spreading over the whole earth, and their correspondence is never interrupted by any civilized nation." [3] His limitations were equally characteristic. In no sense a professional, he gave only sporadic attention to any subject, getting much of his information from popularizers.[4] It was essentially his enthusiasm for science and learning that placed him in the presidency of the American Philosophical Society and gave promise to much more when he became the country's chief executive.

Jefferson's fame as an exponent of science was not entirely an asset. During the election of 1800 especially, the opposition linked this interest with deistic religious ideas and his partiality for the French, even questioning the value of scientific attainments to a public man. "If one circumstance more than another could disqualify Mr. Jefferson from the Presidency, it would be the charge of his being a philosopher." [5] That men like Rittenhouse and Benjamin Rush were ardent leaders in the Jacobin societies indicated to their opponents a sinister connection between science and dangerous ideas. Although this particular election produced a partisan split between Federalist and Republican which seemed to make science in some sense an issue, many leading Federalists were in one way or another contributors to science, especially in New England, while doubtless only a minority of those who wished Jefferson well cared anything about it. That Federalist propagandists should devote space to the charge of dangerous knowledge, however, shows their low estimate of the popular appeal of science and is perhaps a key to the long record of Congressional reluctance to help it throughout this period.

When Jefferson entered the new executive mansion, where he was to use a room for working on fossil bones, he continued to throw his weight for science whenever he could without raising fundamental questions about the relations of science and state. Charles Willson Peale's first thought on discovering a deposit of mammoth's bones at Newburgh, New York, was to notify Jefferson, who offered the use of navy pumps and army tents.[6] He aided the surveyors in the North-west by purchasing instruments from the contingent fund, so that they could make astronomical determinations of base lines to be used in regular surveys.[7] He encouraged his Minister to France, Robert R. Livingston, in the introduction of Merino sheep.[8] Whether he appointed Benjamin Waterhouse to the charge of the marine hospital in Boston because of his republicanism or because the "New England states are indebted to him for introducing vaccination into them," the scientific attainments of the applicant did no harm.

But such genial gestures could not hide the dilemma of Jefferson's position on the Constitution and its relation to science. His views on the strict construction of the powers granted the central government and his fear of its growth were fresh in the minds of everyone so soon after the Kentucky and Virginia Resolutions. When Charles Willson Peale, who was trying to get the legislature of Pennsylvania to support his museum, wrote that before "making my application I wish to know your sentiments on this subject, whether the United States would give an encouragement and make provision for the establishment of this Museum in the City of Washington," [9] Jefferson faced a difficult choice. The chance to aid science by bringing a notable collection to the barren capital collided with his constitutional principles. He replied to Peale that nothing could be done without an amendment and, having no desire at that moment for such a test of strength, he advised Peale to go ahead in Pennsylvania, suggesting also the possibility that Virginia might be interested when she began her university.[10] In this early phase of his great revolution in the government, the need to extirpate the Federalists' debt and reduce expenditure on military preparations precluded a positive program for science.

By the time of his second inaugural, Jefferson considered the errors of the Federalists at least partially rectified. The revenue, under the management of Secretary of the Treasury Albert Gallatin, seemed to insure the disappearance of the debt, allowing the President now to raise the question of what to do with the prospective surplus. He suggested that "by a just repartition of it among the States and a corresponding amendment of the Constitution, [it] be applied *in time of peace* to rivers, canals, roads, arts, manufactures, education, and other great objects within each State." [11] The sale of federal lands and the federal tariff would finance this state-administered program of internal improvements.

In December 1806, when for a few months the United States seemed clear of involvement in the Napoleonic Wars, Jefferson presented to Congress a more detailed program for the use of federal funds when the debt and "the purpose of war shall not call for them." Instead of lowering imposts to the vanishing point, Jefferson suggested "public education, roads, rivers, canals, and such other objects of public improvement as it may be thought proper to add to the constitutional enumeration of Federal powers." Roads and canals to open "new channels of communication" between the states, making "lines of separation disappear," could cement the union. Education, also an internal improvement, would not compete with "private enterprise," for "a public institution can alone supply those sciences which though rarely called for are yet necessary to complete the circle, all the parts of which contribute to the improvement of the country and some of them to its preservation." Jefferson considered this an "extension of the Federal trusts," and he hoped to have a constitutional amendment in hand by the time the debt disappeared. He further urged that the "national establishment for education" be endowed by those lands likely to be "among the earliest to produce the necessary income," a foundation especially advantageous because "independent of war." [12] Jefferson's dream of a country whose political power produced enlightenment here became a concrete proposal. The conjunction of highways and canals with education is significant. All roads led to Rome, where the national university by advanced research created the new knowledge essential to continuing freedom. The old Virginia plan of George Washington for a capital on the Potomac and for a Chesapeake and Ohio Canal as the cultural and commercial brain and spine of the nation had in Jefferson's words its most elevated statement.

What kind of advanced educational institution to "complete the circle" of the sciences Jefferson wanted he did not say, but he had at hand a blueprint worthy of his dream. Joel Barlow, onetime Connecticut poet and later a friend of Tom Paine and the French Revolution, had just returned to the United States after seventeen years in Europe and established himself near Washington as a savant and patron of arts and literature. In a plan for a national university he proposed both "*research* and *instruction*," combining the advancement of knowledge with the "dissemination of its rudiments." After surveying the extensive services of the French government to science and education, he

proposed a chancellor with broad powers, traveling professors, printing presses, laboratories, libraries, scientific apparatus, and botanical
gardens, also suggesting that the university take over the functions of
a military and naval academy, the patent office, the mint, and "a geographical and mineralogical archive of the nation." Such an institution, a real university in a land where none had ever existed, was also
a central scientific agency of greater scope even than Barlow's French
models. Showing a certain awareness that all these things were not
immediately attainable, he hoped that "the legislature, as well as our
opulent citizens, will assist in making a liberal endowment," because
the institution would begin on a suitably small scale with several men
of science giving their services without compensation.[13]

In February 1807, Jefferson talked over with Barlow the draft of
a bill to establish a "National Academy and University," making some
corrections which were "chiefly verbal." At the same time the President expressed a wish for a "Philosophical society" which would have
members spread over the country as well as a central academy at
Washington, perhaps a federation of "the great societies now existing." [14] By the end of 1807, however, all the schemes — Barlow's and
Jefferson's — had collapsed. The President complained that Congress
was balky about amending the Constitution and that "people generally have more feeling for canals and roads than education," only
casually mentioning an even more immediate deterrent when he said
that "the chance of war is an unfortunate check." [15] The attack of the
Leopard on the Chesapeake had shattered even an optimist's hope that
an entanglement in the European war could be averted, and the bitterness that greeted his embargo darkened the closing days of Jefferson's
presidency. Had foreign problems miraculously vanished in 1807, so
many difficulties for a national university still remained — lack of
experience and a dearth of qualified personnel as well as constitutional
doubts, congressional ridicule tied to party bitterness, and general
public indifference [16] — that its success seems unlikely. But the rumors
of war did their work sufficiently.

Explorations

This elaboration of an unrealized dream, however much it en-
nobled Jefferson's concept of an ideal republic guided and permeated
by knowledge, does scant justice to his ability to enlist the government in the cause of science. Even while his frontal attack to get a

constitutional amendment was failing, he proved a gifted organizer with a remarkable sense of political opportunity in coupling science with larger and more definite forces. The greatest of these opportunities was the presence on the North American continent of one of the few remaining gross geographic mysteries which had resisted the efforts of the foremost scientific explorers of the last third of the eighteenth century. Both Cook and Vancouver had failed to bring the Columbia River out of the shadows, its confluence with the unknown upper reaches of the Missouri offering the last possibility of a Northwest Passage, a water route to the Pacific. The cession of Louisiana by Spain to France in 1800 and the explorations of British trading interests approaching the Rockies from Canada stirred up the imperial rivalries in this region. The United States, with its vital interest in the mouth of the Mississippi and its vague claim to the Columbia through Captain Robert Gray's discovery, found here a critical diplomatic problem.

Jefferson, long aware of the enigma up the Missouri, fully appreciated the very definite uses of exploration and geographic knowledge in contests for empire. As early as 1783 he distrusted British schemes to "promote knowledge" west of the Mississippi when, in spite of fears that Americans did not have "enough of that kind of spirit to raise the money," he asked George Rogers Clark about leading an expedition.[17] While Ambassador to France he had given some encouragement to John Ledyard's wild scheme of reaching northwest America by way of Siberia. Doubtless he knew of a secret War Department attempt to send an officer up the Missouri in 1790. In 1793 he coöperated in a scheme to send André Michaux, the French botanist who was harvesting a rich collection of new plants even in the eastern states, to the northwest coast. Although as secretary of state Jefferson tendered official coöperation, the money and the sponsorship came from the American Philosophical Society. He contributed $12.50 and drew up Michaux's instructions to "take notice of the country you pass through, its general face, soil, rivers, mountains, its productions — animal, vegetable, and mineral." Stressing the need for astronomical observations, studies of the aborigines, and a lookout for fossil bones, he specified the types of reports.[18] The principal objective of a route to the Pacific and the subsidiary aims of collecting information on natural history were clear and in the best tradition of eighteenth-century exploration. Yet other ingredients — adequate support to supply a considerable body of men skilled in wilderness ways, and a modicum of military strength — were lacking. Jefferson as a public official seems to have made no serious effort to supply these deficiencies, the whole plan coming to nothing when Michaux became involved in the intrigues of Edmund Genêt.

In the fall of 1802 President Jefferson asked the Spanish minister in Washington "in a frank and confident tone" whether Spain would object to a group of travelers exploring the Missouri River, who would really "have no other view than the advancement of geography." But he would "give it the denomination of mercantile, inasmuch as only in this way would Congress have the power of voting the necessary funds; it not being possible to appropriate funds for a society, or a purely literary expedition, since there does not exist in the constitution any clause which would give it the authority for this effect." [19] When he sent a secret message to Congress in January 1803, to ask for money for the expedition, the President reversed the emphasis. After urging the wresting of the Indian trade in the upper Missouri region from the British, he made a glancing reference to the "Western Ocean" and then reassured the Congress that Spain and France would consider the expedition a mere "literary pursuit." [20] Thus science as an objective was for foreign ears; commerce as an objective was for Congress; and the real purpose, which had to do with the claims of empires, was carefully screened by silence, secrecy, and an ambiguous title to the act.

The amount requested by Jefferson in this bill "for the purpose of extending the external commerce of the United States," was only $2500 — large as compared with Michaux's $128, but small for the considerable party now suggested. However, this modesty did not indicate bad planning, for the entire pay of the company and all their rations while on United States soil came from the War Department.[21] Thus three important precedents at once arose. The Congress, usually so reluctant to grant a penny for science, opened the purse on authority of the commerce clause. The members must have realized that by using army funds they were also blessing scientific exploration under the military powers. In addition, they authorized the expedition to leave the boundaries of the United States to explore foreign soil, a precedent not changed by the news of the purchase of Louisiana which arrived soon afterward.

With the momentous appropriation passed, Jefferson had to organize his expedition both to gather information effectively and to survive in the wilderness. Knowing the scientific personnel resources of the country better than anyone else, he had to admit that "We can not in the U. S. find a person who to courage, prudence, habits and health adapted to the woods, and some familiarity with the Indian character, joins a perfect knowledge of botany, natural history, mineralogy and astronomy, all of which would be desirable." His secretary, Captain Meriwether Lewis, had all "the first qualifications" and although his knowledge of the three kingdoms of nature was not "under their scientific forms," he would "readily seize whatever is new in the country he passes thro', and give us accounts of new things only." [22] Lewis went to Philadelphia for several weeks, where the worthies of the American Philosophical Society taught him to make celestial observations, to collect plants and animals, and to study the Indians. Thus the captain, although not accomplished as a scientist even for that time, had an orientation within the general framework of organized science. [23]

Jefferson's instructions, in their detail, their insistence on astronomical observation, attention to natural history and the Indians, and above all his reiterated admonition to keep every possible record, set a scientific tone for this expedition and for the many that would later copy the pattern he set. He was transplanting to America the idea that exploring was not just going somewhere but was a highly technical enterprise. Although in the seventeenth century a buccaneer like William Dampier could explore fairly well, by the time of Cook and Bougainville expeditions carried expensive equipment and had a professional scientific corps. Jefferson, unable to manage a full company, did inject the ideal into American exploration to stay.

The most striking thing about the Lewis and Clark expedition was its complete success. It replaced a mass of confusing rumors and conjectures with a body of compact, reliable, and believable information on the western half of the continent which caught the imagination of the country. Besides opening trade in the far Northwest and establishing an American corridor to the Pacific, it had made significant findings in botany, zoology, and ethnology. But an expedition's usefulness to science itself depends on classifying the collections and reporting the results. The library, museum, and herbarium were as much a part of scientific exploring as were keelboats and trinkets for the Indians. Since the government had nothing to offer — neither experience nor institutions — both the papers and the collections ended up in Philadelphia. The journals did not appear in any form until 1814, nor in a faithful reproduction until 1904.[24] The botanical collection, which was outstanding even though much was lost in the Columbia, had an unfortunate history, finally being described by a German in England in 1814.[25] The Lewis and Clark expedition, a jewel in the history of exploration, left a precedent reinforced by achievement, but it found no supporting institutions in the government and created none.

Jefferson now mounted a comprehensive effort to bulwark the American diplomatic position in the West by systematic exploration. Dr. Samuel Latham Mitchill, patron of culture in New York City, ardent Jeffersonian in science as well as politics, and chairman of a key committee in the House of Representatives, collected and collated the known information about the trans-Mississippi West.[26] He introduced legislation providing for the exploration of the southwestern border of Louisiana, stressing that "important additions might thereby be made to the science of geography."[27] Hence when Jefferson communicated to Congress the first results of Lewis and Clark, he also included reports from Dr. John Sibley on the lower reaches of the Red River. In 1804 William Dunbar, "a citizen of distinguished science" who had a Scottish education, tried to penetrate the same region.[28]

On orders from General James Wilkinson, governor of Louisiana Territory, Lieutenant Zebulon M. Pike first ascended the Mississippi to explore its source and then made his famous journey up the Arkansas to the Rockies. Pike in the field found himself in the common predicament of the explorer of that day, who "had no gentleman to aid me, and I literally performed the duties . . . of astronomer, surveyor, commanding officer, clerk, spy, guide, and hunter." Confessing that he had no taste for botany and zoology, he complained that he had too much to do to look at the country "with the eye of a Linnaeus or Buffon."[29] Nevertheless, when the expedition returned after a captivity by the Spaniards, who kept all of Pike's notes, Jefferson suggested a protest to the Spanish government "because the object was the advancement of knowledge," and "it is not in the nineteenth century . . . that science expects to encounter obstacles."[30] These activities of only moderate success emphasized exploration's place as a recognized part of the activity of the government and especially of the Army.

When military officers took on exploring and surveying, they needed much more extensive technical knowledge than ordinary garrison duty required. One of the lasting contributions of Jefferson's presidency was the establishment of the Military Academy at West Point in 1802. Although he had opposed a service school as unconstitutional when Washington had advocated the idea back in 1793,[31] he now accomplished the same end by creating a Corps of Engineers which "shall be stationed at West Point, in the State of New York, and shall constitute a Military Academy." [32] This small beginning mixed instruction with engineering duties of all sorts, making cadets more apprentices than students. But the presence of a United States Military Philosophical Society, which lasted about ten years, showed a certain interest in science even at the start,[33] and Jefferson gets credit for beginning the tradition of French mathematics there.[34]

The Coast Survey

Nothing emphasized so clearly the pervading lack of qualified people that Jefferson faced in trying to launch new schemes as the sudden appearance of a real scientist. The range of activity open to the government widened markedly when Ferdinand Rudolph Hassler wandered into its service, bringing from Switzerland and France the skill that in itself suggested a coast survey. The government had inherited from its predecessors an insatiable need for topographical information which had produced various projects. Jefferson, apostle of the rectilinear survey, had brought good mathematicians to the General Land Office.[35] The pressing demands of commercial interests for better charts and navigational aids had produced spasmodic efforts by Congress to survey limited areas of the coast, such as Long Island.[36] The Treasury Department had been concerned with lighthouses and the choice of sites for them since 1789.[37] But the idea of a general and comprehensive survey of the coast, employing the latest scientific methods, came to a head not in maritime New England but among the members of the American Philosophical Society, stimulated by the arrival in their midst of a trained and experienced geodesist who possessed a scientific library, some fine instruments, and a set of French standard measures and weights.

Hassler came to the United States from Switzerland not as a scientist but as a part of a land-speculating agricultural colony, which promptly collapsed. Getting in touch with the savants of Philadelphia and especially Dr. Robert Patterson, Director of the Mint, Hassler applied to Jefferson for some sort of work with the government, for in his Swiss career, he had been not simply a scientist but also a civil servant. The canton of Bern, which had supported much of his topographical work, made appropriations on a four-year basis, allowing his project to plan far ahead. Jefferson, although skeptical of Europeans, who "bring with them an almost universal expectation of office," admitted that some jobs called for "meritorious foreigners . . . of particular qualifications" and allowed himself to go along with the idea proposed by his Philadelphia friends to put Hassler to work on a general geodetic survey of the coast.[38]

Two other forces joined the fortuitous presence of Hassler early in 1807 to make a coast survey possible. First was the brief hiatus in international crisis that had produced Jefferson's proposal for a national university. Second, the commercial interests of the seaboard states were now ready to exert more pressure for correct charts of "every part of our coast," with which "our seamen would no longer be under the necessity of relying on the imperfect or erroneous accounts given . . . by foreign navigators." These proponents pointed to "the lives of our seamen, the interests of our merchants, and the benefits to the revenue" as "ample compensation for making a complete survey . . . at the public expense." Possible military uses added to the commercial appeal, giving enough constitutional dressing for Congress to raise none of the doubts that clouded the national university. What is more remarkable, the usually penurious legislators authorized expending $50,000, twenty times the appropriation for Lewis and Clark.[39]

In a government that had little experience with the necessities of administering a scientific institution, authorization carried with it no guarantee of any result. The act itself mentioned no department but placed the Survey directly under the President. The Treasury Department actually took up the work of organization because of its concern with lighthouses, because it was the most highly developed executive department, and because Secretary Albert Gallatin was both Swiss and a man of scholarly tastes. Hassler later explained that the survey did not go to the military services because the "moral organization of such work has peculiar difficulties" which make all countries who have faced the problem "give their direction to some man of science under a department unconnected with either army or navy." [40] Certainly civilian control was congenial to Jefferson.

Gallatin immediately issued a circular letter soliciting plans for a survey and recommendations on personnel, to which Hassler replied in detail. The other answers were quite short, most recommending him as the person best qualified. Director of the Mint Patterson seems to have become the head of a kind of advisory committee of outside scientists to examine the replies. They selected Hassler's plan,[41] thus uniting the man and the job from the beginning.

Meanwhile Hassler went to West Point to teach mathematics at the Military Academy, and soon Patterson was writing that the Survey was suspended because of the disturbed state of affairs. As with the national university, the *Chesapeake* affair stopped Jefferson's programs and led to the complete concentration of the administration on the embargo. With action in science depending entirely on the enlightened help of a few top officials, anything that diverted them was disastrous. Only in 1810, after Hassler had been forced out of West Point by a ruling against civilian instructors, did Gallatin, now serving under President Madison, take up the Survey. Since clearly nothing could go forward without instruments and since neither they nor their makers existed in the United States, the first order of business was a journey to Europe. Hassler left for England, then the great center for the making of the theodolites and other precision instruments, in the summer of 1811, and stayed until 1815.

The War of 1812, which for the United States was far from total in its mobilization of the country, its diplomatic break with the enemy, or its results, took place in an age when a little more than lip service was still paid to the idea that science was above wars and that belligerents should not interrupt the flow of information and investigators. Only in such a climate of opinion was it possible for the major activity of the United States government during the war years to take place in the enemy's capital. Ironically, while Washington itself lay captured and Madison's government had almost ceased to function, Hassler was supervising with great care the building of a set of instruments especially designed for the problems of our Atlantic coast and for field conditions in the United States. By 1815 he had spent about $5000 more than the $50,000 appropriated. Forced to pay his own way home, Hassler arrived back in the United States poorer than he went, the extreme pains he had taken with the instruments acting only as a source of a general feeling in the government that he had stayed too long and spent too much.[42]

Nevertheless, Hassler, back in Washington, submitted to President Madison a plan embracing both the Coast Survey and the establishment of two permanent observatories which he had mentioned in his earlier prospectus but which were completely unauthorized as yet by Congress. Madison, who like Jefferson could appreciate the scientific worth of the eccentric Swiss, gave the project support. When an appropriation had gone through Congress, Hassler stood for two principles in his negotiations of a new contract. First, the salary of a scientist should be "more than the mere decent existence of my family." Second, he should have complete control of his accounts, for the "subjection of my expenditures to the control of a final account, would subject my whole existence . . . to the control of the accounting officers . . . who, by their absence from the work itself, cannot possibly have any idea of its incidences." [43] Madison, allowing Hassler a munificent salary of $3000 plus $2000 for expenses, conceded the main points of the new superintendent's desires.[44]

Beset by difficulties, delays, and at times lack of coöperation by minor government officials, Hassler finally got into the field during 1816 and 1817. With work preliminary to setting up a base line fairly well along, he received in February 1817 a letter from the Treasury Department asking when he would finish. The question served notice that most people in the government expected quick, tangible results and immediate relief from the necessity of spending money on the business. Furthermore, President James Monroe and William H. Crawford, now Secretary of the Treasury, lacked the exquisite sensitivity of Jefferson, Madison, and Gallatin. Disdaining to keep in touch with a grumbling Congress, Hassler was still going ahead on a scientific survey which would outdo the best of Europe, when in the spring of 1818 he suddenly found that an act had passed providing that only military or naval officers could be employed in the Survey. This seemingly innocent measure had a certain surface appeal in the name of native American skill and reflects the steady pressure to turn scientific jobs over to those already on the payroll. The chaplain of the Navy, one Cheever Felch, had been busy telling Congress he could do a better survey in less time. The argument of usefulness to commerce, which had done good service in 1807 in pressing the first act, now came back in the demand for immediate practical results regardless of scientific problems and opportunities.[45] Without debate and without hearings, Congress by shutting out Hassler had killed the Survey.

After a brief and stormy service for the government in ascertaining the boundary line with Canada, Hassler retired to an unsatisfactory private life to wait out his best years in the hope of getting back into the Survey. Meanwhile, if any alternative talent existed in the country, the government failed to turn it up in the Army and Navy. Money continued to go into surveys, but, with no comprehensive plan and no continuity of personnel, the results were immediately recognized as worthless.[46] Thus for the period before 1830 the government failed to demonstrate that it had the administrative resourcefulness and understanding to conduct a long-range scientific program. It also failed to recognize, granting that the Coast Survey had an immediate and practical objective, that fundamental science is sometimes a safer and cheaper road to usefulness than are hard-headed economy and direct methods, here represented by Chaplain Felch.

Attempted Activities in the Era of Good Feelings

The elevation and subsequent humiliation of the Coast Survey was part of the ambiguous position of scientific institutions in the government in a period usually described as nationalistic and blessed with the designation "Era of Good Feelings." While increased scientific activity was a glory to the nation and hence received some support from patriotic fervor, it shared with other internal improvements a doubtful place in the Constitution and was an exposed target in the bitter factional quarrels that substituted for party alignments. At the same time the old patricians who had supported science both in and out of the government, Federalists as well as the Virginia dynasty, were rapidly losing political power and position.

The national university felt these contributing forces. Madison, always its friend if not always consistent on other internal improvements, asked for it again in his last message to Congress. Enactment in 1817 would have made it a contemporary of the Second Bank of the United States. Although the committee reported favorably, suggesting that it be endowed by lots in the District of Columbia, two drawbacks appeared. One was that the institution had dwindled from the great research center of Barlow's plan to a seminary not unlike the private colleges of the day.[47] The other was that a Congress in deep trouble with its constituents over an act for its own compensation had no stomach for a proposal the merits of which would be hard to explain back home. It much preferred to leave the university alone.

At the same time a movement outside the government combined the urge to glorify the nation with a general scientific institution and the dire need to provide the boardinghouses and war-blackened buildings called Washington with some kind of cultural life. The Columbian Institute, established in 1816, considered itself "a national body . . . embracing every department of human knowledge . . . which will be worthy of the high destinies of the American nation."[48] It proposed to collect plants, minerals, and information on mineral waters, to study agricultural subjects, to prepare a topographical and statistical history of the United States, and to issue publications.[49] The utilitarian cast of these purposes was perhaps necessary in a town much more deficient in scientific men than older centers such as Philadelphia. The membership came largely from local business and professional men, some civil servants and military officers stationed in Washington, and many members of Congress, who were to serve as representatives of the society in their own districts.[50] The federal government was sufficiently coöperative to issue a charter in 1818 and to grant five acres of land near the Capitol for a botanical garden.[51]

The Institute was more or less active through the 1820's, continually petitioning Congress for help that never came. Some draining was accomplished on the botanical garden, but no adequate financial support or any trained scientist as a director seems to have appeared. The real leader, a naval surgeon named Edward Cutbush, left Washington in 1826, already discouraged by the lack of support.[52] All praises of the advantages of the capital were unavailing in the face of lack of resources of every kind to make even a good local scientific society.

Of the technical papers delivered before the Columbian Institute, those on mathematics or astronomy by William Lambert, a clerk in the pension office, had both local and national import. As early as 1809 he had been agitating in Washington the idea of a prime meridian. In determining longitude men had been in the habit of reckoning either from Greenwich or from Paris as zero degrees. A feeling existed that a sovereign nation which was worth anything should extricate itself "from a sort of degrading and unnecessary dependence on a foreign nation by laying a foundation for fixing a first meridian of our own." [53] Lambert industriously made calculations of the longitude of Washington, reading his results at the Institute and submitting them as memorials to Congress, which always drew favorable resolutions but never appropriated any money. For a time he had his instruments set up in the south wing of the Capitol. He soon arrived at the same conclusion Hassler had concerning the Coast Survey, that continuous astronomical observations were necessary to provide accurate bases for such fundamental surveying. Thus the proposal for a first meridian produced a demand for a national observatory, this time in or near Washington.[54] In 1827 Lambert was still pointing to the liberating effects of an observatory where "we might observe and compute for ourselves the right ascensions, declinations, longitudes and latitudes of the moon and such stars or planets as are most suitable for geographical and nautical pursuits, and . . . prepare and publish an Astronomical Ephemeris, independent of the aid of European calculations." [55]

Less directly related to science but much more important for it in the long run was the reorganization of the War Department under John C. Calhoun, whom old Samuel L. Mitchill considered next to Jefferson in his services to science in the government.[56] The secretary's most conspicuous act was to revive the Army's interest in trans-Mississippi exploration. He wished to establish new posts all the way to the Yellowstone River and in 1818 organized a Missouri expedition. Although the devious ways of contractors and the retrenchment forced by the depression of 1819 prevented the main expedition from ever getting fairly started,[57] Major Stephen H. Long was at Council Bluffs with a scientific corps in readiness for plains travel. Abandoning the Yellowstone idea, this segment of the original expedition ascended the Platte and searched for the headwaters of the internationally important Red River, returning by the Arkansas.

The organization of this expedition was considerably more elaborate than that of Lewis and Clark, the instructions to whom Calhoun used as a model. Long had both military and scientific command. The botanist Edwin James, the zoologist Thomas Say, and Titian Peale as assistant naturalist all completed the journey. Both the army officers who handled the topography and the astronomical observations on which their map depended, and the very presence of Say, James, and Peale indicated an increasing supply of men of scientific training. Calhoun in his instructions was fully aware of the value of information on the flora, fauna, geology, and Indians of the plains and Rockies as well as the gross topographical features.[58]

The classifying of collections and the publishing of results after the return of the expedition also showed improvement over Lewis and Clark. A general account of the expedition appeared in 1823, using the main journals kept by Long and others. Thomas Say published zoological papers on his collections. Perhaps the most interesting development was in botany, where James turned over his collections to his teacher, Dr. John Torrey of New York, who brought the superior resources of his private library and herbarium to the determination of the expedition's collection of dried plants. This is the first example of a type of collaboration on which Torrey built a great career as a botanist. The government furnished the military escort and general management of an expedition, while civilian collectors reported back to scientists who maintained privately the headquarters and apparatus necessary for complete analysis and publication.[59]

Long's expedition was only the most dramatic evidence of the new activity in the War Department. The creation of the Office of the Chief of Engineers with a topographer was the beginning of a centrally directed Corps of Topographical Engineers who, in the absence of an engineering profession in America, immediately became prominent in all sorts of surveying work. Despite constitutional scruples, the Congress increasingly appropriated money for roads and for harbor improvements.[60] One offshoot of Monroe's straddling position on the constitutionality of internal improvements was the Survey Act of 1824, under which the Corps of Topographical Engineers made a comprehensive plan for canals between the Chesapeake and the Ohio, along the Atlantic seaboard, and for a road from Washington to New Orleans. This plan, the only one the government ever attempted to make for the country as a whole, required considerable technical competence, and had it been executed would have required even more. Nevertheless, by 1825 army engineers were working on improving the navigation of the Mississippi and the Ohio and on the national road.[61]

The Army could meet these new technical demands largely because part of Calhoun's reorganization put new life into the Military Academy. With the appointment of Sylvanus Thayer as superintendent in 1817, West Point began to take on the characteristics of a college, and in mathematics Thayer modeled instruction on the *Ecole Polytechnique* and other French schools he visited at government expense in two years abroad.[62] Soon the academy became the main source of competent engineers both for the Army and for the burgeoning state and privately financed projects of the country. In the late 1820's the Army sometimes assigned its engineers to surveying railroads and canals, thus giving an impetus to civil engineering, as it was called to contrast it with military. While Major Long himself engaged in surveys for railroads, West Pointers such as G. W. Whistler and William G. McNeill became the leaders of a new profession.[63]

Calhoun's thoroughgoing overhaul of the War Department touched science in other ways as well. The first holder of the new post of surgeon general of the Army, Joseph Lovell, proved to be a man of ingenuity and energy. In addition to varying the soldiers' diet and abolishing the whiskey ration, he began to have his doctors collect weather data, on the theory that some relation might exist between disease and weather.[64] Army medical men before this had served in complete isolation, and the activities of the first surgeon general demonstrate how a central organization can broaden the scope of scientific activity by the government. Jefferson had dreamed of national weather reporting but could not give it the necessary attention. Lovell, a subordinate officer who nevertheless had widely scattered personnel reporting to him, started the system and obtained good results over several years, even without any additional appropriations.

A central officer also had a certain amount of power to aid a subordinate who by chance stumbled on a unique opportunity for scientific discovery. William Beaumont, an army surgeon stationed at Mackinac, had a patient with a wound in the stomach that did not completely close, allowing the doctor to make accurate observations of physiological processes as they occurred. Lovell was able to arrange for publication of Beaumont's first paper and to make various special concessions that allowed him to continue his work. When the woodsman whose unusual misfortune made the experiments possible proved recalcitrant, the United States Government coöperated by enlisting him in the Army, thereby both relieving Beaumont of the expense of paying his laboratory material and making escape amount to desertion. Lovell's successor as surgeon general went out of his way to refuse to help Beaumont's work, ordering him to Florida and provoking his resignation, which overshadowed Lovell's earlier opportune assistance.[65] Thus men with vision as well as improved organization were necessary.

The War Department achieved its aid to science in an atmosphere of congressional hostility which the politics of personal faction and the general suspicion of centralized activity intensified. That this congressional reluctance was not limited to the works of John C. Calhoun is evident in the curious episode of the national vaccine institution. Shortly after the introduction into the United States of the knowledge of cowpox vaccination, one of the few practical medical discoveries of the era, a Dr. James Smith of Baltimore had proposed to distribute genuine vaccine free, through an institute he set up in 1802. In 1813, as a result of his memorials, Congress passed a law naming him vaccine agent and giving him the privilege of using the mails without paying postage.[66]

This arrangement was still in effect in 1820 when a group of Smith's friends petitioned Congress for a federal charter for a "national vaccine institution." [67] But in 1821 a shipment to Tarboro, North Carolina, seemed to preface a fatal outbreak of the disease itself. A congressman from that state immediately demanding an investigation, the House of Representatives doubted whether they could decide the scientific question whether vaccination was effective. But on learning that Dr. Smith had unintentionally sent out smallpox scabs instead of cowpox vaccine, everyone agreed that the practitioner rather than the practice itself was at fault. Although one committee reported in favor of Smith and his retaining the privileges, the North Carolinian was not satisfied, claiming that the life, liberty, and property of his constituents were threatened by this "nuisance of the most dangerous kind." The franking privilege created a monopoly by which Smith could enrich himself, and the whole matter belonged under the police power reserved to the states.[68]

Smith, on the defensive because of his initial mistake despite the efforts of friends in Congress, hotly denied profiteering under the act of 1813. Indeed, he charged that the law had "never contained the provisions I asked," and requested that the law be repealed so that "an end may be put to the prejudices it seems to have so unreasonably and unjustly brought into existence." [69] Thus the connection with the government seemed to him to hurt his reputation more than the postage was worth. Repeal was, however, probably certain whatever his attitude, and his later attempts to get federal aid were unavailing.

In theory, this was a promising start of a public health measure that had a scientific basis. The ideal of free distribution throughout the United States was truly national and one which only the federal government could have easily fostered. In practice, Smith's mistake, a not uncommon one at the time, gave full opportunity for all those arguments designed to limit activity to its narrowest limit.

John Quincy Adams's Program

In such an atmosphere John Quincy Adams became President in 1825, after the bitter fight in the House of Representatives that produced the slogan "bargain and corruption," used effectively by his enemies throughout his presidency. By education and taste the President understood the capabilities of science and knew how to use it. Long head of the American Academy of Arts and Sciences, he had while secretary of state under Monroe prepared a monumental report on weights and measures. Although this masterpiece had no more effect than Jefferson's report of 1790, it showed scientific ability of a high order, and its preparation acquainted Adams with the primitive nature of the scientific resources of the capital.[70]

Adams's first annual message to Congress was the clearest statement ever made by a President of the government's duty toward knowledge, "among the first, if not the first, instrument for the improvement of the condition of men." While calling for roads and canals, he elevated science to the greatest of works to be undertaken by the government, for "moral, political, and intellectual improvement are duties assigned by the Author of our Existence to social no less than individual man." He recalled Washington's support of a national university and a military academy; West Point would have pleased the first President, "but in surveying the city which has been honored with his name he would have seen the spot of earth which he had destined . . . as the site for a university still bare and barren." [71] A national observatory likewise got the emphatic approval of Adams, who was well aware of the relation of astronomy, exploration, national glory, and the improvement of human knowledge. "One hundred expeditions of circumnavigation like those of Cook and La Pérouse would not burden the exchequer of the nation fitting them out so much as the ways and means of defraying a single campaign in war." [72] He asked for a naval academy corresponding to West Point "for the formation of scientific and accomplished officers." He specifically suggested a voyage of discovery to the northwest coast of America, a region where Lewis and Clark had gained so much both scientifically and diplomatically. He asked for an efficient patent office. He even suggested a new executive department to plan and supervise internal improvements generally and science especially.

In requesting these projects and linking them to internal improvements, Adams was but stating the old plan, fostered in turn by Washington and Jefferson, of a cultural capital radiating enlightenment to the entire nation over a connecting network of federally built roads and canals, with the Ohio and Potomac rivers as the main artery.[73] Two decades of experience had sharpened the need for some institutions, such as the national observatory, and Adams's New England background gave a maritime twist to his desire for explorations, but the national university still held the center of this great republican dream. Nevertheless, Adams departed from Jefferson's position when he claimed not simply that scientific enterprise was desirable and worthy of an amendment, but that it was already both constitutional and obligatory. He listed the powers over the District of Columbia, the power to tax "for the common defense and general welfare," to regulate commerce, to fix the standards of weights and measures, to establish post offices and post roads, to make rules concerning territories and other property, to make all laws "necessary and proper" for carrying out these powers. If "these powers and others enumerated in the Constitution may be effectually brought into action by laws promoting the improvement of agriculture, commerce and manufactures, the cultivation and encouragement of the mechanic and of the elegant arts . . . and the progress of the sciences, ornamental and profound," to refrain from exercising them would be to hide "in the earth the talent committed to our charge — would be treachery to the most sacred of trusts."[74]

Furthermore Adams insisted doggedly on a number of comparisons which Americans, committed as they were to the idea that their institutions were the best on earth, found puzzling if not downright undemocratic. Europe, "less blessed with that freedom which is power than ourselves," was nevertheless advancing science at a much greater rate. "Were we to slumber in indolence or fold up our arms and proclaim to the world that we are palsied by the will of our constituents" — a dyspeptic reference to congressional niggardliness — Americans would continue inferior to Europe.[75] In recommending an observatory he taunted his countrymen with Europe's more than one hundred thirty "lighthouses of the skies" to none for the United States. Advocating emulation of the governments of France, Great Britain, and Russia to make the United States "contribute our proportion of energy and exertion to the common stock" was not the kind of talk a man in the street only a decade after the Battle of New Orleans could readily comprehend. And when Adams went on to goad the federal government by referring to the vigorous action of the states in completing the Erie Canal and founding the University of Virginia,[76] he was playing with the sensibilities of the very Jeffersonians who had once advocated his program.

Adams's own cabinet reviewed the draft of the message with shock and horror. Henry Clay, although an advocate of a program of internal improvements as a part of his American system, was enough of a politician to see disaster in his chief's words. He thought the national university "was entirely hopeless" and that "there was something in the constitutional objection to it." While approving the rest of the message in principle, he "scrupled a great part of the details." William Wirt of Virginia, the attorney general, felt the proposals were "excessively bold," that Adams would appear to be "grasping for power," and that "we wanted a great, magnificent government." The reference to foreign expeditions would be "cried down as a partiality for monarchies." The voyage to the northwest coast would make Adams appear "a convert to Captain Symmes," a reference to the fanciful theories of John Cleves Symmes that the earth was hollow and open at the poles. Although only Richard Rush of the cabinet supported the message, Adams went ahead with what he admitted was a "perilous experiment." [77]

No possible motive for thus throwing expediency to the winds exists except Adams's genuine love of knowledge and his desire to have as his monument institutions he knew would live and serve the people. Nevertheless, the results of this first annual message were catastrophic. The administration that should have crowned a half century of independence with bold action in behalf of science produced absolutely nothing. True, those committees in Congress which were in friendly hands labored on favorable reports. They even went against the President's wishes to propose a comprehensive constitutional amendment to place beyond doubt the power of Congress to "make surveys . . . to construct roads . . . to establish a National

University . . . and to offer and distribute prizes for promoting agriculture, education, science, and the liberal and useful arts." [78] But all such gestures were vain. The common man, laughing at the "lighthouses of the skies," deeply distrusted a program of internal improvements which he felt would be financed by the sale of public lands at a high price and which might usher in consolidated government, monarchy, and tyranny.[79] Adams himself soon gave up hope of action from Congress.[80]

The one proposal that received any encouragement at all was the voyage to the northwest coast. A favorable resolution in the House of Representatives gave Secretary of the Navy Samuel L. Southard a chance to use executive authority to get a ship in readiness and to select "astronomers, naturalists, and others who are willing to encounter the toil." [81] Because of pressure from whaling interests, the South Pacific was also included, and an agent named Jeremiah N. Reynolds went to New England to collect log books and data on navigational problems faced by the whalers. A young naval officer, Charles Wilkes, went to New York to buy scientific books and instruments.[82] All this effort on the frail authority of a House resolution drew thunder from the Senate, whose committee feared that the executive preparations were meant to commit them to an appropriation of a large amount of money. They censured Adams's and Southard's use of contingent funds and ordinary naval appropriations as being "to some extent at least . . . an unauthorized application of public funds . . . ." [83] But by this time John Quincy Adams, ridiculed by his opponents and deserted by his friends, had suffered overwhelming defeat for the presidency. On March 4, 1829, with the inauguration of Andrew Jackson, the naval expedition died.

When Adams left the White House, the program of scientific institutions that had developed around the idea of a national university was in complete ruins. The bold attempt to assert the constitutionality of central scientific institutions and to tie them to a vigorous exercise of power by the central government had failed resoundingly. It would have been difficult in 1829 to propose a more unpopular measure than the national university or a national observatory. The generous hopes of Washington, Jefferson, and Madison had after a thorough tarring in the factional strife of the 1820's become political poison. However clearly some statesmen of the early republic saw the potential place of science in the government and however ingenious they were in conceiving institutions for it, in practice they had failed to make their visions reality.

But the almost complete bankruptcy of science in the government in 1829 should not obscure the real contributions of the first forty years. The country that had launched Lewis and Clark had a shining precedent for the blending of science and national interest which could hardly be ignored in the coming decades of rapid expansion to the Pacific. The Coast Survey was quiescent, but poorly charted reefs were still wrecking ships. The patent office languished, but inventors were ever more active. And in some of the few administrative recesses that had developed in the government men were collecting weather data and making astronomical observations. The military had shown a definite capacity for introducing science into its education and in organizing explorations and surveys. All these activities, closer to the needs of specific groups of the people than the ethereal edifice of the national university, still called for science in the government.

In a deeper way John Quincy Adams's departure from the presidency marked an epoch, for he was the last of the politicians with a broad and direct knowledge of science in his own right to gain the highest office. No longer was learning a necessity or even an asset in public life, a fact that added gloom to Adams's own reflections on his administration. Patrician control, through such amateur groups as the American Academy of Arts and Sciences, was having ever greater difficulty in giving effective organization to an American science that was beginning to change rapidly. Both science and the government were seeking new social adjustments in 1829, and they were fortunate in having as building materials the fragments of institutions and ideas inherited from John Quincy Adams.

III

PRACTICAL ACHIEVEMENTS IN THE
AGE OF THE COMMON MAN

1829–1842

WHEN Andrew Jackson came to Washington in 1829 he had
no hostility toward science, but he had no policy concerning it — a
lack which hardly troubled him. The men who put him into office
and swarmed so affectionately around him had only a guffaw for John
Quincy Adams's lighthouses of the skies, when they thought about
scientific institutions at all. With banks, revenue, public lands, and
the surplus occupying the official mind of Washington, scientific activity was a piecemeal response to forces that often seemed unrelated
to any coherent plan. Under the cloud of the failure of Adams's ambitious scheme, science fared well to have the government's actions
based on specific needs rather than larger constitutional and philosophical arguments. For the forces demanding specific actions by the
government were not only still present in 1830, but rapidly changing
and growing.

New Trends

Science itself was departing from the eighteenth-century norm
that had framed the efforts of Jefferson and John Quincy Adams.
The old societies continued on, joined by younger ones such as the
New York Lyceum of Natural History. Here the amateur still was
the typical scientist. The colleges, especially medical schools, as before provided training in most of the sciences and a shelter under
which professors could carry on a modicum of research. But forces
impinging on science from American life made these patrician institutions seem less adequate. Rapidly increasing wealth and economic
activity opened both new potential opportunities and new responsibilities. The many-fingered reform movements helped to bring knowl-

edge to more people. On their extreme fringe Utopians such as Robert Owen and Frances Wright gave a central place to science and did all they could to advertise its virtues. Less spectacularly, the ideal of education for all offered broadened opportunities for both teachers and students of the sciences, at least at an elementary level.

The common man, through universal white manhood suffrage, was beginning to have a real voice in affairs. Manufacturers and businessmen as well as Western farmers and city laborers had new power to make governments do their bidding. In fields where they could see some practical advantage from science, common men began to press legislatures to use this tool. States both north and south, as usual in this period taking the initiative ahead of the federal government, began to authorize surveys of natural resources. In Massachusetts, for instance, a resolution of 1831 called for a report on "botanical and geological productions of the Commonwealth," thus clearly implying a staff of experts, which was indeed the method used.[1] The hope of unsuspected wealth in minerals and other resources spurred on these efforts.

Inside science also forces were working for change in the direction of complexity and specialization. The old naturalist or natural philosopher who aspired to universal competence was being crushed under such a weight of accumulated knowledge that he perforce became a specialist. And even within the great areas of physics, chemistry, and biology smaller subdivisions were attracting the exclusive attention of scientific workers. Increasing competence and continuity in a restricted field replaced the general view.

In the gradations of scientific activity from fundamental discoveries to applications, specialization was also setting in. Jefferson's and Franklin's easy shifts from pure discovery to the most mundane of inventions became much less easy. The abstract scientist who worked on basic problems but had no aptitude for applying his results put an increasing interval between himself and the inventor and entrepreneur who specialized in technology. Young Joseph Henry, who by 1833 had worked out the principles necessary for an electric telegraph, continued his research into electricity and magnetism instead of trying to apply what he had already discovered.

As these changes proceeded, a body of scientific men of a new type grew up in the country. They were more specialized, more nearly professional in their attitude, more willing and anxious to cut the time lag in the flow of information from Europe. Benjamin Silliman was one of the first of the new men, and his *American Journal of Science and the Arts* was their organ. The Military Academy at West Point had played its part, giving employment to Dr. John Torrey and Jacob Whitman Bailey, the microscopist. Graduates of the 1820's such as Bailey and Alexander Dallas Bache were important additions. But as the ranks of young and ambitious specialists grew, a crisis in employment also grew. For the institutions of the country — old-line societies and colleges — offered few opportunities for the scientists to be hired as such, while amateur status no longer proved adequate. In this basic situation, which prevailed until the rise of universities after the Civil War, governments both state and federal had the chance to become great supporters of science. Every specialist they employed had an opportunity he would probably not have had if left to private institutions, and every dollar spent added to the short sum of funds available for support of science. The states, first in the field with their natural resources surveys, were already by the 1830's providing such a volume of support that geologists as professionals were a better organized and more self-conscious group than were the followers of any other science. In contrast to the dearth of personnel in the period of the early republic, a considerable group of men now clamored for scientific place and glory in the government.

The Implications of a Growing Technology

In addition to the political and social changes of Jacksonian democracy, in addition to the subtle but profound shifts within science, technological developments were becoming conspicuous. L. J. Henderson's dictum that before 1850 science owed more to the steam engine than the steam engine owed to science still applied. The day of research technology had not yet come. Inventions sprang directly from the empirical observations of practical men, most of them ignorant of contemporary science. But the application of steam to ships and later to the railroads, the growth of the factory system in textiles, and the application of machinery to agriculture were beginning to have incidental but important repercussions both on science in America and on the government.

The increasing number of inventions made unavoidable a reform of the patent system, vainly urged by John Quincy Adams. In 1826 a better patent office had seemed like a part of a dangerous scheme of unconstitutional centralization. In 1836 it appeared a simple and obvious necessity. The law of that year created the permanent office of commissioner of patents and provided the incumbent with a small staff. It reinstated the principle of an examination to determine the novelty and utility of each patent. It placed an obligation on the Patent Office to test each invention, calling implicitly for the use of some scientific principles by the examiners, who were often men of considerable attainments.[2] To assist them the law tightened requirements for recording patents and provided for a more efficient museum of models. The standards were hardly set as high as Jefferson had originally put them, for in practice patents were seldom refused for their lack of utility,[3] but searches to prove novelty gave the patent a certain face value. They also necessitated a "library of scientific works and periodical publications, both foreign and American." [4]

Henry L. Ellsworth, the first commissioner of patents, was a son of Chief Justice Oliver Ellsworth and a man of large ideas. Believing that the "natural and practical sciences, as well as the arts, have usually found their best patron in the munificence of a wise Government," he wished to make the Patent Office into a central depository not only of the mechanical models of patents but also of unpatented models, specimens of manufactures, and collections of minerals "illustrative of the geology of the country." He also thought that these scientific activities should directly serve the great economic interests. Since more had been done for commerce and manufactures than for agriculture, he attempted to make his office a clearinghouse for seeds and plants, starting a kind of agricultural service which will be discussed in more detail later. Thus on the narrow base of a patent law Ellsworth saw the possibilities of erecting a great scientific bureau.[5]

The reorganization of the Patent Office coincided with the rapid increase of technological innovation that began in the 1840's and reached large proportions in the 1850's. These were the decades of the reaper, the telegraph, and the sewing machine. From 436 patents granted in 1837 the total rose to 993 in 1850 and 4778 in 1860.[6] Whatever the cause-and-effect relation between invention and patent protection, the government had the administrative machinery for the first time in 1836 to make good on its obligations in the patent clause of the Constitution, while the increasing income from fees made a certain amount of money available for Ellsworth's larger scientific objectives.

Of the important inventions before the Civil War only one departed from the general pattern. Most sprang empirically from the traditional technological base, with only incidental and very distant assists from organized science. The electric telegraph, however, followed a new and different pattern, for it emerged from the discoveries of basic science without any supporting technology. As S. F. B. Morse, relying entirely on the discoveries of Joseph Henry and others, developed a workable electrical signaling system, he found that it had no place in the economy of the country. Few precedents indicated who should develop the new utility and who should control it. Morse early turned to the federal government, giving a demonstration in 1838 before the House committee on commerce. He was of course interested in patent protection, but he was even more interested in a direct subsidy which would allow him to develop his inventions.[7] For four years he endured the life of a petitioner and lobbyist. Finally in 1842, Whigs friendly to internal improvements were in control of Congress and they, with Democratic associates who had the ear of President Tyler, passed Morse's bill. It appropriated $30,000 for a test of the telegraph by building a line from Baltimore to Washington. The fact that some uncertainty existed whether the secretary of the treasury or the postmaster general should have supervision indicates that the commerce clause and the post office clause were the main constitutional justifications.

Morse favored not only government investment in experimental lines but permanent government ownership, or at least control of the new device by ownership of the patent.[8] However, powerful opportunists such as F. O. J. Smith, the chairman of the House committee on commerce, quickly recognized the commercial possibilities and fought for private control. Smith had a rather questionable financial stake in the enterprise and eventually left his seat in the House of Representatives to exploit it. In 1845 Congress refused to extend the telegraph from Baltimore to New York and merely appropriated $8000 for the upkeep of the original line.[9] The telegraph as a new type of industrial development stemming directly from science had the possibility of setting a powerful precedent for the entry of the government into the development of new devices. In general the support for Morse came from the North and West, and the opposition from the South,[10] indicating that the uncertain constitutional position of internal improvements generally was responsible for the feeble start and the abandonment of this policy.

In 1849 Congress used the Morse precedent to grant $20,000 to Charles G. Page, an examiner at the Patent Office, to develop an electric motor under the general supervision of the secretary of the Navy. After preliminary tests on types of batteries and model engines, Page built a full-scale electric power plant for a locomotive. The secretary of the Navy confessed he did not "understand the subject or have any notion of its object." Although Page had the use of some facilities at the Washington Navy Yard, he failed to gain the confidence of the authorities sufficiently to become his own disbursing officer.[11] In 1850 Senator Thomas Hart Benton tried to get an additional appropriation, claiming that Page had used $14,000 already and that the engine could develop 7½ horsepower. Scientific men such as Joseph Henry, Benjamin Silliman, and Benjamin Peirce supported the project. But the opposition, voiced by Lewis Cass of Michigan and Jefferson Davis of Mississippi, insisted that "the government should be the last agent to interfere in these matters." Davis attributed the right to make the appropriation to the patent clause, which he interpreted as protective only. He feared that the government would assume the risk of the enterprise and that Congress would become an examining agent. Indeed, the fear of a flood of petitions and the difficulty of recognizing the many charlatans then loose in the country was a genuine check on Congress. Benton's effort failed, and with it the Morse-Page precedent for the development of inventions lapsed. The exhaustion of his funds halted Page's researches after his locomotive had reached a speed of 19 miles per hour on a test run.[12]

Even those technological changes firmly rooted in an artisan tradition sometimes incidentally involved a combination of science and the government in a new role. The steamboat and the steam locomotive, the great innovations of the first half of the nineteenth century, were largely the products of "An Age of 'cut and try' and 'rule of thumb.'"[13]

Yet when the explosion of boilers began to take human lives in a more dramatic way than the inhabitants of an agricultural country were accustomed to, a cry for federal regulation arose. The greatest deterrent to congressional action was the lack of any accurate information on why boilers burst, and in a more general way how steam and iron behaved under high pressures. As early as 1830 the secretary of the treasury began inquiries, following a resolution of the House of Representatives. Finding his own agents unable to get much information from captains and engineers, he approached the Franklin Institute in Philadelphia, a young organization devoted to "the promotion of the mechanical arts," which already had a committee working on explosions. The chairman was Alexander Dallas Bache, a graduate of West Point who had resigned from the Army to become professor of natural philosophy at the University of Pennsylvania.[14] The committee, using government funds for materials required in experiments but not to pay the volunteers conducting them, first circularized practical steam engineers for likely causes of explosions and then built a model boiler with glass windows, a steam gauge, and thermometers.[15]

This very early example of a grant of research funds to a private body to conduct experiments with a specific objective produced a considerable body of data but hardly an answer to the explosion problem. The law of 1838 introducing federal regulation ended neither the explosions nor the disagreements as to their cause and cure.[16] The procedure stands as an isolated example in the 1830's. Nevertheless, the episode is important. In a period of strong states and comparatively weak central control, the federal government proved itself able to go out into the scientific institutions of the country to get the information it needed. It also proved itself able to regulate in the general interest on the basis of scientific data. Although the execution was not outstanding, the fact that the government could do these things at all indicated a capacity which would later bear more fruit.

After 1838 the main interest in explosions in the government centered in the Navy Department, which, beginning to use steam in warships, had a direct interest in finding effective safety devices. Since the Patent Office made no serious effort to test the utility of such inventions, the Navy hired Walter R. Johnson to examine for their actual effectiveness all the schemes proposed. As a professor of physics and chemistry in the medical department of the University of Pennsylvania, Johnson was able to apply scientific principles to his task. This activity, however, did not extend beyond evaluation of schemes worked out by others.[17] Again it was a sporadic response to a little understood need rather than a settled policy. Johnson later showed his ability in a more creative way by investigating for the government different varieties of coal, comparing their value in producing heat.[18] But the Bureau of Mines, which honors him as precursor, was far in the future.

The same sort of pressure and the lack of a comprehensive policy prevailed in the field of public health, which from time to time came into the range of the federal government because diseases, like explosions, are no respecters of state boundaries. The marine hospitals inherited from the earlier period now followed the lines of trade and the steamboat into the Mississippi valley. This medical service, extended to the West in 1830, called for eleven hospitals, but none opened before 1850, and the same lack of coördination and vigorous administration that had always characterized the system prevailed.[19] The immediate spur to this expansion was cholera, which, entering the United States by way of Canada and the Erie Canal in 1832, spread rapidly through the western river system. The Board of Health of the City of New York, pointing out the national character of the menace, requested that "Congress should, without delay, constitute a sanitary commission, whose duty it should be . . . [to go] to the part of Europe and Asia where the disease now prevails, and to collect facts relating to the means of preventing, and remedies applied to it." The House committee on foreign affairs thought they could get the information from consuls, and, considering this a commercial matter, passed the memorial on to the committee of commerce.[20] But in a period when state and municipal authorities were becoming genuinely concerned with public health, the federal government largely stood aside.

The Revived Coast Survey under Hassler

If on the whole technological change and the increasing calls on science for answers to practical problems found the government unprepared with either a basic policy or adequate machinery, exploration and surveying elicited a very different response. These activities were fundamentally in tune with the main task of the American people in this period — the conquest of a continent. The government had urgent need for geographical information in the broadest sense, and the science of the day had reached a stage that could supply it. From the earlier period clear precedents existed in the Lewis and Clark expedition and the Coast Survey. The Army already had experience and through West Point a group of trained officers. The real question was not whether the government could constitutionally undertake exploring and surveying on a large scale, but rather how well its institutions were adapted to carrying it on and to what extent it could see beyond immediate practical ends to general service to science.

The first agency to benefit from the basic congeniality of the exploring-surveying mission was the Coast Survey, personified as it had always been by the redoubtable Hassler. The aging Swiss, after twelve unhappy years out of the government service, had finally found useful work for the secretary of the treasury in examining the weights and measures used by the various customs houses. Hassler was qualified in this field because of his old teacher Tralles, who had participated in the final deliberations on the metric system in France and who had given Hassler a highly accurate standard meter. Uncovering wide discrepancies, Hassler undertook to provide a set of weights and measures for each customs house. A law of 1836 extended this project to a standard set for each state and territory as well. Thus, although the government was still without the policy for weights and measures so clearly suggested in the Constitution, the personal interest of Hassler instituted regular activity in the field. Appropriations before 1836 came from Treasury Department funds and after that from the Coast Survey. Hassler could style himself superintendent of weights and measures even though this task never took anything like his whole time.[21]

In the meantime the pressures working for a survey of the coast continued. The commercial interests needed charts ever more urgently. Hassler himself had never given up hope. The Navy, although it had conducted local surveys between 1818 and 1830, confessedly had no organized policy and correspondingly produced worthless results.[22] In 1832 Congress revived the Survey by authorizing under the act of 1807 the employment of anybody considered desirable. This was, of course, in line with the policies advocated by the Adams administration. Hostility came to the surface in a specific provision that "nothing in this act, or the act hereby revived, shall be construed to authorize the construction or maintenance of a permanent astronomical observatory."[23] An obvious slap at Adams and an advance prohibition of a transformation Hassler would undoubtedly attempt to make, this ill-natured proviso also revealed a distinction important in the minds of congressmen throughout the period. What they feared was not scientific activity as such or the appropriation of money to accomplish a particular end, but rather the creation of a permanent scientific bureaucracy which involved a long-term commitment of funds. The survey of the coast was a specific task which on completion would allow the people involved to disband.

With Hassler's return came his insistence that the Survey be a true contribution to science and not just a compiled map. He also brought his scheme for detailed triangulation from a few accurate base lines, the determination of which actually required astronomical observations of a high order. Also back were Hassler's ideas that a scientist should be paid in proportion to his exceptional abilities and that he should not be subject to dictation, either financial or military.

The attempt to move the Coast Survey to the Navy Department put to its severest test Hassler's ideal of the independence of science in the government service. The 1807 law had given authority directly to the President, who had assigned the secretary of the treasury as administrator. The reasoning which had effectually killed the Survey in 1818 still carried weight in that some people hoped for economy by letting the Navy do the work. In 1834 the switch actually took place, after Hassler had resisted the formation of an interdepartmental committee, under which "all would come under deliberation, with conflicting interests, and the work of course would be lost." [24] He bitterly complained that while both the Survey and weights and measures were mainly concerned with aiding commerce, the change to the Navy Department would put him under two different cabinet officers. After the transfer his accounts came under the scrutiny of Amos Kendall, a power in the Jackson administration and fourth auditor of the treasury, who held up payment because of what he considered irregularities.[25] Hassler was so indignant that he made himself troublesome even to his friends, such as President Jackson, in demanding investigations and redress. The only authority to which he willingly offered to submit was a board of scientists capable of understanding the scientific side of his work.[26] Jackson finally smoothed over the impasse in 1836, returning the Survey to the Treasury Department. Hassler combined this campaign with a successful appeal both to Congress and to the public to get his salary raised to $3000 plus $3000 for expenses.[27] He thus won his main contentions — civilian control and adequate pay for scientific talent — but, since his real demands were so extreme that they bordered on administrative irresponsibility, his troubles were bound to recur.

The Coast Survey was a pioneer in tackling the problem of getting and training scientific personnel. Although the principle of civilian control held for the general direction of the Survey and helped ensure scientifically respectable objectives, Hassler had to draw most of his assistants from the officer corps of the Army and Navy. Seeing the need for completing their education in the use of instruments and in the application of physics and mathematics to surveying problems, he provided books and as much instruction as possible. He tried, usually successfully, to keep the officers assigned to him long enough to give them a real education. Consequently the military services gradually became highly competent surveying organizations.[28] This function was especially important for the Navy, which had as yet no academy to compare with West Point.

Hassler's method was to begin at New York and work north and south. By 1841 the survey covered 11,000 square miles from Rhode Island to the Chesapeake Bay. Fearing that cheap imitators would use his findings, and desirous of getting as near perfection as possible, Hassler steadfastly refused to publish results as he went along, saving everything for a final report "according to good principles of science." [29] However correct in theory, this procedure was not the one to give the Survey a good reputation with the congressmen of a young republic whose economy was just then suffering violent fluctuations. Appropriations continued year after year, always increasing, without charts to show the merchants and cut maritime losses. The $20,000 of 1832 became the $100,000 of 1840 with no end in sight. Indeed, this sufferance came almost entirely because of an early and dramatic find, Lieutenant T. R. Gedney's discovery of a new channel into New York harbor. The release of this single bit of information gave the Survey a dollar value and a talking point.[30]

By 1841 Congress was asking serious questions. The House of Representatives undertook a full-scale investigation, and a serious effort to cut the appropriation began. Those who felt Hassler did not publish fast enough and those who thought he was getting too much money, or who wanted his office, joined the "clique in the Navy" seeking a cheap and easy survey by chronometer — the old logic of 1818.[31] In addition to restiveness at the time and money consumed, a note of resentment crept in because of the "mystery" element in science which Hassler used as a refuge. The select committee had difficulty in judging the quality of the survey work, sending a man who, according to Hassler, "measured the map with a foot rule, like an undertaker, to make a coffin for a dead body." [32] At one point Hassler's English and testimony were so confusing that the committee secured an interpreter.[33] The contrast between Hassler and the politicians with whom he had to deal was clear to John Quincy Adams, who introduced Hassler to John C. Spencer, the new secretary of the treasury in 1843. The old statesman reported that "Hassler, already restive under the yoke fitting to his neck, said that the work, being scientific, must be conducted on scientific principles. The Potentate answered in a subdued tone of voice, but with the trenchant stubbornness of authority, the laws must be obeyed." [34]

The committee finally recommended and the Congress adopted a reorganization in which the Survey kept its $100,000 appropriation, but all plans had to reach the President with the recommendation of a majority of a board consisting of the superintendent, his two principal assistants, two naval officers now in charge of surveying parties, and four topographical engineers.[35] Such a board was hardly a solution to the problem of how a government should keep a check on its scientific enterprises, but its creation emphasized that a link was necessary. The complexities of science made control by nonscientists difficult but did not relieve the despised politicians of their responsibilities as officers of the government. This yoke fitting to Hassler's neck did not weigh him down long, for, already over seventy, he became ill in the field in the summer of 1843 and soon died.

The last tie with the age of Jefferson, the personification of American scientific dependence on Europe for men as well as ideas, the bearer of the standard measure from the original source of the metric system, Hassler had fought and suffered for science in a hostile land. Because of his stubborn devotion the Coast Survey was a scientific enterprise which bore well the comparison he so often made with similar European endeavors. Though his sublime belief in his own mission prevented him from achieving a workable position within the framework of the government, his truculent stand for the integrity and independence of science contributed toward its elevation in a period when emphasis on these aspects was needed and overemphasis was pardonable. The men he trained both in and out of the services were part of the reason why no new Hassler was necessary to take over the Survey in 1843. Alexander Dallas Bache, completely different in background and approach, was the typical leader of science in the government from 1840 to 1865 as completely as Hassler had been earlier.

The United States Exploring Expedition

If the Coast Survey responded to the urgent need for scientific information about the shores best known to Americans, the demand for knowledge of distant regions was also intense. Both expansion westward across the continent and the spread of commerce to all the seas made important groups demand that the government provide them with expensive data on geography and natural resources. These were very practical needs, but their fulfillment had to come within the exploring tradition handed down from Cook and Bougainville by way of Lewis and Clark and Long. The object was no less practical for being the most complete picture of the geography, geodesy, geology, flora, and fauna available. Nor was this picture any the less fundamental to science for having commercial implications. In general, the exploring expedition had much greater affinity to basic science than it did to any form of technology.

Since John Quincy Adams had urged an exploring expedition to the northwest coast and had through Southard tried to accomplish it, the Jackson administration in this field as with the Coast Survey and Patent Office had to reconcile itself to its despised predecessor's proposal. The abortive Navy Department attempt to mount an expedition in 1828 had lined up an unlikely coalition of whale fishermen with the followers of an eccentric ex-army officer who thought the earth was a series of concentric hollow spheres open at the poles. Captain John Cleves Symmes had toured the country in the 1820's lecturing on his theory of the earth and trying to arouse interest in an expedition to either the North or South Pole which would, on reaching a high latitude, be able by gradual stages to pass over the verge into the interior of the earth. A government voyage to the Pacific could easily be diverted into a polar expedition that would validate Symmes's ideas. The most zealous of the captain's followers was the same Jeremiah N. Reynolds who collected whalers' logs in New England for Southard in 1828. He sensed the potential commercial backing for a voyage of exploration to the whole Pacific Ocean area and became its leading advocate.[36]

After the Adams expedition collapsed with the inauguration of

Jackson, Reynolds left the United States in a whaler for high southern latitudes. When he returned in 1834 aboard the U.S.S. *Potomac*, the exposure had cured him of Symmes's theory and shifted his interest somewhat from the Antarctic to the tropical Pacific, but he was now more than ever a zealot for an expedition. Voluntarily assuming the role of lobbyist, he marshaled petitions from the whaling states at the same time that he used his acquaintance among Ohioans for inland votes. He adroitly blended several themes to make a compelling practical and constitutional case. His emphasis on whaling and the new Latin American trade found a sure basis on the commerce clause. The charting of islands and hydrographic studies helped to gain the support of practical sailors both commercial and naval.[37]

While he did not hesitate to cite Lewis and Clark as a precedent for this commercial and military combination, Reynolds also appealed to national glory in another way. This was the decade of the *Beagle* surveys, the Antarctic expedition of Sir James Clark Ross, and the voyage of Dumont D'Urville. The rivalry here was nationalistic but also largely scientific, extending beyond gross geography to all the standard departments of research on an expedition — hydrography, magnetism, meteorology, and natural history. Prominent in the list of supporters for Reynolds's scheme were the great scientific names of the country, including Benjamin Silliman and Peter S. Du Ponceau, president of the American Philosophical Society.[38] For the United States to enter the Pacific, the classic ground of Cook and La Pérouse, was a mark of her growing scientific stature.

Significantly, action on the expedition came on the report of Samuel Southard, now back in public life as a Senator. But the years of quiet from 1829 to 1834 had allowed the old idea, in spite of the continuity of the interested backers, to appear as a new and nonpartisan enterprise. Andrew Jackson enthusiastically signed the bill authorizing the United States Exploring Expedition on May 14, 1836. All doubts about constitutionality were smothered under the appeal to aid for commerce. Besides the $150,000 directly appropriated, the President could "use means in the control of the Navy Department, not exceeding $150,000." [39] Although this method of depending in part on funds of the military establishment had been used before by Jefferson in the case of Lewis and Clark, the size of the enterprise and the appropriation was so unprecedented that people generally considered this the "first National Expedition." [40] For the executive to organize an exploration on this scale put great strain on American experience and ingenuity.

Three points of view almost immediately developed. The first was that of Reynolds, who, securing a position as the naval commander's secretary, wished to head the scientific corps. He rounded up some of the best of the young scientists who were beginning to appear in the country, including Charles Pickering, James Dwight Dana, and Asa Gray.[41] These men had in common a competence in particular branches of science superior to that of the older generation of naturalists, coupled with a lack of secure employment in private institutions. Reynolds had large ideas about the size of the fleet and the amount of equipment needed, arguing for at least one frigate, and he wished the civilian scientific corps to do all the important work, leaving the Navy only the task of transportation. The naval officers on the other hand naturally wished to control the entire expedition, including the science, but found themselves hampered by the lack of training which a service academy such as West Point might provide. They also suffered from jealousies over rank in the near-stagnant officer corps.[42] The third point of view was that of the secretary of the navy, Mahlon Dickerson, who by dissipating the ships and equipment of the expedition continually disrupted outfitting. Applying the spoils system to the selection of personnel, he mixed with Reynolds's qualified candidates several political appointees. At one time the scientific corps numbered thirty-two. In general his actions gave the impression of outright hostility to the very existence of the expedition.[43]

At first Reynolds had his way with the appointment of his scientists and the fitting out of the frigate *Macedonian*. He hoped for a time to send part of the scientific corps to Europe to buy books and instruments. One of them wrote enthusiastically that "with a cabin in a firm frigate fitted up for our especial accommodation, with a better collection of books and instruments than any we shall leave behind us in this country . . . we may . . . hope for a pleasant time." [44] But Reynolds, being untrained, had trouble maintaining a position of authority over the scientists, and together they clashed with the secretary. The purchases in Europe turned into a Navy enterprise, with Lieutenant Charles Wilkes as agent. Dickerson also delayed calling the scientific corps together, refused to put them on the payroll, and allowed the enlistment of the crew to lag.

When Martin Van Buren became President on March 4, 1837, he

continued Dickerson in office. Although Reynolds had hopes from the new administration for a little while, he and Dickerson soon began to blast each other in New York newspaper articles signed with only slightly disguised pseudonyms.[45] The summer and fall of 1837 saw no improvement, and as the sailing date receded indefinitely the expedition became a symbol of folly, mismanagement, and extravagance. One commodore reputedly said to his gunner, "If you have any gold guns on charge, send them to the Exploring Expedition." [46] In November 1837, Captain Thomas ap Catesby Jones fell ill and resigned the command, which Dickerson was unable to fill as one officer after another turned it down. With the appropriation nearly gone, the expedition seemed doomed. So far the government had proved unequal to the job of mounting an exploring enterprise that would do it credit.

Early in 1838 a convenient breakdown in his health removed Dickerson from the scene, allowing the ruins of the expedition to fall into the control of Joel Roberts Poinsett, Van Buren's secretary of war. As a genteel naturalist with a European education and at home in the cultural circles of Charleston, Poinsett fully appreciated the scientific goals of the expedition and felt no need of Reynolds's advice. While following Dickerson's lead in cutting the number of ships and civilian specialists, he still tried to keep the real scientists, weeding out the political appointees. His experience with the Topographical Engineers and the Army exploring tradition led him to appreciate scientific attainment in the services. Thus he assigned to the Navy the hydrography, mapping, and magnetic and astronomical observation. Making science the prime qualification in choosing a commander, he went far down the list of lieutenants to select Wilkes.[47] Then to placate William L. Hudson, whom he wished for second in command but who stood some numbers higher than Wilkes, Poinsett issued a general order stating that "the objects . . . [of the expedition] being altogether scientific and useful, intended for the benefit equally of the United States, and all commercial nations of the world, it is considered to be entirely divested of all military character." [48] Although precipitated by a specific and even trivial consideration, this general order throws much light on the balance of aims of the expedition. It was to be above all for science — in Poinsett's words, "to extend the bounds of human knowledge" — and so clearly was the military aspect depreciated that in case of war before the return of the squadron "its path upon the ocean will be peaceful, and its pursuits respected by all belligerents." [49] Although not an explicit admission that the federal government can constitutionally support science, this general order shows the change in emphasis since Jefferson's stratagem for the approval of Lewis and Clark.

Poinsett now had a winning combination. Charles Pickering wrote that he "had gone over to the enemy . . . I had the honour of dining with our new Commodore about a week ago, and think many of his ideas not bad, and though rather hasty in his conclusions and in danger of running against a few posts . . . he will bring up about right." [50] Wilkes fulfilled this description, and with great energy had his flotilla at sea in August 1838, to be gone four years. Having surmounted the difficulties of getting started, he would be equal to his task at sea, and in spite of all would redeem the honor of the nation. Reynolds, a sacrifice to harmony, remained fuming on the beach. A workable if not perfect organizational form had emerged from the two years and more of quarreling and confusion, and exploration had gained a secure place as an activity of the government.

The results of the expedition came from Latin America, the Antarctic, the Central Pacific Islands, and the western coast of America. It touched the sciences of ethnology, anthropology, zoology in all its major branches, geology, meteorology, botany, hydrography, and physics. In addition, the surveys resulted in large numbers of charts,[51] some of which were not the worst used by the Navy at Tarawa in 1943.[52] The collections in sheer bulk were the largest scientific treasure in the country, fully worthy of the struggle that later took place over their control. The total cost ran to $928,183.62, some three times the specific original appropriation.

Since the actual voyage is such a small part of exploration as a scientific investigation, the story of the Wilkes expedition after its return in 1842 significantly shows the attempt of the government to find proper means to accomplish the vital clean-up chores. The very attempt to solve the problem that had overwhelmed Lewis and Clark was a marked advance. After a confused struggle, which is to be recounted in the story of the National Institute,[53] the responsibility for publication fell to Congress's joint committee on the library. Thus the greatest scientific publishing program undertaken by the government before the Civil War was directly under Congress and provides a test of the efficiency of the legislative branch in the detailed administration of scientific affairs. A pale reminder of this experiment is the

United States Botanic Garden, founded with the seeds and live plants brought home by Wilkes, and still under congressional control.[54]

Wilkes himself administered the publication program for the committee.[55] Enlisting an able group of specialists to work on the various volumes, he tied up a measurable portion of the available scientific manpower in America for thirty years. The design finally grew to a series of some twenty-four great volumes. Many decisions, petty in themselves, had to be made which vitally affected the usefulness of the work for science. How many copies should be printed? Should any part be elaborated by foreigners? Should systematic treatments include any material not actually collected on the expedition? Should botanists he required to make their page format conform to the style of the zoologists? Each of these questions and many more were the subject of acidulous and sometimes angry debate among the committee, Wilkes, and the scientists. Unquestionably, the final result embodied many mistaken decisions. Especially the limitation of the official printing to one hundred copies permanently hampered the usefulness of the whole effort. The volumes that remained unpublished, especially those of Gray and Agassiz, were an unrequited loss. The appropriations for publication, continuing year after year through the 1840's and 1850's, had reached a total of $329,578.21 by 1856,[56] when even the members of Congress most closely associated with the work heartily wished to transfer it to some agency better qualified than the committee. Thus Congress proved to itself and to the scientific community that its committees were not the proper bodies in the government to supervise detailed and complex scientific undertakings. But the experience gained here served the country well. By the time of the railroad surveys of the middle 1850's the organization both of expeditions and of the following scientific work was reasonably smooth.

The Beginnings of Permanent Agencies

The United States Exploring Expedition, although in itself an *ad hoc* organization, immediately began by its needs to put pressure on the government to establish permanent institutions. When Wilkes was about to sail, orders from the Navy Department requested Lieutenant James M. Gilliss of the Depot of Charts and Instruments and William Cranch Bond of Dorchester, Massachusetts, to make astronomical and magnetic observations at home for comparison with those made on the expedition.[57] Bond went on to become director of the

Harvard Observatory, while Gilliss had by 1842 virtually converted an obscure navy storehouse into an embodiment of the most hated of John Quincy Adams's proposals.

The Naval Observatory is the classic example of the surreptitious creation of a scientific institution by underlings in the executive branch of the government in the very shadow of congressional disapproval. No more hated proposal existed, and nowhere had more pains been taken to prevent the creation of a new agency. Yet despite this vigilance the forces that required an observatory gained their ends. Before 1830 each ship of the Navy obtained charts, chronometers, and instruments individually, with no tests before purchase and no responsibility for what became of them at the end of a cruise. In that year the secretary, on the advice of interested officers and the Board of Navy Commissioners, issued an order setting up a Depot of Charts and Instruments which would take care of all nautical instruments, books, and charts when they were not in actual use.

Lieutenant L. M. Goldsborough, the first to be in charge, at a very early date mounted a transit instrument for the purpose of determining accurate time for the rating of chronometers. His successor in 1833 was Wilkes, who got most of his scientific background while in this job. In 1834 he built an observatory sixteen feet square on the Capitol grounds and mounted one of the transits Hassler had brought back from England in 1815. Thus the practical need for taking care of chronometers opened the way to astronomical observations, just as questions concerning the accuracy of the charts led to a certain interest in hydrography. When Gilliss took over from Wilkes in 1836, he found his opportunity broadened by the expedition, which needed observations in the United States of culminations of the moon and stars, eclipses, falling stars, and meteorological and magnetic phenomena. An intense young man who wished to prove that naval officers could accomplish good work in science, he labored hard under many difficulties and produced very accurate results.[58]

By 1841 Gilliss felt bold enough to agitate for an adequate building. After he convinced the House committee of the need for studies in hydrography, astronomy, magnetism, and meteorology for the practical business of the Navy, an act of August 31, 1842 authorized $25,000 for a building. Although the name was still the Depot of Charts and Instruments, Gilliss clearly had more in mind, as the architecture of the new building made obvious.[59] It even mollified John

Quincy Adams, who was "delighted that an astronomical observatory — not perhaps so great as it should have been — had been smuggled into the number of institutions of the country." [60] Although Gilliss was detached in the hour of his victory and his successor chose to stress hydrography and meteorology over astronomy, a permanent organization had arisen without benefit of specific legislative blessing.

While the Wilkes expedition was the most dramatic single enterprise, the area that most persistently attracted governmental efforts after 1830 was the trans-Mississippi West. The drive of Americans into the region and the fact that the unsettled public domain was a federal rather than a state responsibility challenged the government to take affirmative action. In the area between the Mississippi and the Great Plains the action was incidental to some other function. It was also the territorial counterpart of state activity in the burgeoning geologic and natural history surveys. When H. R. Schoolcraft went to the Lake Superior region in 1831 as a part of an effort to make peace between two Indian tribes, he turned the journey into a kind of natural history expedition.[61] David Dale Owen undertook a survey of mineral lands in Illinois, Iowa, and Wisconsin for the federal land commissioner for the Treasury Department.[62]

The Army, however, had the largest responsibility west of the Mississippi. Even in the 1830's it carried on the tradition of Lewis and Clark and the Long expedition by giving general aid and comfort and some positive support to scientific exploration.[63] Officers such as Dr. M. K. Leavenworth who were stationed at frontier posts contributed their collections.[64] Dr. G. W. Featherstonhaugh's visits to the Ozarks and the St. Peter's River region on the Missouri were slightly different in that he was a civilian specifically sent by the Army to make geologic surveys. His importance lies not in results — described as "worthless rubbish" by a later geologist [65] — but in the fact that the Topographical Engineers who sent him out were even then in the process of becoming a kind of scientific corps for the Army with special training and aptitude for exploration.

Behind the Topographical Engineers was West Point and the Army's extensive experience in helping private enterprise with engineering problems. The practice of lending army engineers to railroads for surveying reached a peak in the early 1830's,[66] as did the river and harbor improvements undertaken by the Army. In 1831 the Topographical Engineers became a separate department,[67] and its head, J.

J. Abert, moved into construction as well as surveying duties. The result was more engineering work than the Army could handle. Presidents Jackson and Van Buren both asked for a reorganization, which finally came in 1838 under the strong guiding hand of Poinsett. All fortifications for defense went to the Corps of Engineers while the Topographical Bureau got civil improvements.[68] This clarified its status, and in authority over road surveying and mapping it had the key to the trans-Mississippi West. It is too much to say that the Topographical Bureau actually directed all the Army's western explorations. With the repeal of the general survey act in 1838 no overall plan for surveying the routes of the country — west or east — existed. Rather, the Topographical Engineers served as a pool of trained officers from which individual expeditions could draw their leaders.

The first survey after the reorganization was that of Joseph N. Nicollet to the region between the Mississippi and Missouri Rivers.[69] A Frenchman who had been working privately on a physical geography of the Mississippi valley, Nicollet is a transitional figure, a civilian who took over as commander of an expedition. His assistant was also cut from an unusual pattern. Lieutenant John Charles Frémont, who had received his commission and his assignment because of a friendship with Poinsett, learned mapmaking and exploring from Nicollet. By the time the 1840's opened, the Army was ready to take up the heavy demands which that stirring decade placed upon it.

The scientific activities of the 1830's must not be discounted simply because they all sprang from definite needs and because the subject of the ultimate nature of scientific institutions in the government was usually avoided. All advocates of scientific enterprise learned to side-step constitutional issues and discussions on the ultimate responsibility of the government to aid science. The really striking thing is that, given these limitations, so many activities sprang up and so much was accomplished. Except for a national university, John Quincy Adams's program approached realization.

Certain general trends emerge from these activities so consciously kept piecemeal. First, practical problems tended to reach out to ever-widening circles of theoretical considerations. Chronometers and mapping led to celestial observations, which led to astronomy. Compasses led to terrestrial magnetism. Natural resources led to geology and natural history. Second, and relatedly, *ad hoc* organization tended to become permanent or require permanent services. The Coast Survey and the Wilkes expedition had set off chains of scientific demands the end of which, much to the consternation of Congress, was not even remotely in sight in the early 1840's. Third, although technological change was beginning to create a sporadic need for government science, it was surveying and exploring that set the tone of activity in a period of rapid geographical and commercial expansion. The Coast Survey, the Wilkes expedition, and the Topographical Engineers were much better staffed, more continuously supported, and closer to the best elements in American science than were the projects of Morse, Walter Johnson, and Charles G. Page. Fourth, in spite of the great forces working, the course of development measurably felt the imprint of men and their visions. Why the Coast Survey and weights and measures go together is incomprehensible if Hassler is left out of account. Why a Depot of Charts and Instruments became an observatory is a mystery without Gilliss. Even poor Reynolds kept alive the idea of an exploring expedition through bitter years. Poinsett left the imprint of his large vision on many activities he touched. Finally, the general supply of scientific men and behind them the level of science in America, were rising, an increase the government was stimulating as well as using. This increase meant new opportunities for both American science and the government.

IV

THE FULFILLMENT OF
SMITHSON'S WILL

1829–1861

FUNDAMENTAL discussion of the nature of scientific institutions and the relation of the government to them was no more really dead in the 1830's than was the rejected John Quincy Adams himself. It was revived by an Englishman of whom virtually no one in America had ever heard. James Smithson, a British chemist who had spent much of his life in France and whose considerable inherited wealth from noble parents contrasted with his illegitimate birth, died in 1829.[1] Since he was a bachelor and had no very obvious heir, his leaving a fortune to some scientific institution was by no means unnatural. But among the legal provisions of his will appeared this one: "In the case of the death of my third nephew . . . I then bequeath the whole of my property . . . to the United States of America, to found at Washington, under the name of the Smithsonian Institution, an Establishment for the increase and diffusion of knowledge among men." [2] In one sense this bequest was clear and complete. The recipient, the location, the exact name, and the purpose of the projected institution were stated unequivocally. Yet the highest officers of the government furrowed their brows for years over each phrase. The answers depended not so much on logic as on Americans' understanding of their own institutions. Was the United States of America the people, the government, the Congress, or the executive? Could it take title to a bequest or be a trustee? Was Washington to be simply the physical headquarters of the Institution, or was this to be the local society for the coming metropolis of the country? How did one go about increasing and diffusing knowledge? Which was more important? What is knowledge, anyway? Science, or science and something else? And does the "among men" make the establishment universal?

The answers to these questions were sought in the intentions of the founder in vain. Perhaps the copy of a travel book by Isaac Weld in his library gave him the idea that Washington would become a great center of culture. Perhaps he knew Joel Barlow in Paris, catching from him the dream of a national university. Perhaps he had republican sympathies. Perhaps his illegitimate birth and a snub in the Royal Society gave him a motive for posthumous vengeance.[3] But these were at best largely speculations. John Quincy Adams had returned to public councils as a member of the House of Representatives, chastened by his presidency and convinced that national scientific institutions could only come "in an after-age . . . when the sciences shall be more ardently cultivated than they are . . . at the present time." [4] But the Smithson bequest quickened his old desire. He saw in it "the finger of Providence, compassing great events by imcomprehensible means," an explanation no one has improved upon.

The Acceptance of the Bequest

In 1836, after word had seeped through diplomatic channels that the last of Smithson's heirs had died and the United States had inherited the equivalent of $500,000, the first test came in Congress on the question of accepting the bequest. The old arguments about the constitutionality of the national university came up again. John C. Calhoun led the opposition in the Senate, claiming that "acting under this legacy would be as much the establishment of a national university as if they appropriated money for the purpose; and he would indeed much rather appropriate the money, for he thought it was beneath the dignity of the United States to receive presents of this kind from anyone." [5] Calhoun's fellow South Carolinian W. C. Preston, who later talked in a different vein, put this austere doctrine somewhat more racily when he claimed that "every whippersnapper vagabond that has been traducing our country might think proper to have his name distinguished in the same way." [6] Against the clear states' rights position that this was a national university which the government could not constitutionally establish, two main lines of argument appeared. The committee report claimed that no question of a national university was involved because the government was simply acting as a trustee for the District of Columbia.[7] But the faithful Southard, Robert J. Walker of Mississippi, and John Davis of Massachusetts went farther to argue that Congress did have the power

to found a national university and called "the establishment of institutions for the diffusion of knowledge a vital principle of republican
government." [8] Not often directly expressed but oviously present was
the fact that the taxpayer had to take no loss. The final vote on accepting the bequest, 31 yeas and 7 nays, measures the strength of the
pure states' rights argument on one side and a congeries of motives
on the other. With the House concurring, Jackson sent Richard Rush
— another shade from the Adams administration — to England to
claim the legacy.

A common assumption during the first debates was that the
Smithsonian Institution would be, in fact, the national university.
That the donor had had in mind some kind of school as the proper
means of increasing and especially diffusing knowledge continued to
dominate discussion when in 1838 Rush returned to the United States
with the legacy in the very tangible form of gold, as well as with
James Smithson's library and collections of minerals.[9] President Van
Buren instructed the secretary of state to "apply to persons versed in
science and familiar with the subject of publication for their views." [10]
The list to which the secretary sent the letter notably includes the
heirs of the Jeffersonian tradition in Thomas Cooper and Albert Gallatin, and the variant Adams tradition in John Quincy Adams himself
and Richard Rush. Conspicuously absent was any professional scientist such as Silliman.[11] The replies of Cooper and of President Francis
Wayland of Brown University developed in great detail schemes of
education differing only in degree from the contemporary college.
Wayland wished the classics to remain the center of the curriculum,
while Cooper wanted "No Latin or Greek; no mere literature . . .
Things, not words." [12] Others not invited by the secretary made haste
to proffer their pet educational projects for the munificent aid of the
windfall bequest. The president of Columbia College in the District
of Columbia naturally saw it as his salvation. Horatio Hubbell of
Philadelphia knew something of German universities and wished to
recreate one in America.[13] Charles Lewis Fleischmann, "Graduate of
the Royal Agricultural School of Bavaria, and a citizen of the United
States," wished studies in agronomy, chemical and mechanical agriculture, vegetable and animal productions, and agricultural economy
taught to boys of ordinary education at least fourteen years old.[14]

The main legislative proponent of a university was Senator Asher
Robbins of Rhode Island, who introduced a series of resolutions that

"the Smithsonian Institution should be a scientific and literary institution." Calhoun naturally opposed the whole business, and when Jacksonians such as Thomas Hart Benton joined him because of the analogy of a national university to a national bank, the proposal met defeat by a vote of 20 to 15.[15]

Even at first some people saw a different role from that of education for the fund. Richard Rush wanted a building in Washington with grounds for seeds and plants, and lectures in physical and moral science with their publication as a kind of prize.[16] More specifically, Professor Walter R. Johnson, who tested coal and safety devices for the Navy, wanted "an institution for experiments in physical sciences," pointing to the Royal Institution and, in France, the Polytechnic School and the School of Mines.[17] James F. Espy of Philadelphia, whose interest in weather led John Quincy Adams to call him the "storm-breeder," sought part of the fund for "simultaneous meteorological observations all over the Union." [18]

But the two men with the largest and most statesmanlike views and the greatest power to accomplish their ends were Adams and Joel Poinsett. Each in his different way sought to make the Smithson bequest do more than poorly endow another college. Although neither entirely saw his design fulfilled, their efforts did much to change the course of discussion.

Adams from the beginning, besides insisting upon the preservation of the principal and the use of the interest only, stood firmly against "the endowment of any school, college, university, or ecclesiastical establishment." He considered the "education of youth . . . a sacred obligation, binding upon the people of the Union themselves, at their own expense and charge, and for which it would be unworthy of them to accept an eleemosynary donation from any foreigner whomsoever." [19] The finger of Providence seemed to him to point directly toward using the income for several years for his cherished astronomical observatory. Carefully investigating the organization of the Greenwich Observatory, he used his position as chairman of the House committee on the bequest to forward his scheme.[20]

The confusion of voices on what to do with the money is clear simply from the array of Adams's real and fancied adversaries. Everywhere, of course, was the remembered aversion to his policy as President. One politician thought an observatory impossible because, as Adams testily put it, "I had once called observatories light-houses *in* the skies. My words were light-houses *of* the skies." [21] Van Buren seemed "ostensibly neither to favor or oppose it, but . . . he will underhandedly defeat it, taking care to incur no personal responsibility for its failure." [22] Among the advocates of a "school of education for children" [23] he saw many foes, such as the "English atheist Cooper, a man whose very breath is pestilential to every good purpose." And on all sides he saw the danger of sinecures, "jobbing for favorites," [24] and "monkish stalls for lazy idlers." [25] Espy was seeking a comfortable salary. Asher Robbins wanted to become Rector Magnificus. Even Hassler, who wished to use the fund for an astronomical school before the erection of an observatory, seemed to Adams to covet a position as the head, "and would continue to absorb the whole fund in the management of it." [26] Against all the old ex-President saw "the finger of John C. Calhoun and of nullification" opposing "the establishment of the Institution in any form." [27] As late as 1842 little sign appeared that opinion had jelled in any one mold.

The National Institute for the Promotion of Science

Meanwhile the other statesmen who had a better than common grasp of scientific affairs approached the use of the fund from a vantage point different from Adams's. Joel R. Poinsett as secretary of war had organized the Wilkes expedition and the Topographical Engineers and had served widely as a diplomat with an eye for natural curiosities (witness *Poinsettia*). Hence he saw the need of the museum as an essential part of the exploring enterprises he had done so much to launch. As early as December 1838, he was looking forward to the collections to be sent home by the Wilkes expedition, and also had his eye on the Smithson fund. For the time being he agreed with Adams on the use of the income for an observatory, which was also an auxiliary for exploration, but he went ahead with plans for a more comprehensive organization.

In 1840 some eight people gathered at Poinsett's house and formed a National Institution for the Promotion of Science.[28] Among those present were Colonel J. J. Abert of the Topographical Engineers, Colonel Joseph G. Totten of the Corps of Engineers, and Francis Markoe, a clerk in the State Department.[29] A short time later the number of resident members had reached 84 and the corresponding members 90.[30] The imminence of the arrival of the first advance collections from the Wilkes expedition made some such organization almost necessary, and within a few months Poinsett was publicly insisting on a connection with the Smithson fund.

But the National Institution was something else in addition. It was the heir of the Columbian Institute and the old desire to give Washington a scientific society worthy of the capital. As Poinsett put it, "the lovers of science, literature, and the fine arts, residing in this district . . . were mortified to perceive that . . . at the seat of the government . . . there existed fewer means than in any other city of the Union of prosecuting those studies, which, while they impart dignity and enjoyment to existence, lead to the most useful practical results." [31] Members of the Columbian Institute received an invitation to join, and its records were turned over to the new organization. Dr. Edward Cutbush rather ruefully hoped that the objects he and Thomas Law had sought back in 1816 "will *now* meet the approbation and support of the Government, and of the scientific men of the District of Columbia." [32]

The ninety or so members of the National Institution were mainly representative citizens of Washington — congressmen, clergymen, scientific men, and others. Those living outside Washington were corresponding members, with the management in the hands of a president, a vice-president, a treasurer, a corresponding and a recording secretary, and twelve directors. The members of the cabinet of the President of the United States automatically became directors. The membership was divided into eight classes: (1) geography, astronomy, and natural philosophy; (2) natural history; (3) geology and mineralogy; (4) chemistry; (5) the application of science to the useful arts; (6) agriculture; (7) American history and antiquities; (8) literature and the fine arts.[33] In short, Poinsett was attempting to cope with a new and rapidly changing situation by creating an organization almost completely similar to the private societies of the period of the early republic. He was trying to create another American Philosophical Society for Washington just when the amateurs of Philadelphia, Boston, and New York were proving unequal to the task of supporting American Science in an era of increasing specialization and professionalization.

Like all local societies of a country's capital, however, the National Institution had an additional dimension in that it hoped to include, at least indirectly, the whole nation. These aspirations were the reason and the hope that "the government might extend its patronizing hand . . . in erecting a temple to National fame." [34] When incorporating it in 1842, Congress had a chance to discuss the national character of the new organization. On this occasion the name was changed to National Institute to assert its superiority over the proposed Smithsonian Institution. When the inevitable cry arose that Congress had "no constitutional power to incorporate an institution of this kind," the advocates of the measure hastened to give assurances that it "was only intended to operate in the District of Columbia" and that the national name was "a mere matter of taste and fancy." [35] But the Institute usually acted as if it represented the nation and in return deserved support from the general government.

The National Institute based its activity on a drive to get collections. David Dale Owen sent in his geological specimens. A circular letter to consuls and army and navy officers brought curiosities of all sorts, from animal skins to hieroglyphics.[36] The control of the Wilkes expedition collections was the real test of whether the National Institute might become a national museum and headquarters for the follow-up activities of exploration. Success in administering this public scientific property would in effect make the Institute an official agency of the government and give it a commanding claim either on the Smithson bequest or on large appropriations.

In 1841 the Institute made progress. By an agreement with the secretary of the Navy it was given the care of the collections, and Congress appropriated $5000 for the purpose. When the first advance boxes arrived in Philadelphia, an order from the secretary had them transferred to Washington, and the new Patent Office building provided space. The Reverend Henry King, who had done some geology in Missouri, became curator, undertaking the unwelcome but nevertheless important chore of unpacking the boxes.[37]

Almost immediately, in spite of the promising beginning, two fatal internal flaws appeared in the Institute. First, Poinsett left the War Department with the defeat of Van Buren and the overturn of the Democratic party, retiring to South Carolina early in 1841. Thereafter, although he retained the presidency, the Institute lacked both its ablest administrator and its shield and sword in the political arena. Colonel Abert and Francis Markoe, who remained as leading spirits of less than outstanding vision, were handicapped by their subordinate positions in the executive branch of the government. The second flaw was the lack of scientifically competent people to conduct the museum operations. The Reverend Henry King turned out to be one of the unsung villains of the history of science in America. He dried pickled specimens, ran them through with pins, made references to catalogues impossible by taking off labels, and allowed "a general scramble for curiosities by irresponsible members of the Society." [38] In addition, he antagonized his friends in the Institute, lacking, according to Abert, the judgment and intellect necessary to sustain himself in a position of consequence. By dabbling in politics he managed to get John Quincy Adams dropped from the board of directors, thus for a time alienating the Institute's most powerful potential friend in the House of Representatives.[39] None of the younger generation of competent scientists of the country had any say in the management of the collections before untold damage had occurred.

Just after the Wilkes expedition itself returned home in 1842, Congress appropriated $20,000 for the care and preservation of specimens that came to the National Institute. But on August 26 the act providing for the publication of results specified that the "objects of natural history . . . shall be deposited and arranged in the upper room of the Patent Office, under the care of such person as may be appointed by the Joint Committee on the Library." [40] This meant that custody shifted from the executive to the Congress, and the informal agreement with the secretary of the Navy no longer gave the Institute any legal control.

For a time a new working arrangement seemed possible when King left, to be replaced as curator by Charles Pickering, who had done good work on the expedition. The joint library committee also appointed Pickering, allowing coöperation to continue. But a coalition against the Institute was in the making as, quite probably for different reasons, three men in authority questioned the right of the National Institute to government property. Henry L. Ellsworth, the commissioner of patents, with his own ambitions in the direction of a National Gallery, questioned the use of the hall in the Patent Office for any collections belonging directly to the Institute.[41] Wilkes himself refused to recognize their legal right to the collections of the expedition.

The most formidable opponent of the Institute emerged, however, when Senator Benjamin Tappan of Ohio became chairman of the joint committee on the library. When the friends of the Institute tried to get special legislation giving them title to the collections, Tappan not only handled their efforts very roughly but raised the issue of the responsibility for government property if it were let out to a "private corporation" — an epithet which in the 1840's raised all the passions of the war on the United States Bank.[42] In the summer of 1843, Pickering resigning, the joint committee made Ellsworth the custodian of the government collection and put Wilkes in special charge of both the expedition's collections and their publication. This move launched the ambitious attempt of Congress to administer a scientific enterprise directly, and the organization of the museum function in the government followed these lines for several years. When the National Institute lost control of the Wilkes expedition collections, it had lost its chance to be a true national museum.

The failure of the National Institute was not at once apparent. Resident membership reached 232 and the list of corresponding members went up to the large total of 1148, but a special category labeled "Paying corresponding members" numbered only 40.[43] In 1844, along with an attempt to get an appropriation, the Institute held a great national meeting with the coöperation of the American Philosophical Society, the Association of American Geologists and Naturalists, and others. In this great effort a number of respectable scientists gathered in Washington to give papers and hear encomiums of the Institute from Adams, Richard Rush, President Tyler, and other politicians who could lose nothing by oratory.[44] But when Congress adjourned without an appropriation and Poinsett declined the presidency, it was clear, in spite of further efforts and temporary revivals, that the National Institute was without a sound foundation for existence.

The appeals of 1843 and 1844, which had yielded so little in money and political influence, had been much more successful in bringing in a flood of donations of all kinds. An attempt to show that the collections were not "of very trifling extent and value" only revealed them in very poor condition. Members themselves pointed out that private munificence in the form of dues (even if they were paid) could not cope with the flood. By November 1846 one thousand boxes and barrels were on hand unopened. Much of this material, stowed in an open passageway in the Patent Office, either disappeared or was stolen.[45] Not the least legacy which the National Institute left to its successors was a horrendous example of how not to run a museum.

The influence of the National Institute in its effort to become a general agent for the government in scientific affairs was both positive and negative. It stressed the museum function at a time when explorations were making that the greatest single activity of the government. It gained for the cause of science at least lip service from most of the high officers of both the executive and Congress. By the device of using the President's cabinet as part of the board of directors, it made a start toward bridging the gap between the government and a private society of the model of the American Philosophical Society.[46] This was the only organizational novelty of importance. The National Institute did not provide a pattern for either the Smithsonian or the American Association for the Advancement of Science, but in a diffuse way its aspirations to national significance acquainted many people with the idea of national action in scientific matters.

The negative side of the Institute's record was the general inadequacy of the whole organizational concept for the heavy tasks that the government's burgeoning scientific activity on the practical level placed upon it. The local base of authority in the District of Columbia was too narrow both in membership and in law to make good a claim as a truly national organization. More seriously, the Institute was a group of amateurs in an age when professional competence in science was becoming both possible and necessary. Even though the names of most of the new generation of professionals appeared in the list of corresponding members, neither the management of policy nor the care of collections was in their hands.

Most serious of all was the Institute's ambiguous relation to the government, a cloud not dispelled by the presence of cabinet officers and members of Congress in its counsels. For behind all the disparaging remarks about the "trash" collected by the Institute, which seemed to reveal such critics as Ellsworth and Tappan as unappreciative of science, lay the real question of the democratic control of the government's growing stake in science. After all, the National Institute was a private, closed, self-perpetuating corporation with no responsibility to the people and with no effective control by their elected representatives. Appeals to Poinsett's democratic principles and the love of science that motivated such lesser members as Francis Markoe did not alter this objection. And in the mistakes of the period of King's curatorship there is real evidence that the public's interest in the scientific usefulness of the Wilkes expedition collections suffered severely. A solution that safeguarded the people's legitimate stake, while recognizing the hard complexities imposed by science, was yet to be found.

The Creation of the Smithsonian Institution

About the same time the National Institute was making its great effort of 1844, Congress again turned to the Smithson bequest, which had become something of a scandal both because of the years of delay and because of the depreciation of the state bonds in which the Treasury Department had had the bad judgment to invest the money. The Senate began action in June 1844, when Tappan of Ohio, the foe of the National Institute, reported out a bill which reflected the decline of most of the earlier proposals. Although silent on many important features, Tappan's bill did provide an emphasis on useful sciences, which seemed to mean agricultural experiments, plus the usual sciences of natural history, chemistry, geology, and astronomy. The creation of professorships might have meant either some kind of instruction or a research and lecture staff. The board of managers was to be elected by joint resolution of Congress. The museum function entered with the provision that "all objects of natural history belonging to the United States which may be in the city of Washington, in whosesoever custody the same may be, shall be delivered to . . . the board of managers." A significant advance over the National Institute was provision for a superintendent, who was to be also the secretary of the board.[47]

The debate on the Tappan bill revealed a many-cornered fight on somewhat other lines than the discussions of the late 1830's. The observatory idea was now dead and the pure university no longer prominent. The National Institute made a last effort to get control of the fund when Levi Woodbury, who had succeeded Poinsett as president, offered an amendment to vest control in its officers. In reply Rufus Choate of Massachusetts, who now emerged as one of the principal figures interested in the question, directly attacked the Institute as antidemocratic and antirepublican, placing the Smithson trust in the hands of "a close body . . . wholly irresponsible to either Congress or the people." [48] The best that Woodbury could settle for was to have two places on the board of managers saved for members of the Institute, who would also be residents of the District of Columbia.

Choate's appearance introduced the new element, for he led a group passionately in favor of a "grand and noble public library," one that "for variety and extent, and wealth, shall be . . . equal to any now in the world." [49] This appealed especially to those devoted to the classics and to those who stressed the moral and political sciences as proper objects for the Smithson bequest. A note of condescension toward the natural sciences also sometimes entered, as when G. P. Marsh, leader of the library faction in the House, said, "Sir, a laboratory is a charnel house, chemical decomposition begins with death, and experiments are but the dry bones of science. It is the thoughtful meditation alone of minds trained and disciplined in far other halls than can clothe these with flesh, and blood, and sinews, and breathe into them the breath of life." [50] The library adherents in the Senate proved stronger than Tappan, forcing his bill back into committee. It emerged with a provision that $20,000 of the approximately $30,000 annual income be spent for books. The board of managers also had a sharper definition, consisting of the vice-president, the chief justice, three members of the Senate, three members of the House, and seven other people, of whom two were to be members of the National Institute and the rest from different states. Except that the individual seats were reduced to six, this arrangement was destined finally to prevail.[51] Cumbersome as it was, it recognized directly the problem of control by the government, including officers of all three branches. Congress, having the choice both of its own and of the public members, tended to be dominant. The Tappan bill, thus altered by the rise of the library group, passed the Senate early in 1845.

In the House of Representatives John Quincy Adams, standing alone for an observatory and in part soothed by the development of one within the Navy,[52] became obsessed with the project of saving the investment of the principal from the "fangs of the rattlesnake" — the state bonds — leaving leadership on the bill itself to Robert Dale Owen. Son of Robert Owen the Utopian of New Lanark and New Harmony, and brother of the geologist David Dale Owen, this free-thinking representative from Indiana had a very different approach to the whole problem.[53] A veteran of movements for the education of working men by cheap tracts and popular lectures, he saw the true function of the Smithson bequest as direct diffusion on the broadest scale. "The People govern in America. Ere long the people will govern throughout the habitable earth. And they are coming into power in an age when questions of mighty import rise up for their decision. They who govern should be wise. They who govern should be educated." [54] Instead of introducing the bill passed over from the Senate, Owen substituted one of his own. Putting the emphasis on cheap popular publications, he cut the annual expenditure for books to $5000 and added a normal school for the graduate instruction of teachers, "a far higher and holier duty than to give additional depth to learned studies, or supply curious authorities to antiquarian research." [55] This effort to revive the school idea met immediate attack from all sides, led by Adams, and a vote of 72 to 42 struck it out. But Owen's opposition to the library idea was resolute, as was also his insistence that after ten years some sort of legislation was absolutely necessary.

Before action was possible, however, the long-delayed bequest had to undergo an attack by those who still wanted to send it back to England. Among these, of course, were the Southern followers of Calhoun. Others, like John Chipman of Michigan, who had fought the corporate monster of the United States Bank, asked "what distinction was there between a corporation in the form of a United States Bank and a corporation intended to elevate humanity in close approximation to the throne of Heaven?" [56] Andrew Johnson, the self-educated representative from Tennessee, feared that the "extravagance, folly, aristocracy, and corruption of Washington" would demoralize the Institution.[57] G. W. Jones of Tennessee claimed "there is too great a centralization in this government already." [58] But these sentiments, threatening as they were to all government action in science, could muster only 8 votes to return the money against 115 opposed.

After W. J. Hough, a New Yorker with a preference for "regents" instead of "managers," had redrafted Owen's amended bill, the House was at last ready to pass on a measure whose corners were so rounded by compromise that it bound the new Institution to the program of no one party. The library provision had been softened to give the regents power to make an appropriation "not exceeding an average of twenty-five thousand dollars annually, for the gradual formation of a library composed of valuable works pertaining to all departments of human knowledge." [59] Whether this meant spending the $25,000 every year — five-sixths of the income and a victory for the library party — or whether it was a ceiling that need not be closely approached was a question for the board of regents to decide. The museum function, still very much in evidence, also had a loophole in that the regents were to take over government collections "in proportion as suitable arrangements can be made for their reception." [60] Most of the other functions were implied only by the facilities of the building, which besides a library and natural history cabinets was to house a chemical laboratory, a gallery of art, and lecture rooms. The secretary would have charge of the buildings, record proceedings, serve as keeper of library and museum, and, most importantly, employ assistants. A paper "establishment," consisting of the President, the vice-president, the cabinet, the Chief Justice, and the mayor of Washington, never became active. The board of regents, whose design came over from the Tappan bill, was the effective governing body.

The final vote in the House of Representatives, 85 to 76, reflects more than anything else the desire to get some bill passed, even one not very satisfactory. The Senate vote was 26 to 13. Party lines meant little, and perhaps the most indicative name in the list of opponents besides the always consistent Calhoun was that of Thomas Hart Benton, a Jacksonian and a part-time friend of science whose fear of a corporation here was paramount.[61]

As soon as Congress had done its work, the board of regents found they could mold the Institution's very nature without leaving the law behind. Rufus Choate, now a public member, still tried for his library. Robert Dale Owen, nursing his desire for a school, devoted most of his energy to selecting plans for a building, which in the end embodied his singular theories about Norman architecture as most appropriate for the American spirit. Colonel Joseph Totten, representing the National Institute, could be expected to uphold the museum point of view, as could William C. Preston of South Carolina. Richard Rush's presence was a reminder of the continuity from 1825, the time of John Quincy Adams's first annual message, to 1846.

Joseph Henry as the First Secretary

By far the most important member of the board was the youngest. Born in 1806, Alexander Dallas Bache knew his way around in two groups who heretofore had had almost no communication. As might be expected of a descendant of Benjamin Franklin, he was one of the new generation of professional scientists at the same time that he was intimate with important politicians. In 1836 and 1837 he had gained the European dimension so necessary to the new generation by examining educational institutions in several countries as a prelude to the establishment of Girard College. While in London he became the friend of Joseph Henry, America's leading physicist and professor at Princeton, who was also getting his European initiation. Bache devoted his attention to Girard College, including an observatory for terrestrial magnetism, until 1843, when he entered government service as superintendent of the Coast Survey.[62]

In 1846 Bache as a civil servant was quite different from the unpopular Hassler. The vice-president, George Mifflin Dallas, was his uncle. Robert J. Walker, senator and then secretary of the treasury, was his brother-in-law. He bore the name of the secretary of the treasury in Madison's cabinet, Alexander J. Dallas. And behind them all hovered the ghost of Benjamin Franklin. The political connections implied in this pedigree made Bache an unusual scientist indeed, especially on the board of regents, presided over by his Uncle George, the vice-president.

With Bache the professional scientists for the first time were getting a voice in the planning of a central scientific organization. The evidence of his part in the deliberations is largely the results themselves — two entirely new developments which had appeared in no earlier organization or discussion. One was a paid, full-time executive officer with ample power; the other was a real scientist to fill the post. At the meeting on December 3, 1846, the regents adopted a resolution, probably prepared by Bache, calling for the secretary, whose job as described in the law might be quite without distinction, to be a man possessing "eminent scientific and general acquirements . . . capable of advancing science and promoting letters by original researches and effort." He should also command respect in foreign countries "and, in a word [be] a man worthy to represent before the world of science and letters the Institution over which the Board presides." [63] Such qualifications restricted the post to one of the new generation of scientists, restricted it, in fact, very nearly to the candidate Bache had been working for ever since his appointment to the board of regents. He had already secured support in behalf of his friend Joseph Henry from such scientists as Michael Faraday abroad and Benjamin Silliman and Robert Hare in America.[64] In the final vote the tally stood: Joseph Henry, 7; Francis Markoe, 4; Charles Pickering, 1.[65] Pickering had support from New England and probably was Choate's choice. Markoe as a leading spirit in the National Institute reflected the museum interest, but he would have had difficulty measuring up to the resolution. Henry, with his years as professor at Princeton and his great discoveries in electromagnetism behind him, stood high as an investigator. Only the stern concept of duty emanating from his Presbyterianism could have persuaded him to take up at the age of fifty the unaccustomed burdens of administering an ill-defined and unformed institution. Because Henry had submitted an outline of his ideas on the nature of the Institution before the vote,[66] his election meant an immediate and deliberate orientation of Smithsonian policy. The plan of organization already prepared for the regents by Robert Dale Owen was sent back to committee, to reëmerge with many marks of Henry's influence. As he finally put it down late in 1847, Henry's plan for the Institution proved to be much more influential than the organic law itself.

In his interpretation of the Smithson will Henry made sharp distinctions where earlier discussion had often been confused. In the first place, the Institution was to benefit all mankind. Hence it "is not a national establishment, as is frequently supposed" and "unnecessary expenses on local objects would be a perversion of the trust." The new Institution thus sharply differed from its local-national predecessors. It was a universal body, tied to the District of Columbia only by its physical existence there and to the United States only by the trusteeship. In the second place, Henry clearly separated the increase of knowledge from its diffusion. He thought the Institution could increase knowledge by "facilitating and promoting the discovery of new truths" while it "can be most efficiently diffused among men by means of the press." Thus he put the whole emphasis on research or the publication of its results — an idea that had been secondary in most of the plans put forward by politicians and amateurs. An enthusiast for popular education like Robert Dale Owen had no feeling for such a program.

In the third place, while taking full responsibility for all branches of knowledge, Henry insisted that in proportion to this vast field "the funds are small." A comparison with the Wilkes expedition, whose operating cost far exceeded the total value of the Smithsonian endowment and whose annual appropriation for publication usually approached the Institution's income, indicates that Henry was quite correct even by the standards of his own day. His operating rules stemmed from this estimate. He would seek to do only the things "which cannot be produced by the existing institutions in our country." His organization would be small, allowing him to take up projects provisionally, modify them easily, and abandon them "in whole or in part, without a sacrifice of funds." [67]

Finally, Henry felt that the Smithson will envisaged neither a library, nor a museum, nor a gallery of art, these objects being added by Congress. This announcement drew the line between him and those of the regents, such as Choate, who favored a big library. Henry bound himself to resist the expenditure of a large part of the Smithson money for books and for the care of collections. Research had made its appearance as a scientific activity rivaling the older and dominant interests.

Henry saw two main ways to sponsor research. If a man capable of discovering scientific principles has fallen on a new vein, the Institution should aid him with grants and provide a channel for publication. A commission of specialists would choose the projects and judge whether the results should be published. Henry's own methods of physical research showed through in the much-discussed requirement that "all unverified speculations" must be rejected. Out of these recommendations grew the *Smithsonian Contributions to Knowledge*.[68]

But since scientists of genius were rare, Henry provided also for projects in which the end in view was defined first and a competent person then chosen to accomplish it. As a precedent for this type of research he cited the steam-boiler experiments performed for the secretary of the treasury by the Franklin Institute.[69] In modern jargon this was programmatic research. Henry offered a number of examples — meteorological observations, explorations in natural history and geology, new determinations of physical constants, chemical analyses, statistical inquiries into moral and political as well as physical subjects, historical researches, and ethnological studies of the races of North America.[70] Meteorology and ethnology became great fields of activity within the Institution, while some of the other suggestions were still-born.

To diffuse knowledge Henry proposed a series of reports on "new discoveries in science, and . . . the changes made from year to year in all branches of knowledge not strictly professional." He also proposed the occasional publication of popular treatises on — with a faint bow to Owen — the statistics of labor, the productive arts of life, and public instruction.[71] The exquisite fairness Henry here showed scarcely masked his lack of real interest.

Early Policies and Activities of the Institution

With a high likelihood of friction between Henry and Bache on one hand and Choate the library advocate, Owen the popularizer, and various museum men on the other, the board of regents avoided a showdown by reaching a compromise. Early in 1847 they agreed that of the $30,000 annual income half would go toward Henry's program and half for the museum and library. Charles C. Jewett became assistant secretary in charge of the library at the same meeting.[72] Since the law gave the regents the discretion on when they should take over the collections in the Patent Office, including the Wilkes expedition objects, Henry adopted the policy of refusing to receive them as long as the Institution's building was unfinished. Finally in 1850 he appointed young Spencer F. Baird of Carlisle, Pennsylvania, as assistant secretary for the museum, but its contents at that time consisted almost entirely of Baird's own personal specimens which he brought with him. With this precariously balanced organization Henry carefully steered his course until the great storms of 1854.

By circulating his program to all the scientific men in the country, Henry collected an impressive array of testimonials. The committees of local societies and the professors of natural history and natural philosophy at colleges great and small all added their praise.[73] Back in 1838 they had not even been consulted. Henry had plenty of use for this support, for the infant Institution was immediately beset by all sorts of attack both from within and without. Some people still did not like the idea in any form. Andrew Johnson in 1848 proposed in the House to change the name to "Washington University for the benefit of the indigent children of the District of Columbia."[74] Every time the question of printing the annual report came up in Congress some opposition appeared. For instance, in 1858 the vote in the House for printing was only 84 to 50. Some congressmen never tired of remarks such as Simon Cameron's, "What do we care about stuffed snakes, alligators, and all such things?" On another occasion he lamented, "I am tired of all this thing called science here."[75] Robert Dale Owen as an early regent drew criticism to the Institution because of his religious views, and the effort to get him reappointed after his term in the House expired drew the charge that the Institution was "a hospital for destitute politicians."[76] At the same time Henry detested the monstrous building forced on him and doubtless

rejoiced at the departure of its enthusiast. Politics indeed was a vex-
ing factor in the selection of assistants as well as regents. James D.
Dana warned Spencer F. Baird, who was applying for a place, that
"a word from a Political man is perhaps quite as important as from
Scientific, since much depends on favor in all Washington appointments." [77]

The demands to use the supposedly ample funds for all sorts of
projects for which congressmen hated to take the blame of appropriating money were even more troublesome. Henry claimed the income was not enough to carry out one fourth of the plans already
mentioned in the act of Congress or in actions of the board.[78] Powerful politicians such as Stephen A. Douglas nevertheless contemplated
throwing the money away as a sop to special groups — for instance,
the farmers.[79] Diffusion of knowledge, and plans with the word "practical" prominently displayed had a surface appeal to most Americans.
That Henry resisted the temptation to resign and return to physics
at the same time that he stood uncompromisingly for original research as the basis of Smithsonian policy is the true core of his
greatness.[80]

The most formidable single threat to his plan was the desire to
turn the Smithsonian into a library. Since the Library of Congress
was not then what it is today, and since Choate, George P. Marsh,
and Jewett were friends of learning after their own fashion, the final
outcome was long in doubt. Henry had at first gone to considerable
lengths to coöperate under the half-and-half compromise agreement,
and Jewett had acted with great energy to amass a collection of
some 32,000 volumes and to gather a great deal of bibliographical
information. However, in 1854 the secretary felt strong enough to
attack the compromise openly, and the board of regents sustained
him by a vote of 8 to 6. When Assistant Secretary Jewett tried to
appeal directly to congressmen, Henry dismissed him. Immediately
Rufus Choate resigned from the regents, touching off a debate in
the Senate and an investigation in the House of Representatives. Up-
held by a majority of the board, by Congress, and by a great outcry
from the scientific community of the country, Henry emerged with
his plan vindicated and the powers of his office confirmed. With a
rare policy of self-denial, he then withdrew the Institution from the
library business as completely as possible. In 1857 he had the original
act amended to repeal the provision making the Smithsonian a de-

pository of all copyrighted material. In 1866 he sent to the Library of Congress for safekeeping some 40,000 volumes, a collection that has continued to grow.[81]

Henry's attitude toward a museum was much the same as toward a library. It did not become a source of conflict only because Spencer Baird was an outstanding naturalist with a benign disposition and because exploration (and hence the museum) was the dominant scientific activity of the period. Heeding the example of the National Institute sunk by specimens before his very eyes, Henry came out against the permanent inclusion of the National Museum in the Smithsonian even though the law clearly provided for it. But at the same time Baird began his work as assistant secretary in a quiet way, building up his own holdings and serving as adviser to the government concerning the many explorations then under way. In his hands the collections became real research tools which in their way contributed to the advance of knowledge as well as to the entertainment of the public. This modest "Museum of the Smithsonian Institution" was a living enterprise under control both scientifically and financially, in contrast to the "National Cabinet of Curiosities" moldering in the Patent Office.[82]

Although under heavy pressure from both Congress and the Patent Office to take over the older collection, Henry for a long time refused either to use Smithson income for the purpose or to make himself dependent upon an annual appropriation, with all it entailed in political uncertainty and, he feared, control.[83] But as Baird held his peace and kept working during the library uproar of 1854 and 1855, and as the building neared completion, Henry finally agreed to come down from his lofty perch to the extent of accepting the collections and an annual appropriation from Congress to care for them. "While, on the one hand, no appropriation should . . . lessen the distinctive character of Smithson's bequest, on the other it is evident that the government should not impose any burdens upon the Institution which would impair its usefulness or divert its funds from the legitimate purpose."[84] A kind of buffer was arranged by having the $4000 appropriated to the secretary of the interior, "thus obviating the necessity of an annual application to Congress by the Institution itself."[85] Nevertheless, this annual appropriation brought the Institution closer to the government and made the subsidiary museum national in a way the Smithsonian itself was not. That Henry re-

mained jealous of the money spent on the collection is indicated by
the fact that until 1872 he delegated to Baird no control over National
Museum disbursements.[86]

Nevertheless, the annual appropriation for the National Museum
added a new dimension to the Smithsonian's relation to the government. Before this, the Institution had received its support entirely
from its endowment; the government's control was limited to its
position as trustee of the fund and to the officers who served ex
officio on the board of regents. After receiving the Wilkes expedition collections from the Patent Office, the parent Institution, with
its own relation unchanged, took on a direct subsidiary, the National
Museum, which was dependent on public money for its support. The
force of events and Baird's patient work turned the National Museum
into the most significant branch of the Institution, despite the doubts
of the secretary. And yet Henry's very laggardness was of great
positive importance, for he insisted on the Museum's being financially
able to take care of the collections, and prevented the dead weight of
the National Institute jumble from sinking the new Museum at the
outset.[87]

To evaluate the work of the Smithsonian in the 1850's outside of
the museums and exploration, it is necessary to weigh basic structural
limitations as well as the decisions of the management. With an income of $30,000, fixed not by its opportunities but by its endowment,
the Institution's ability to take a commanding place in the American
scientific community was sharply limited. In 1854 the Coast Survey's
annual appropriation of $489,000, coming very close to equaling the
Smithsonian's endowment, was 16 times the income Henry had at
his disposal. By adhering to this inelastic structure, the secretary may
well have kept the Institution from fulfilling many of the more elaborate hopes for it, but the reasoning behind his actual program is only
apparent in the light both of the amount and the rigidity of his income. Even so, the record of the Institution changes in aspect if different questions are asked concerning it. To ask, for instance, which
activities became the basis for permanent agencies makes its appearance less favorable than to ask to what extent it stimulated science in
its own day. Perhaps it is most legitimate to ask how the secretary's
own plan was actually executed.

Henry made a real mark in American science by publishing costly
research works in the Smithsonian *Contributions to Knowledge*. The

first volume, E. G. Squier's and E. H. Davis's *Ancient Monuments of the Mississippi Valley*, began a lasting tradition of the study of the natives of America by the Institution. Succeeding volumes presented the researches of the best of the new generation of professionals in many fields — Sears C. Walker, Jacob Whitman Bailey, Louis Agassiz, Asa Gray, Joseph Leidy, Jeffries Wyman, John Torrey, Wolcott Gibbs, Frederick A. Genth, J. L. Le Conte, and by 1860 S. Weir Mitchell. One European, W. H. Harvey, published his great work on North American algae in the series. The eminence of these men insured the importance of the *Contributions*, and the fact that they published there indicates the need for such support.

The results of Henry's progress reports on the various sciences and on special subjects were perhaps less happy. Although translation of Johannes Müller's "Report of Recent Progress in Physics" appeared from time to time, no Smithsonian publication even remotely rivaled Silliman's *American Journal of Science and the Arts* as the great source of current scientific news. The difficulties involved in commanding the talent necessary for an authoritative survey are evident in Henry's experience with a "Report on Forest Trees," of which he expected great things because its author was to be Asa Gray, the foremost botanist in the country. But before Gray could begin work in 1849 he became involved in the publication of the Wilkes expedition botany. Other research problems and opportunities crowded so hard on Gray that the report never reached completion.

Often Henry invested in research to serve going agencies, notably the Coast Survey. For instance, the Institution prepared tables of the occultations of stars for the determination of longitude until the *Nautical Almanac* began to appear.[88] For a time the Institution ran a magnetic observatory, first in Washington and then in Key West, even though Henry bluntly stated that "this establishment ought to be supported by the government." [89] Through most of the 1850's the Institution maintained a chemical laboratory, but this activity was evidently choked off by the development of agricultural activity in the Patent Office.[90]

Henry never engaged in empire building for its own sake. He dropped an activity as soon as someone else either inside or outside the government seemed willing to do it. For instance, among the projects he initially encouraged were some researches by Louis Agassiz. During the early years after the arrival of the Swiss naturalist, the Institution put up funds for the preparation of the plates for various memoirs. But when Agassiz proved to be the ablest money raiser yet to appear in American science, getting $360,000 in subscriptions for his *Contributions to the Natural History of the United States*, the Smithsonian pulled out, "though it may lose the honor of a more permanent association of the name of this celebrated individual with its own publication." [91]

The establishment of an international exchange service was both enduring and indicative of the universal ideal of the Institution. Henry used the experience of the older societies and of individuals such as the botanist Asa Gray, who had for years been struggling with the problem of establishing a flow of publications between Europe and America. Besides trading its publications with European societies, the Institution hired agents in various key cities abroad and offered its facilities to scientific organizations in the United States. By handling customs clearances and often getting free transportation, the Institution rendered a real service to the American scientific community. [92]

Meteorology was a prominent part of the program all through the 1850's. Calling upon the agencies already collecting weather records, notably the surgeon general's office, the regents of the University of the State of New York, and a Pennsylvania group under the auspices of the American Philosophical Society and the Franklin Institute, Henry organized a corps of observers, adapted the telegraph for simultaneous reports, and published a number of volumes of observations. In 1856 he arranged with the Patent Office, which was then interested in agricultural statistics, to use some of their funds for extending the Smithsonian system. [93] In attracting to the service a scientist of the caliber of Arnold Guyot, in organizing the corps of observers, in distributing instruments, in utilizing new techniques such as the telegraph, the weather map, and the newspaper forecast, the Smithsonian made for itself a prominent place in meteorology in the 1850's. [94]

Joseph Henry so dominated the early Smithsonian that his own interests and researches formed a part of the Institution's activities. Although he made a sharp distinction between pure and applied science, gaining his most enduring fame in research in physics, his versatility came out in a large number of practical interests. After 1852, when the Treasury Department established a Light House Board, Henry became a member. He not only recommended the adoption of the Fresnel lens but also tested sperm oil, rapeseed oil, lard oil, and kerosene.[95] Because he was on a commission to help the army engineers design new rooms in the Capitol, he took up the study of acoustics as applied to public buildings.[96]

On one famous occasion Henry used his board of regents as a kind of supreme court to decide a personal scientific controversy. Although S. F. B. Morse had used freely Henry's discoveries and advice while he was developing the telegraph, he forgot his debt and claimed all credit for himself in the course of his later litigations. Henry chose as his method of answering a full report, accompanied by documents, to the board of regents, who passed unanimously a resolution stating that Morse had not refuted Henry's statements and that their confidence in the secretary was unimpaired.[97] Here for an instant the Smithsonian moved toward the kind of judicial function performed in France by the *Académie des Sciences.* However, the uniqueness of this event and the use of Henry's official position as an excuse to give the regents jurisdiction only serves to underscore the incompatibility of the Smithsonian with the role of a central scientific organization on a European model.

Perhaps Henry's greatest contribution to the Smithsonian was his own austere personality. With a fanatic contempt for the common run of politician, he presented to the country the very picture of the new scientist, above passion and emotion, judging all things in the cold light of fact. This detachment enhanced his reputation with the very politicians he despised. For instance, he was in the course of time able to win over the Institution's old enemy Andrew Johnson.[98] Party lines made no difference with him. When in the library fight of 1855 it appeared that Jewett might line up the Know-Nothings against the antilibrary policy, Henry promptly countered with "a number of warm friends in the House who belong to that party and will defend the Institution." [99] So much of his time and energy went into the Institution that he was never again able to take up abstract physics.

One service he steadfastly performed was to battle in the name of inductive science the army of quacks who infested the lower fringe of science in this country. Believing that the wide diffusion of knowledge and the freedom of thought and discussion in the United States bred a lack of regard for "the present system of science," he adopted the rule of stating "candidly and respectfully the objections" to the communications received by the Smithsonian, hoping "to convince their authors that their ground is untenable." Thus America's leading physicist gave his time in the name of the Institution to writing patient answers to letters on "the quadrature of the circle, the trisection of the angle, the invention of self-moving machines, the creation of power, the overthrow of the Newtonian system of gravitation, and the invention of new systems of the universe." [100]

While the Smithsonian did not dominate the scene in the 1850's, it gave American science badly needed support in many projects, strategically selected to be of most immediate help. But it made its greatest contribution by raising the banner of original research. In his annual report in 1859 Henry pointed out that in the early discussions of Smithson's will even "prominent and enlightened men" had confused the aims of the donor. In contrast, the Institution had "succeeded in rendering familiar to the public mind in the United States the three fundamental distinctions in regard to knowledge, which must have an important bearing on the future advance of science in this country: namely, the *increase* of knowledge, the *diffusion* of knowledge, and the practical *application* of knowledge to useful purposes in the arts." Although the legislatures of states had provided generously for the diffusion, and although practical applications received encouragement from the government in the patent laws, few people realized that "the advance of science or the discovery of new truths, irrespective of their immediate applications, is also a matter of great importance, and eminently worthy of patronage and support." Henry felt that "this Institution has done good service in placing before the country the importance of original research, and that its directors are entitled to commendation for having so uniformly and persistently kept in view the fact that it was not intended for educational or immediately practical purposes, but for the encouragement of the study of theoretical principles and the advancement of abstract knowledge." [101]

V

THE GREAT EXPLORATIONS
AND SURVEYS

1842–1861

THE late 1840's and the 1850's are a new period in government scientific activities in other ways than the rise of the Smithsonian Institution. The rapid expansion of the country westward set the tone for American life and raised problems for a government almost immobilized by sectional strife over the slavery issue. One facet of this expansion was scientific. As in the days of Lewis and Clark the great imperial need was accurate information. The advance of settlement into the area of the northern and western reaches of the Mississippi Valley, the Mexican War with a vast new territory as a result, the settlement and acquisition of Oregon, the gold rush, all brought Americans into new and strange environments. That expansion, which took the form of acquiring and settling public domain, brought in the federal government as the foremost information-collecting instrument. With previous experience on which to build, the government developed its surveying and exploring functions to fill the needs of expansion. As earlier, the work was done *ad hoc*, with the basic unit of organization the survey or expedition, higher administrative groupings playing only a secondary role.

Attempts at Land Classification

Along the western edge of the timbered country, in lands now rapidly filling up with settlers, the surveys tended to originate in the General Land Office, which was interested in the problem of reserving valuable mineral lands still in the public domain. David Dale Owen, who had done this work back in the 1830's, undertook a geological reconnaissance in Wisconsin and Iowa in 1847.[1] With additional assistance he did a more thorough job between 1848 and 1850. Owen's survey covered federally owned lands either in organized territories or in newly created states. It was therefore an example of the federal government's carrying on an activity which in more settled regions was done entirely by the states.[2]

More directly concerned with resources of great potential value was Dr. Charles T. Jackson's survey of the United States mineral lands in Michigan. The act of Congress provided for a geological examination and authorized the President to sell lands which "may contain copper, lead, or other valuable ores" with a "brief description of the lands to be offered." [3] This implied a classification of lands by scientific methods as to suitability for agriculture or mining, and the reservation of the mineral lands for special sale. Jackson himself felt that a geological survey might be useful for giving the public information, but for the government to use it as a basis for land policy would "most seriously embarrass . . . the settlement of newly acquired territory." [4] On Jackson's resignation in 1849, the survey under J. W. Foster and J. D. Whitney completed an extensive series of geological studies of the region.[5]

After 1850 this beginning at scientific land classification dwindled away as the various new states began surveys of their own territory. Mississippi in 1850, Illinois in 1851, California in 1853, Missouri in 1853, Wisconsin in 1853, Iowa in 1855, Arkansas in 1857, Minnesota in 1859,[6] captured the geology of their regions from the federal government. Owen himself entered the service of Kentucky. Thus a potentially important movement toward the use of science as a regular basis of public policy decisions was temporarily checked.

The Army in the Trans-Mississippi West

The great theater of action in the West lay beyond the organized territories, where the Army was both occupying new regions and attempting to control those already won. In the Topographical Engineers it had at hand a trained body of officers to command expeditions and to make the surveys and astronomical observations necessary for accurate mapping. Among the earliest and most famous of these Far Western expeditions by the Topographical Engineers were those of Frémont, who set out to examine the route of the Oregon trail until he linked with Wilkes's survey of the Columbia River region.[7] But Frémont attempted to keep up with all departments of science by himself — geology and natural history as well as topography and astronomy.

The more usual practice, and one that yielded much greater results, was to depend on the surgeons and on civilians for the geology and the collections of plants and animals. The expeditions varied greatly in organization, mission, and quality of scientific results, but taken together they add up to an impressive picture of the nature and resources of the trans-Mississippi West. These collections were the raw materials that Spencer F. Baird at the Smithsonian turned into a real National Museum. By encouraging naturalists, getting them places on expeditions, furnishing them with equipment, in some cases making small grants, and above all by receiving and studying the collections, Baird with the coöperation of a few other strategically placed scientists welded the separate expeditions into the scientific conquest of half of the continent. He was tireless in working both with collectors and with the officers themselves. "When Capt. Marcy comes on," he wrote to Captain George B. McClellan, "I will endeavor to inoculate him with the Natural History virus." [8] Between 1850 and 1860 about thirty expeditions sent their results to the Smithsonian. [9]

Often the explorations took place as a part of regular army operations. The Mexican War, for instance, occasioned Lieutenant J. W. Abert's examination of New Mexico and Lieutenant W. H. Emory's reconnaissance from Fort Leavenworth to San Diego. [10] A more usual occupation was the building of wagon roads. [11] Although in the later 1850's some of these were under control of the Interior Department, the Topographical Engineers did all the important surveying. Large collections were often made as well. Other expeditions investigated the possibility of artesian wells along the thirty-second parallel, the navigation of the Colorado in steamers, and the locating of lands for the Indians of Texas. [12]

As one example of this exploring activity, a typical expedition was that of Lieutenant G. K. Warren, who was to find the best route from the Missouri River to South Pass, to explore the Black Hills, and to examine the Niobrara River with an eye to opening a road between Fort Randall and Fort Laramie. The appropriation of $25,000 came under the head of those "for surveys for military defenses, geographical explorations and reconnaissances for military purposes." [13] Warren could hire a topographer, an assistant topographer, an astronomer, a physician, a geologist with an assistant, and a meteorologist at salaries from $130 to $60 per month.[14] The geologist, F. V. Hayden, was destined for a long career in the government service — an example of how the surveys were the great graduate schools for a whole generation of naturalists. In addition to the official reports covering climate, physical geography, medicine, geology, paleontology, botany, mammals, birds, fishes, reptiles, and molluscs, the expedition collections were the basis of fifteen articles in various society transactions.[15]

Diplomatic expansion westward also provided opportunities for the soldier-civilian teams of scientific explorers. The Mexican Boundary Survey resulting from the treaty of Guadalupe Hidalgo in 1848 had parties in the field for several years. Although the experiment of a civilian commissioner was tried, a topographical engineer, Lieutenant W. H. Emory, who had served as astronomer, pushed the survey to completion.

Because of the delays and confusions of authority in the early years of this survey, an extraordinary number of naturalists, sometimes rivals, got into a new region and underwent an arduous apprenticeship. Thus the Mexican Boundary Survey proved unintentionally preëminent in training the rather peculiar equivalents of graduate students. A collector had to know some science and also know how to live on the frontier, an incompatible combination which attracted and produced strange men. C. C. Parry, Charles Wright, and John M. Bigelow, all of whom collected plants in the Boundary Survey, were sui generis, equally unlike civilized botanists and ordinary frontiersmen. Later and more efficient, the Northwest Boundary Survey made a less interesting scientific record.[16]

The great efforts of the decade of the 1850's, in which the soldier-civilian teams operated on the largest scale and with the greatest efficiency, were the famous railroad surveys seeking a transcontinental route. With an appropriation of $150,000, Secretary of War Jefferson Davis sent out expeditions along six potential routes from the Mississippi to the Pacific. The forty-seventh parallel just below the Canadian border was the northernmost, with the thirty-second parallel west through Texas and along the Gila River the southernmost.[17] Davis instructed the parties to "observe and note all objects . . . which have an immediate or remote bearing upon the railway, or which might seem to develop the resources, peculiarities, and climate of the country." He emphasized topography, meteorology, magnetic surveys, geology, zoology, botany, and statistics of the Indian tribes.[18]

In instructions and organization these surveys followed the patterns already laid down.[19]

In one sense the railroad surveys were a failure. The driving force behind them was the bitter sectional conflict in which both North and South were seeking advantage by securing the terminus of the transcontinental railroad which everyone envisaged. Some conciliators had the hope that the surveys by scientific methods could solve a problem muffed by politicians and businessmen. No clear-cut answer of the superiority of a northern or a southern route emerged from the reports. Amid all the statistics of grades, timber, and water supply it was clear that the country could build a railroad originating in either section if it were willing to pay for it. Hence the decision went right back to the embattled politicians. The Topographical Engineers could not make basic policy decisions for Congress.

In a less spectacular way the railroad surveys were a marked success. The scope, covering the whole West, and their simultaneous execution gave a great body of fairly comparable data on a whole empire. The entire process from collecting to publication took only about seven years. In contrast to the Wilkes expedition, the twelve giant volumes of reports appeared promptly, the various sections bearing the names of the best civilian specialists in the country. Editions, printed in large quantities, circulated the results widely to the scientists, both at home and abroad, who could use them. To the student of the flora, fauna, and geology of the West, the volumes still seem as live and as important as they seem futile to the political historian.

Overseas Exploration

Just as American expansion overseas in the 1850's was a part of the same general movement as the push westward, the exploration of distant lands was closely related to the scientific activity in the continental territories of the United States. Naturally the Navy, following the pattern of the Wilkes expedition, played a predominant role, but the problems and even the personnel were often much the same as those working in the West. For instance, the greatly heightened interest in the various possible routes for either railroads or canals across Central America after the gold rush stemmed from the very same sources as the desire for a transcontinental railroad.

Alexander von Humboldt had postulated as a necessity for the

building of an interoceanic canal a thorough exploration of all possible routes, using similar methods to gather comparable data.[20] Although this condition was not met until the 1870's by the Navy Department, the government made some efforts in the 1850's to get accurate information on the various routes. In 1856 a coöperative expedition under Lieutenant Nathaniel Michler of the Topographical Engineers and Lieutenant T. A. Craven of the Navy examined a route
for a canal between the Atrato and Truando Rivers, just where the
Isthmus of Panama joins South America.[21] In 1860 the Navy examined
the Gulf of Chiriqui both for a coaling station and as a terminus of a
railroad. The geologist who went along, John Evans, was a veteran
of the railroad surveys.[22] Although these efforts were small compared
to those of private concerns in the area, they were part of the American dominance in surveying that laid the basis for the American engineering triumph in Panama.[23]

The demands of commerce and communications which called attention to Panama also sent Americans into other areas. Lieutenant
W. F. Lynch, U.S.N., leading an expedition to the Dead Sea in 1847,
collected geological, zoological, and botanical data as well as making
topographic and hydrographic observations. Lynch saw the Middle
East as strategic for world trade with the coming of steam.[24] Similar
views seem to have motivated his reconnaissance of the coast of West
Africa in 1853.[25]

South America attracted a good deal of attention. Perhaps most
revealing of the forces behind American overseas imperialism in the
1850's was the exploration of the Amazon by Lieutenant W. L.
Herndon of the Navy. Under the sponsorship of his brother-in-law,
M. F. Maury, Herndon ostensibly sought to report on the navigabil-
ity of the river system and the possibilities for commerce "waiting
for the touch of civilization and the breath of the steam engine." [26]
But as a Southerner Herndon also dreamed of a slave and plantation
civilization transferred from the United States by "planters . . . who
. . . [looked] with apprehension (if not for themselves, at least for
their children) to the state of affairs as regards slavery at home." [27]
Just as the railroad failed to settle a burning sectional issue in the
West, overseas exploration failed to discover a way out by a settlement in the tropics. South America also attracted an expedition under
Captain Thomas J. Page, U.S.N., to La Plata River estuary between
1853 and 1856.[28]

The Naval Astronomical Expedition to Chile was more strictly scientific in motivation and more considerable in its results. The impetus came from the American Philosophical Society and the American Academy of Arts and Sciences, both of which recommended astronomical observations in the Southern Hemisphere.[29] The primary object was to determine the sun's parallax by observations of Venus and Mars from stations in both northern and southern latitudes. Lieutenant James M. Gilliss was the real instigator of the enterprise. Forced to make a career for himself outside the Naval Observatory he had done much to create, he lined up support and coöperation for the expedition in Europe as well as in the United States. Spending four years in Chile, he extended the range of inquiry far beyond his primary purpose. Besides working on observations for a catalogue of stars visible from the Southern Hemisphere, he also directed the collection of data on earthquakes, weather, magnetism, and natural history. These results did much to redeem the expedition, for the observations of the planets were compromised by the failure of the home observations.[30]

In the Far East, science also rode along with expanding American commerce. Japan had long been closed to Western science as well as trade. Commodore Matthew C. Perry, almost alone of the high officers of his day, saw his mission to Japan as a sensitive one in which security was important. Accordingly he breasted the general practice of the services to exclude all civilian scientists. He felt his expedition was "not scientific, but naval and diplomatic; to attempt both would probably be to succeed in neither." He could not see civilian scientists under naval discipline, thought they would take up too much space for quarters, and feared their correspondence would reveal secrets to his rivals, especially the Russians. This attitude was reinforced by another, the desire to make naval officers as proficient in science as the Topographical Engineers of the Army. He wanted them to record facts, even if "they might not always, in their early efforts, be able to account philosophically for what they saw." [31] Thus Perry was at once the hardest enemy of civilian specialists and the warmest friend of science in uniform that the Navy produced before the Civil War. Ironically, the pattern of civilian-military coöperation in exploration was so strong that despite the commodore's orders the botanical collections, among the most significant made on the expedition, came from civilians who had managed to go along. Back in the United States, their publication found civilian channels.[32]

Closely following Perry's visit, Secretary of the Navy John P.

Kennedy, who was an enormously energetic friend of science, sent the North Pacific Exploring Expedition to Asiatic waters. A botanist and a zoologist selected by the Smithsonian Institution went along, outfitted with "nets, kettles, dredges, etc., amounting to near $2,000, all of which were authorized and paid for without flinching. They go much better prepared than the old expedition (Wilkes), although with few hands." [33] Formally, the expedition's instructions called for a survey "for naval and commercial purposes" of the Bering Straits and the ports of the North Pacific Ocean and China Seas frequented by American whaleships and by trading vessels. Kennedy more rhetorically wrote the commander that the expedition was "not for conquest but discovery. Its objects are all peaceful, they are to extend the empire of commerce and science; to diminish the hazards of the ocean." [34] After a bad start Cadwallader Ringgold, a veteran of the Wilkes expedition, had to be relieved by Captain John Rodgers. This distinguished and intelligent officer was careful to give to the secretary of the Navy the names of Bache, Agassiz, C. H. Davis, and Maury as references for his fitness to command. [35] The good work done by the expedition has necessarily suffered from the fact that the Civil War stopped the preparation of an official report, and only scattered papers carried its results before the scientific public. One of these, however, by clarifying the relations of the flora of Japan with that of eastern and western North America, was a pioneer work in the new science of dynamic plant geography and the first major scientific publication in America to bear directly in favor of Charles Darwin's theories. [36]

By going through Bering Straits Rodgers engaged briefly in Arctic exploration as the Wilkes expedition before him had penetrated the Antarctic. But in general, polar exploration by Americans in the 1850's followed a somewhat different pattern from the dominant type of expedition that flourished elsewhere. The pressing and immediate need for expeditions to the far north came from the disappearance of Sir John Franklin. During the ten years after 1847 forty search parties combed the regions north of Canada in search of the British ships. [37] The American government, for all its experience in exploring, proved ill-adapted to the quick action necessary for a relief expedition. When Lady Franklin appealed to President Zachary Taylor, he could only spread the news of British rewards and ask Congress for an appropriation. [38] Had government effort been the only possibility, the year 1850, very late for any real hope of finding any of Franklin's party alive, would have gone by without action. But the merchant Henry Grinnell put up the money for two small ships which the Congress accepted for the Navy by joint resolution. The command went to Lieutenant E. J. De Haven, who had accompanied Wilkes.[39] Thus began an enduring pattern of government participation by providing military personnel but excluding ships and equipment. Science appeared here as a secondary object. For instance, the expedition had some idea of testing Maury's theory that beyond a rim of ice lay an open polar sea. Perhaps the greatest accomplishment of the expedition was the introduction of its surgeon, Elisha Kent Kane, to the Arctic.

In 1853 Kane headed his own expedition, again under navy orders, and with the private support of Grinnell and George Peabody. The Geographic Society of New York, the American Philosophical Society, the Smithsonian, Bache of the Coast Survey, and Maury of the Naval Observatory, all helped. By early 1855 it was clear that a relief expedition would be necessary — an expense which Congress shouldered. In a harrowing retreat over the ice Kane gave high priority to his records. His results led Bache to remark that "Dr. Kane appreciated highly all the relations, direct and indirect, which science has to an exploring expedition." [40] Through Kane's surgeon, I. I. Hayes, who returned to the Arctic in 1860, and through Charles F. Hall, the American exploring activity in Arctic regions continued in a trickle into the post-Civil War period.

In all their variations in organization and results, the exploring expeditions of the 1840's and 1850's had certain common features. They were almost always military in organization and command. They almost always depended upon civilians either at the Smithsonian or elsewhere for working up and publishing the results. Although showing continuity in personnel and purpose, they were usually *ad hoc* missions, flourishing for a while and then vanishing.

Together the explorations were an area of activity in which the federal government found a steady scientific occupation where its participation was indispensable. They made up the great bulk of the research carried on and paid for before the Civil War. They reflected the needs of the basic forces of commerce and westward expansion that dominated the period economically. They largely determined the sciences to be encouraged by the government. Astronomy, hydrography, terrestrial magnetism, meteorology, topographical mapping, geology, botany, zoology, and anthropology were the exploring and collecting sciences, the ones that received government support in the period. Sciences in which laboratory work predominated, where discoveries were made in test tubes rather than in distant mountains, were notably absent from the federal government's interest. And the Coast Survey — the organization for domestic exploration inherited from the Jeffersonian past — felt the same forces and developed the same sciences as the trans-Mississippi and overseas explorations.

The Coast Survey under Bache

When Alexander Dallas Bache took over the Coast Survey on the death of Hassler in 1843, he found it in theory an *ad hoc* task which was permanent only in that completion seemed to most congressmen to be impossibly far in the future. Hassler's regime left to him important assets — a truly scientific approach to the problem, a tradition of civilian control in a civilian department, a coupling with the weights and measures problem, data on a small stretch of the coast, a body of officers in both the Army and the Navy trained in the work, and an annual appropriation which against continuous opposition had reached the order of $100,000. The redoubtable Swiss had also left a reputation for extravagance and lack of results that had kindled the enmity of economy-minded congressmen.

Bache, being at once a capable scientist in civilian life, a graduate of West Point, and a familiar in high political circles, was admirably situated to make the Coast Survey a "triple organization" which "brings the scientific and practical training of civilians and of officers of the army and navy to its aid." [41] He used a small number of army officers, topographical engineers when he could get them, for triangulation work on land, and a larger group of naval officers in hydrographic work offshore. The civilian element provided "a more permanent nucleus . . . than the wants of either the military or naval service could yield." For "concert of action" he relied on "a central authority, the department in which matters pertaining to the trade, commerce and navigation of the country centre." [42] This meant civilian control under the Treasury Department.

Mindful of the criticism that had hampered his predecessor, Bache, after biding his time for a year, divided the Atlantic and Gulf Coast into eight sections and placed parties in all of them simultaneously. In addition to gaining support from congressmen of all the coastal states both on the Atlantic and the Gulf, Bache could turn out immediately useful charts much more rapidly. He himself admitted he could expand in this way only because colleges and West Point "had been pouring out educated men . . . The science of the country was altogether upon a different level in 1845 from what it was when the survey was proposed in 1807, or when it was commenced in 1816." [43] The increased number of parties also took more money, and Bache proved a master in relations with Congress.

By explaining technical terms clearly, stating his needs and how he intended to use his money, by playing up projects of immediate usefulness, Bache avoided petty irritation of legislators' tempers. Watching closely the propensity of Congress for economy, he continually presented statistics to show that the Survey "not only furnishes accurate scientific details and practical results, but it affords them at a very moderate cost." [44] He was careful to cultivate congressmen from inland districts, some of whom proved willing to get publicity into the newspapers for him.[45] Through the new American Association for the Advancement of Science he mobilized scientific opinion in his behalf and used it skillfully.[46] His extensive personal friendships made it easy to mold opinion among scientists and to shape a common course with Joseph Henry.

Bache needed all this diplomacy, for, besides chronic attempts to cut his salary and his appropriation, enemies in Congress again made a determined effort to give the Navy control of the Survey. The ostensible and oft-advanced reasons were the large use of the Navy's personnel and ships as well as its lively interest in the hydrographic side of the results. In 1851, for instance, eleven army officers were assigned to the Survey while as many as sixty-six naval officers were serving.[47] With Annapolis in its infancy, the Coast Survey was still the real counterpart of West Point in giving naval officers advanced scientific training. Bache could answer the arguments by citing the high turnover of naval personnel, the need of operations on land as well as on sea, and the need of a civilian "permanent nucleus."

But the real drive for navy control arose out of the fact that the growing Western and overseas explorations were under military control and appropriations. Since the standing military establishment cost the government in these years more than all civil and miscellaneous appropriations combined, it seemed attractively simple to let the

Navy, whose equipment and officers had to be paid for whether used or not, do the work, thus eliminating a conspicuous item from the civil expenditures. Senator Thomas Hart Benton, for instance, who considered the Survey a sop to the commercial interests anyway, claimed that the shift would save $400,000 annually.[48] The friends of Bache, led by Jefferson Davis in the Senate, answered that to maintain the Survey at its present efficiency no saving would be possible in the Navy, which had failed miserably in the 1820's. Denying that civilian control reflected on the capacity of military officers, they pointed to the advantages of a chief who was neutral between the two services. With scientific support from the AAAS, Bache managed to weather the storm. By preserving civilian control he maintained the Coast Survey with a continuity and a relative permanency that the explorations to which it was so closely akin were unable to achieve.

The Coast Survey's tendency toward permanence received encouragement from territorial expansion. Bache, who continually had to answer the question of when the Survey would end, pointed out that Texas had added two years. "Since then Oregon has been made a territory, and California acquired, and thus the limits of our coast have been greatly extended, and . . . the importance of the survey has greatly increased." [49] He lost no time in getting operations under way on the Pacific coast, producing preliminary sketches by 1850.[50] Amid charges in Congress that the Survey was creating a "new corps of officers" responsible to the Treasury Department and that idle employees had used government ships as hotels and "headquarters for frolicking," Bache's friends were able to get through an addition to the appropriation to take care of the increased operations on the West Coast.[51]

Besides the simple geographical extension, the Coast Survey moved beyond its narow functions of chart-making to a wider range of scientific inquiries. Bache saw that the practical studies necessary to make the Survey accurate had broad implications. His own specialty, terrestrial magnetism, he considered "eminently . . . practical . . . though reached by a scientific discussion which seems to pass beyond the bounds" of applied science. Even with this emphasis he claimed that the Coast Survey table for magnetic variation was "one of the contributions to general physics." [52] Thus, while never losing touch with its practical mission, the Survey energetically developed techniques and explored phenomena of basic scientific interest. Magnetism became a continuing study.[53] By 1858, 103 stations were spread over the Atlantic, Gulf, and Pacific areas.[54] The advantage of civilian personnel is illustrated by the career of Charles A. Schott, who served in the computing division of the Survey for 52 years beginning in 1848, contributing much to the science.[55]

When Bache inherited the office of superintendent of weights and measures from Hassler, he continued the distribution of standards and also carried out experiments on the accuracy and construction of standard bars.[56] The Coast Survey thus kept alive a continuing tradition concerning a subject explicitly provided for in the Constitution.

Accurate measurement of time was also a legitimate interest of the Survey. The telegraph obviously had possibilities for the simultaneous comparison of chronometers in widely separated places. This finally provided the basis for accurately determining longitude, a classic problem of navigation and surveying. Sears C. Walker of the Survey, coördinating the private efforts of several scientists, worked out the mechanical and organizational details necessary to put the telegraph to this use. He even made experiments on the human factor involved in observing and recording signals.[57] The progressive, experimental temper of the organization in seeking improved tools also showed itself in demands for steam surveying vessels and in the application of photography and electrotyping to the production of charts.[58]

Since astronomical observations are the raw materials for an accurate survey, Bache took a steady interest in them. He tended to look for his astronomy, however, not to the Naval Observatory under Maury, but to Harvard College, where Professor Benjamin Peirce served as the Survey's consultant. When Benjamin Apthorp Gould became head of the longtitude department of the Survey, he made his headquarters at the Harvard Observatory. Cambridge thus became a center of astronomical research in the 1850's in part because of Coast Survey support. Harvard, although still primarily a college, was beginning to develop research facilities in a few departments. The strong personal and scientific ties established between Bache in Washington and Peirce and Gould in Cambridge made an axis of power within the scientific community.

In other ways the Coast Survey scientists showed vision and energy in attacking problems that came in their way. They worked out the pattern of tides in the Gulf of Mexico.[59] In the Atlantic they extended their interest sufficiently offshore to include the Gulf

Stream, which was fairly definitely in the Navy's sphere.[60] Bache enlisted the greatest microscopist of the day, Professor Jacob Whitman Bailey of West Point, to study for organic remains samples of sea bottom collected on the Survey.[61] Later, Louis Agassiz became an adviser on these studies. This enthusiastic newcomer, who stirred up activity in every organization he touched, soon extended his work for the Survey to the origin, growth, character, and probable future progress of the coral formations of the Florida keys.[62] Hence biological research developed out of the Survey's mission.

While its tendencies in the direction of basic research never got completely away from the practical business of the Survey, they were sufficiently extensive to give some validity to the claim that this potentially perfunctory operation was in reality the general scientific agency of the government. Bache's personality, his ties with Cambridge and with the Smithsonian, aided its preëminence. In sheer size if nothing else the Survey dominated the scene. Its appropriation for 1854, which so dwarfed the Smithsonian, was $489,537.20, the highest single one but not atypical. The appropriation in 1853 was nearly three times that for the railroad surveys. In the two years 1854 and 1855, Congress made available more money than the Wilkes expedition cost in four years' operation.[63] Before the Civil War the Coast Survey was the best example of the government in science.

Two important handicaps kept the Coast Survey from becoming a fully developed national scientific institution. In the first place, for all the continuity of policy and personnel, and even though its job lay swelling in front of it to give an illusion of permanency, the Survey remained in theory an *ad hoc* enterprise. If the day ever came when Congress considered that the coasts had been adequately mapped, the Survey's reason for existence would vanish. This distinction, academic as it was in the 1850's, tended to discourage the development of the long-range objectives that are necessary in planning scientific policy. In the second place, by being tied to a specific mission, the Survey, however deeply it went into theoretical aspects of the science within its scope, could not easily break out into disciplines that had no connection with its work. All the sciences taken up by the Coast Survey were for obvious reasons on the list of those stimulated by the exploring expeditions. Astronomy, topography, hydrography, and terrestrial magnetism received most emphasis, with a minor measure to meteorology and natural history. As with the expeditions, the laboratory side of science received little encouragement.

Thus several of the general tendencies recognized elsewhere emerge in the history of the Coast Survey. A temporary agency tended to become permanent. A practical agency tended to go into the theoretical aspects of the sciences it used. However, even when great generality was reached in this trend, a practical agency did not wander through those scientific disciplines too far from its area of primary responsibility. Finally, the expansiveness of an agency depended largely on the quality, drive, and continuity of its leadership. The Coast Survey, under two chiefs for the sixty years from 1807 to 1867, reached its pinnacle of influence under Bache.

The Naval Observatory and the Nautical Almanac

The history of the Naval Observatory from 1842 to 1860 is a contrast and also a complement to that of the Coast Survey. The Navy's Depot of Charts and Instruments, which became in fact a naval observatory in 1842, moved into its new building in 1844. It was more than the Coast Survey in that it was tied to no single project which would some day be completed. It was less than the Coast Survey in that it had no clear mission or even existence in the statutes, which allowed its program to swing through a rather wide arc according to the whim of its director. Gilliss had been an astronomer primarily, who had made both his and the institution's mark with the home observations of the Wilkes expedition. Matthew Fontaine Maury, who took over in 1842, not only had different interests but also a great intensity about his ideas.[64]

A Southerner and an officer who had distinguished himself largely by his anonymous but trenchant criticisms of the Navy, Maury in 1842 turned his attention to the oceans rather than the heavens. His great discovery was that the ships, always making their way around the world, provided a ready-made corps of observers, with ships' logs past and present as a continuous record of the face of the sea. He shifted the main emphasis of the observatory from astronomy to collecting and collating data on winds, currents, and the nature of the ocean generally. Out of this grew a world view of meteorolgy, the first comprehensive picture of the currents of the ocean, and a new organization to the science of marine physical geography. He issued six series of Wind and Current Charts — Track Charts, Trade-Wind

Charts, Pilot Charts, Whale Charts, Thermal Charts, and *Storm and Rain Charts*. Accompanying these he published *Sailing Directions*, expanding with each edition, which translated the new information into practical guides by which a mariner could choose the most favorable route.

Quick to see that Navy ships were few in number, Maury made a firm alliance with the merchant owners and captains, trading his publications for logs kept on a standard form. Since this was the heyday of the clipper ship and the American merchant flag on the high seas, since the passage around Cape Horn to California was, after the gold rush, a national route, Maury's directions paid off in cash to the merchantmen. By 1848 ships using his instructions made Rio de Janeiro in 35 and 40 days as compared to a former average of 55. By 1855 he had reduced the time of passage from New York to San Francisco from 180 to 133 days. Maury could take some of the credit for the *Flying Cloud's* record of 89 days and 21 hours.[65] Such figures broke through even to the consciousness of the President of the United States,[66] an official who usually paid little attention to science in these years. Maury's hydrographic work paid off in an equally spectacular way when he advised Cyrus Field on a path for the transatlantic cable.

By 1851 Maury began to work on a universal system of meteorological observations, leading up to an international Congress at Brussels in 1853. Here he achieved the adoption of a universal abstract log and established coöperative relations with such European scientists as the great statistician, L. A. J. Quételet. Among the commercial nations of the world, at least, Maury may well have been the most famous of American scientists. In a measure this fame has continued. A note at the top of each chart issued by the Hydrographic Office crediting Maury's research is a kind of living monument.

One group, however, that had little use for Maury was the American scientific community. The impartial Joseph Henry, who made a policy of stretching the meager Smithsonian funds by coöperating with everyone possible, came into direct conflict with Maury over meteorology. The Navy man wished to extend his observations inland, especially to include the Great Lakes, where the Smithsonian had the coöperation of a Topographical Engineers' survey under Lieutenant George Gordon Meade.[67] Although Maury was never able to carry his system of observations inland, using farmers in place of sea captains, his failure to coöperate with the Smithsonian kept Henry from coördinating all the meteorological work that had sprung up in several branches of the government.

With the Coast Survey Maury's relations were much slighter than would be expected from their community of interest. The astronomers in Bache's organization were especially bitter, feeling that Maury in replacing Gilliss had stifled their science at the observatory.[68] The *American Journal of Science* had precious few references to Maury's work, and those few stressed the belief that some of his theories were "unsustained by facts." [69] In 1858 James D. Dana, editor of the *Journal*, wrote to Bache, "Where is the review of Maury that Gould was to furnish us? . . . I should like to have his theories sifted in an article in the Journal, and the chaff shown up. The only restriction I suggested . . . was that it should be scientific instead of personal." [70] Thus the network of friendship centering on Bache deprecated Maury and his works.

The isolation of Maury from the rest of the scientific community both in and out of the government accounts for the way in which the Nautical Almanac emerged. The idea of an astronomical almanac for the use of mariners was at least as old as John Quincy Adams's time and was tied to the same nationalistic motive that had led to early work on an American prime meridian. The avowed aim of an American publication was to outdo in accuracy and convenience the *British Nautical Almanac*. When Congress made provision for one in 1849 it was in a paragraph dealing with the Hydrographic Office, administratively part of Maury's Observatory. However, the wording made possible the setting up of a separate office directly under the secretary of the Navy, and the personal connections and the ability of the first appointee emphasized its independence.[71]

Lieutenant Charles Henry Davis, the first head of the *Nautical Almanac*, was an officer who had utilized the long years ashore which marked every naval career at that time first to get a Harvard education and then to get a great deal of scientific experience with the Coast Survey. By the late 1840's he had become an intimate friend and trusted subordinate of Bache. He was also the brother-in-law of the Harvard mathematician and Coast Survey consultant, Benjamin Peirce.[72] Thus when he set up headquarters for the *Nautical Almanac* he chose Cambridge, Massachusetts, instead of Washington. Although formally correct relations with Maury prevailed, the affectionate and continuous intercourse with astronomers necessary for a good almanac took place with the Harvard Observatory, where the Coast Survey work on longitude was going on. Benjamin Peirce, already working for the Survey, became chief mathematical consultant of the new work. Benjamin A. Gould also appeared on the payroll.[73] This arrangement, besides interlocking the *Nautical Almanac* closely with the Coast Survey, emphasized the role of Harvard as a center of government scientific work and the position of Peirce as expert adviser from the academic world. The publication itself, which first appeared in 1852, was an outstanding success, continuing to the present. It established the government as the proper and most efficient source of scientific publications for the navigator. After the Civil War the Navy absorbed the two classics of the field which had earlier been published privately — Nathaniel Bowditch's *Practical Navigator* and Blunt's *Coast Pilot.*

With the rise of the Naval Observatory and the *Nautical Almanac* headed by regular officers of the line, the Navy faced a problem on how much scientific attainments would weigh in the scales of promotion. Maury was still a lieutenant in 1855, after 13 years at the Observatory, when a selection board put him "on leave of absence pay," which cut his salary and froze his rank.[74] In the popular outcry and congressional debate that followed, the naval affairs committee of the Senate proposed a scientific corps of one captain, two commanders, ten lieutenants, and seven masters who would rank with other officers but exercise no command.[75] These officers would be free to develop science in a concentrated way and would staff both the Observatory and the *Nautical Almanac.* But congressional opposition was ready with arguments against those who are "officers and not officers" at the same time. One senator felt the Navy could find "men much higher in science" outside the service and should hire them rather than maintain a "corps of savans." [76] Although the Senate rejected the proposed corps, pressure for Maury and others was so great that he was restored to the active list and promoted to commander. He stayed at the Observatory until 1861, when he went over to the Confederacy. In the fall of 1856 Davis left the *Nautical Almanac* for a sea command to protect his place on the active list.[77] Thus the Congress faced for a moment the problem of a close relation of naval officers to science and decided instead in favor of well-rounded professional competence as the prime measure of an officer in all cases.

The budding specialization demonstrated by the Observatory and the *Almanac* had raised a fundamental problem, but adjustments to meet it were far away.

Meteorology and Agriculture

Persistently cropping up among these activities in the 1850's was the reporting of weather data. Although often simply a recording of facts, meteorology became potentially a new science with the invention of the telegraph. Immediate communication of data gathered over a large region made possible both predictions and the formulation of a theory of storms. The necessity for widespread observations almost irresistibly suggested that the federal government take a hand. Efforts in the surgeon general's office, the Smithsonian Institution, the Patent Office, the Topographical Engineers in the Great Lakes Survey, and Maury in his worldwide observations on the high seas show the interest that the subject could arouse. In addition, James F. Espy, who was working on storms for the American Philosophical Society and the Franklin Institute as early as the 1830's, came to Washington in 1842 hoping to become the national meteorologist. This "Storm Breeder" was at times attached to the Navy as a professor of mathematics and to the surgeon general's office in the Army. In some form or other his friends in Congress managed to appropriate $2000 for him each year until 1859. Espy developed a network of observers and also published extensively on the theory of storms.[78]

Thus with meteorology the government faced for the first time a situation where coördination among existing groups was obviously the most acute need. In his ambition to bring together this whole complex of activities, Henry at the Smithsonian was undertaking a timely and important work. His success, however, was limited by his inability to devote much money to the project and by the partial escape from his influence of Espy and Maury. Although Espy's name appears often in the Smithsonian reports, he himself attached little significance to Henry's aid and instructions.[79] When Congress refused the inland extension Maury wished, he charged that the "Smithsonian Institution and the Agricultural Bureau of the Patent Office stole this idea and attempted to carry it out, but with what success let silence tell." [80] Thus coördination of weather work under the Smithsonian was far from complete, and Henry, with his usual stern self-denial, eventually advocated an independent meteorological department inside the government.[81] Such an agency became a reality only after the Civil War.

In terms of the economic interests of the country, the exploring and surveying activities so extensively supported by the federal government told almost entirely in favor of commerce. A counterdemand for scientific aid to agriculture was the natural response to such a one-sided program. In a diffuse way such ideas went back to the Pinckney plan in the Constitutional Convention, to the abortive American Agricultural Society of the 1790's, and to the energetic if random activities of statesmen-farmers such as Washington and Jefferson. By John Quincy Adams's time this activity took the form of circulars to government representatives abroad and lengthy investigations of the silk industry.[82] The common purpose of all such efforts in this period was to import either from Europe or from some recently explored land either a new breed or an entirely new crop which would put money into the pockets of American farmers. In its extreme form this hope amounted to the search for a panacea. At the time of their settlement the colonies had built their economies largely around plant and animal introductions,[83] and the great expeditions of the nineteenth century encouraged the hope that new crops of unprecedented yield would be found to succor the worn-out areas of the eastern seaboard.

Outstanding in the search for new agricultural products was Henry Perrine, U. S. Consul at Campeche, Mexico, from 1827 to 1837. Taking advantage of John Quincy Adams's circular letter, he flooded the State, Treasury, and Navy Departments with reports of tropical plants of medical and economic importance. In 1832 he proposed a station in Florida to work on the naturalization in the southern United States of chocolate, coffee, tea, sisal, and plants producing a wide variety of other tropical products. He hoped that these would revitalize the areas of the Southeast, ruined and then abandoned by cotton and tobacco farmers. By 1838 Congress had granted him a township on Biscayne Bay, but before he could produce any results he was killed in the Seminole War.[84]

When Henry L. Ellsworth became commissioner of patents, he almost immediately moved to make the Patent Office a central clearing house for plants, seeds, and agricultural information. He claimed that for "commerce and manufactures, much has been done; for agriculture, the parent of both, and the ultimate dependence of the nation, much remains to be done," [85] but this activity stemmed more from Ellsworth's personal interest than from its tenuous connection with the patent law. In 1838 a bill suggested by Ellsworth to appropriate $5000 for a seed depository and an agricultural clerk in the Patent Office failed in Congress. The appropriation act of the next year granted $1000 from the Patent Office fund for statistics and other agricultural purposes. Most of this was used to distribute seeds free to farmers — some 30,000 packages in 1840.[86] This practice, while pleasant both for farmers and for their representatives in Congress, did little to improve the quality of crops. Indeed, the disappointing results of this enterprise extended to the vast quantities of plants and seeds sent home by the burgeoning exploring expeditions of the 1840's and 1850's. However much they added to the knowledge of botany, they put few dollars into the pockets of farmers. No revolutions occurred, and few glimpsed the amount of complex experimentation necessary to select varieties that would really prove beneficial. No scientists of the day, much less any clerk in the Patent Office, had the knowledge at his command to make the program a genuine success. Nevertheless, the collection and distribution of seeds remained, like efforts to introduce silkworms and tea plants, a prominent part of the agricultural program of the Patent Office until the Civil War.

Ellsworth's large plans, however, included two strains of science that had little to do with the exploring activities of the government. One was the gathering of statistics on the country's agriculture, a necessary prelude to dealing with problems on a large scale. Statistical methods, while long used in astronomy and certain governmental operations, were only in the 1840's finding widespread use in the social and economic sphere in the hands of men such as Quételet. Although part of this new interest concerned agriculture, so much had to be done that the Patent Office had essentially a clear field up to the Civil War. The report of the commissioner of patents for 1843 contained a large amount of statistical information.[87]

More important was Ellsworth's call in 1841 for the application of chemistry to agriculture. Chemical analysis approached more closely the great problem of soil exhaustion than did importation of new plants. The timing was most important, for 1841 was the year after the German chemist Justus Liebig's work, *Chemistry in Its Applications to Agriculture and Physiology*, was published in America. By placing the knowledge of the composition of soil on a chemically sound basis and by connecting this with the nutritional needs of plants, Liebig pointed to a general and direct application of science to agriculture. More important than his actual scientific work was his great ability and zeal as a publicist. His works carried the gospel of soil analysis far beyond scientific circles. His laboratory at Giessen became the mecca for young American chemists, who worshiped at the feet of the master and then returned to preach the word to their native land. The agricultural press went wild about soil analysis and promised direct and easy results which could not possibly follow, so that by 1860 a general disillusion prevailed about the immediate usefulness of soil analysis; but the larger validity of chemistry as a weapon against soil exhaustion remained.[88]

John Quincy Adams, who declared that Ellsworth had "turned the Patent Office from a mere gimcrack-shop into a highly useful public establishment," attributed the stiff opposition to his program to John C. Calhoun and other Southerners. Since Ellsworth was the kind of supporter of science in the government of whom Adams approved, he felt that the commissioner's leaving office soon after Polk's inauguration was the result of iniquitous political pressure.[89]

Although no commissioner of patents following Ellsworth showed his driving interest, the annual reports continued to swell with agricultural information along the lines he established. The agricultural function slowly became differentiated and more definite. After 1844, appropriations gradually increased to $5000 in 1850 and $10,000 in 1854. Between 1856 and 1862 they varied from $30,000 to $75,000.[90] At their peak these figures do not approach an average appropriation for the Coast Survey. In 1848 Congress specifically granted $1000 for "chemical analyses of vegetable substance produced and used for the food of man and animals in the United States." After the shift of the Patent Office to the new Interior Department in 1849, a kind of agricultural division was set up, headed first by a farm journalist named Daniel Lee. He stressed the need for "institutions designed expressly to develop new truths in agriculture." Under his successor, Daniel J. Browne, also a journalist, reports continued to touch scientific subjects — for instance, the first entomological paper by Townend Glover, which appeared in the Report of 1854. In 1858 a five-acre propagating garden was laid out in Washington. The purpose was not experimental, but to serve as a place to grow plants for distribution — 30,000 tea plants and 12,000 grapevines in 1859. In 1860 the personnel of the division included the superintendent, four clerks, a curator or gardener, and some assistants.[91]

The very inadequacy of the Patent Office program for agriculture led to two movements outside it in the 1840's and 1850's toward a more adequate federal agricultural establishment. As usual, the governments of the several states were already active, granting bounties to agricultural societies and appointing boards which sometimes had close relations with scientists.[92] One great push in the federal government was for a separate agricultural bureau, and the other was for a system of agricultural schools. Both Zachary Taylor and Millard Fillmore urged a bureau, placing a good bit of emphasis on the importance of chemistry.[93] By the middle 1850's the United States Agricultural Society, a group whose aim was to secure the establishment of some sort of national bureau, had given up the idea of doing anything with the Patent Office and was agitating for a separate department.[94] Although nothing tangible happened before the end of Buchanan's administration, the idea of a department had developed with adequate political backing from pressure groups who had a general understanding of the possible help science could be to the farmer.

The use of federal public lands to support agricultural and mechanical colleges also became a practical and important proposal by the middle 1850's. Many schemes appeared in different parts of the country. For instance, Jonathan B. Turner, a professor of Illinois College, began a crusade for "Industrial Universities for the People" and established the Industrial League of Illinois. The essential features emerging from these plans were the donation of land instead of money by the federal government, and the emphasis on practical agriculture and mechanics, in the way advocated earlier by some of the memorialists concerning the Smithson bequest.

In 1857 Justin S. Morrill of Vermont introduced a bill for land-grant colleges into Congress. Against constant opposition, largely Southern in origin and states' rights in argument, the bill struggled through both houses by 1859, only to founder on the constitutional scruples of President Buchanan, whose veto stood. Significantly, Jefferson Davis of Mississippi, staunch friend of the Smithsonian and the organizer of the railroad surveys as secretary of war, led the opposition in the Senate. A potential means of federal aid to agricultural research thus had reached definite legislative form when the crisis of the Civil War engulfed the nation. It is hard to avoid the conclusion that sectional strife held both the Department of Agriculture and the land-grant colleges in abeyance in the late fifties, and that their passage awaited only the release of the tension between North and South.[95]

When in the fall of 1860 the election of Lincoln produced the long-lowering storm, the United States still hesitated to embrace the theory that the government should have a permanent scientific establishment. The concept of the Union as a federation of states was still a powerful argument against a forthright commitment to the support of science. Hence the Smithsonian Institution arose from an accident and maintained an element of universality and independence. The explorations and the Coast Survey were in theory temporary task forces directed at specific goals. The Naval Observatory and the *Nautical Almanac* grew up under the guise of caring for charts and instruments. Sciences that only the government could easily coördinate, such a meteorology and aid to agriculture, failed to find adequate organizational expression. The attempts of the government to use science in regulation and aid of technology had been timid and intermittent.

In practice, however, the force of American expansion and the growth of science within the country had made these theoretically temporary and disguised institutions into an establishment which along some lines had impressive strength. The Smithsonian, its National Museum, the Coast Survey, and the Naval Observatory were active organizations with continuity of policy and of personnel. They have endured to the present with their form and organization of 1860 unaltered in essentials. The great exploring expeditions of the period laid the basis of the National Museum and of several important private collections in natural history. Taken altogether, if states are included, government activities were possibly the largest and most important source of funds and employment for science in the country. The federal government surpassed both the private societies and the colleges as a patron of science.

VI

BACHE AND THE QUEST FOR A
CENTRAL SCIENTIFIC ORGANIZATION
1851 – 1861

AS the federal government became more deeply involved in scientific activities in the 1840's and 1850's, the problem of a central organization took on a new form. The national university had never been born. The Columbian Institute and the National Institute had failed to flourish. The Smithsonian Institution had purchased survival by trimming its activities and influence to conform to the smallness of the income from its endowment. Yet the need for some kind of coördinated effort became more apparent with every accretion to the government's scientific establishment.

Meanwhile, the profession as a whole was reaching a level of maturity that allowed it to organize itself. As early as 1840, ten geologists, mostly connected with the state surveys, met in Philadelphia to form the Association of American Geologists and Naturalists. In 1848, this organization, with a widened and open membership, adopted a new constitution based on that of the British Association for the Advancement of Science. On the drafting committee were the geologist Henry Darwin Rogers, Benjamin Peirce, and Louis Agassiz. This American Association for the Advancement of Science immediately became the main meeting place of the scientists of the United States. in 1848, it had 461 members, and by 1854, the number had risen to 1004.[1]

Although the new organization had no direct ties to the government, it was a means by which the scientists could express their views on public policy. As early as 1849, the Association induced the State Department to make representations in behalf of a Professor Schumacher, whose work was impeded by unsettled conditions in Schleswig-Holstein.[2] By 1851, of nineteen special committees, eight aimed directly at the federal government. The subjects included Johnson's experiments on coal, Maury's wind and current charts, the prime meridian, the Coast Survey, uniform standards of weights and measures, the use of public lands to aid Missouri in a geological survey, scientific exploration, and the corps of observers on the Mexican Boundary Survey. Scientists in the government service seemed to dominate the organization. Henry appeared on five committees, Baird on three, C. H. Davis on two, and Bache on six. Of the officers in 1851, Bache was president, Baird corresponding secretary, Henry and Wilkes members of the standing committee. Even the most prominent college scientist, Benjamin Peirce of Harvard, was a consultant on the Coast Survey and the *Nautical Almanac*.[3]

The significance of the connection of the government to the new AAAS was not lost upon Bache, the superintendent of the largest and strongest segment of the federal scientific establishment. When he delivered his address as retiring president at the Albany meeting in 1851, he showed that he recognized the problem in its broadest dimensions. He saw that of the older societies only the American Philosophical Society and the American Academy of Arts and Sciences had "struck very deep roots," and that neither was well endowed. He saw that American science had labored under the evils of "the prevalence of general lecturing on various branches, the cultivation of the literature of science rather than of science itself." Pointing out that the Institute of France paid its members "a moderate support, that the country may have the benefit of their labors," he insisted that researches such as those of the Franklin Institute on steam boilers fell short because "the laborers were without hire, though neither they nor their works were deemed unworthy of it." Indeed, he reckoned as one of the largest "obstacles to the progress of science with us" the want of "direct support for its cultivators as such." He saw too that the Smithsonian, "had it fivefold its present endowment . . . would not be able to meet the actual demands upon its funds for purposes in its 'active operations.' " He saw that the organization of a scientific association had to wait on the development of professional standards, with the geologists leading the way because the state surveys gave the basis for testing competence by "positive work." He saw that the AAAS, now that it had come into existence, had its role to play in advising the government, which was "called upon often to decide questions which belong rather to scientific than to political tribunals.

A timely recommendation by a scientific congress would frequently be a relief from serious embarrassment." However, both lack of money and of "that working spirit . . . which alone could bring experiments to a working conclusion" severely limited its committees.

Beyond this general pattern of the institutions of the country in 1851, Bache saw that the government could be a positive force in the advancement of science. For instance, he attributed the relatively advanced state of astronomy to its being "chiefly at first from its connection with navigation . . . the science which all governments, our own inclusive, have selected to encourage." The same thing could happen to meteorology if it had the patronage. "The results of even the partial effort made in behalf of magnetism and meteorology" were hopeful, with materials "gathered, or gathering, from which important conclusions are daily derived, and which await the master mind to weave into [a] new 'Principia,' a new 'Mécanique,' or a new 'Theoria.'"

Bache then declared that *an institution of science, supplementary to existing ones, is much needed in our country, to guide public action in scientific matters.*" He disclaimed any antidemocratic bias in the suggestion. "It is . . . a common mistake, to associate the idea of academical institutions with monarchical institutions," and he warned against "the two extremes of exaggerated nationality and of excessive imitation: let us modify each by the other, and be wise." He also fended off the inevitable states' rights argument, for "the idea of a necessary connexion between centralization and an institution [does not] strike me as a valid one." The proposed body would have members who "belong in turn to each of our widely scattered States, working at their places of residence, and reporting their results; meeting only at particular times, and for special purposes; engaged in researches self-directed, or desired by the body, called for by Congress or by the Executive, who furnish the means for the inquiries."

Avoiding the mistake of the old National Institute, Bache would not depend solely on the "men of science who are at the seat of the government," as they were "too much occupied in the special work which belongs to their official occupations." The heart of his proposal was appropriations from the "public treasury," which "would be saved many times the support of such a council, by the sound advice which it would give in regard to the various projects which are constantly forced upon" the notice of government officials, "and in re-

gard to which they are now compelled to decide without the knowledge which alone can ensure a wise conclusion." The spheres of activity were already quite clear. "Without specification, it is easy to see
that there are few applications of science which do not bear on the
interests of commerce and navigation, naval or military concerns, the
customs, the lighthouses, the public lands, post-offices and post-roads,
either directly or remotely." This new institution of science would
step into an area otherwise "left to influence, or to imperfect knowledge." [4]

Two assumptions underlay all that Bache said that day in Albany.
One was that only through the professionalization of scientists and
the "minute subdivision" of their efforts in specialties could real research go forward. The other was that "science" meant to him essentially those branches which the surveying and exploring enterprises
of the government had stimulated. Because of his own specialties he
put great emphasis on mathematics, physics, and astronomy. Admitting a good deal of prominence to geology, he gave a little grudging
recognition to descriptive natural history. But chemistry, laboratory
biology, and the application of these fields to agriculture do not
enter into his scheme at all. An institution that sprang from his ideas
might be expected to be a group of professional specialists whose interests heavily favored the physical sciences.

Although Bache's institution got no overt support in the 1850's,
the idea never died. It could live a kind of subterranean existence in
high places because of the extraordinary importance of a small group
of men. Scientific organization had reached a new state not only in
the AAAS, but in small self-conscious gatherings of professionals who
recognized their common goals and their differences from the older
generation. The body of scientific men was now numerous enough for
comradeship and still small enough not to be impersonal. As early as
1847, one scientist discovered that the "fewness of men well-grounded
in science, and the disparity that exists between those claiming to be
adepts" made especially likely in America "the formation of predominant cliques." [5]

As Bache himself stood before the AAAS in Albany to give his address as retiring president, he had another title also, the "Chief" of the
Scientific Lazzaroni. The membership of this group, whose stated purpose was to "eat an outrageously good dinner together," centered in
Bache, Henry, Peirce, Louis Agassiz, James D. Dana, C. C. Felton of

Harvard, John F. Frazer of Philadelphia, the astronomer Benjamin Apthorp Gould, and the chemist Wolcott Gibbs.[6] At that very meeting, the Lazzaroni were pushing for the establishment of a private national university in Albany. Later they tried something of the same thing in New York, and for a time backed the Dudley Observatory in Albany.[7] Their enemies, including some eminent scientists, conceived of them as a clique either connected with the Coast Survey, because of Bache, or located in Cambridge, because of the towering eminence of Peirce, Agassiz, Felton, and Gould. Although all these men had important friends outside the group, the fact that they knew each other, saw each other regularly, and often coöperated was a condition of some importance on the scientific scene of the 1850's. The Lazzaroni took Bache's ideas of an institution of science seriously, and the time would come when they would do something about it.

VII

THE CIVIL WAR

1861–1865

THE Civil War, which broke out among a completely unprepared people in 1861, was unique in its technological problems and their impact on the government's ability to use science. In the "last of the old wars and the first of the new," [1] the battles on land were not impossibly different from those of the Napoleonic era, which served as models for the officers of both sides. And yet, the introduction during the years of peace of the railroad, the telegraph, and innovations in the construction of guns had changed the pace of warfare and upset the balance of offense and defense, producing a new pattern in which the old rules, based on a relatively static technology, offered poor guidance. At sea, where technology had affected the methods of warfare more deeply, the change was dramatic. In addition, the very size of the Union war effort called for organization and efficiency in fields in which the federal government had done very little before.

The Application of Science to Technology

The use of railroads illustrates how the technological problems of the Civil War were ones of organization and administration. Herman Haupt and D. C. McCallum, the leading figures in adapting the railroads to military purposes, were both experienced engineers and managers. Their essential contribution was to apply the railroads as they knew them to the urgent war problems of supply and transport. In the course of their work, many incidental innovations appeared — special equipment, such as hospital cars, more efficient methods of construction of both track and bridges, barges for ferrying freight cars, and techniques for destroying track and locomotives effectively. But this kind of experimentation was largely done on the spot, when the men were actually confronted with a specific and pressing problem to solve.[2] Most of the roads remained in private hands, and, although McCallum at one time had some 10,000 men in his construction corps,[3] no federal organization emerged with a creative attitude toward railroad technology. For example, the large-scale manufacture of steel was very close to practicality at the outbreak of the war, and, with the heavier traffic wearing out iron rails faster than they could be replaced, the Union could have gained important advantage by a research and development program in steel. Instead, the experiments on steel rails that took place in 1863 and 1864 were at the Altoona yards of the Pennsylvania Railroad. Only gradually did steel rails begin to be used, and all installed before the end of the war were imported from Europe.[4]

A Civil War "Manhattan District" to reduce a theory to military practice, as later with the A-bomb, would have had great opportunities. Steel, high explosives, and the germ theory of disease come immediately to mind. However, this was unthinkable because no one could conceive of the government in the active role of first selecting research problems and then hiring and organizing scientists to pursue them. Rather, the responsible heads of the military departments were swamped with proposals and inventions of all sorts, some with merit and many more without, some by men unknown and others by those who could exert powerful political pressure. An observer from Europe, the Comte de Paris, early in the war described the process by which "every inventor who had any patronage could easily manage to have a few of his guns recommended to the principal of some foundry, who was generally his partner." Research was simply a "few shots fired in the neighborhood of the factory," and "if chance favored them, the piece was immediately received and added to the diversified assortment which already existed in the Federal artillery." [5] Thus, the most pressing need was for experiments to determine the value of inventions already developed and presented to the government for adoption. It was the war and the peculiarly public and centralized nature of military problems that ineluctably forced the government into the business of using experimental science to evaluate technology.

In these research efforts, spasmodic and on the fringes of the great organizational accomplishments of the war, a consistent relation between science and technology appears for the first time. This new union, of course, far exceeded the bounds of the war effort, because it was one of the most significant changes of the nineteenth century

— the one which in the popular mind eventually linked science with the production of material wealth and with the enormous additions to the power of man over his environment. Sometimes technology benefited directly by the application of a principle discovered in the general pursuit of knowledge. More often science entered technology in the guise of replacing cut-and-try with more orderly forms of experiment. Rule of thumb no longer sufficed as it had even in the early days of the steam engine. A new research technology was growing up which brought science into all forms of engineering. Whether or not adequate guiding principles were at hand, a well-ordered empiricism replaced the hit-or-miss technical knowledge of the earlier age. The government, in its war effort of the 1860's, had to grapple with this new research technology for the first time.

The Navy was most intimately involved in technological change because of the sevenfold revolution in sea warfare which in 1861 was well under way in Europe. The introduction of steam, the screw propeller, the iron ship, and armor completely altered naval construction, and these changes were forced by new developments in ordnance — the fuzed shell with explosive charges, the built-up rifled gun, and slow-burning powder.[6] By 1860, the French navy especially and the British navy partially had gone a sufficient way with these changes to render completely obsolete the wooden sailing ships of the line of Nelson's day.

The United States Navy, removed from the neighborhood of the European armament race and chained by the theory that its mission was to raid commerce on the high seas and to defend the coasts, had fallen appreciably behind European powers, notably France. Although it had no consistent policy toward the revolution, it did have a considerable record of technological innovations. Robert Fulton's *Demologos*, launched in 1815, was a false start for the application of steam to warships, but from the 1830's onward, with prodding from such enthusiasts as Matthew C. Perry, the Navy built steam warships and experimented with the serious problems of protecting vulnerable machinery. For instance, the *Princeton*, commissioned in 1844 under Captain Robert F. Stockton, had a screw propeller and an engine below the waterline designed by John Ericsson. The explosion of a 12-inch wrought-iron experimental gun aboard her not only killed two members of Tyler's cabinet but set back the cause of technological change in the Navy considerably.[7] One experiment with an iron-hulled warship on the Great Lakes made no permanent impression.

The most important lash to change in naval warfare was the development of a more powerful ordnance. Major George Bomford of the Coast Artillery had developed a shell gun during the War of 1812, and by 1820 the French Paixhans gun had definitely upset the balance between offense and defense at sea.[8] From the 1830's on, each international crisis produced its spate of proposals from inventors hoping to design some kind of adequate armor protection from shells. The most famous of these was the "battery" of Robert L. Stevens. A fast ironclad with long guns for both shot and shell, its plans were approved by Congress in 1842. In spite of $500,000 spent by the government, the ship was still incomplete in 1855 and the contract still pending in 1861.[9]

Much more significant to the Navy that actually fought the Civil War was the career of John A. Dahlgren. After a scientific initiation in the Coast Survey under Hassler, he was assigned to his first ordnance duty in 1847. By the 1850's, Dahlgren was continuously experimenting with guns. Paralleling the work of Captain T. J. Rodman of the Army, he gathered accurate data on the pressure at various points along a barrel as the projectile passed. Based on this information were his bottle-shaped cast-iron guns with the greatest reinforcement at the points of greatest stress. In 1857, the Navy got an appropriation of $49,000 for an experimental gunnery ship on which Dahlgren proved his 9- and 11-inch guns in service afloat.[10] In 1859, as head of the ordnance department of Washington Navy Yard, he supervised the founding of all light guns and the handling of shells, fuzes, and the like. In addition, his shop tested iron and powder. Since he could deal with the Navy Department only via the Yard Commandant, he aspired to make ordnance an independent organization with the status "enjoyed by the Naval Observatory and the Naval Academy." [11]

In the months immediately preceding the outbreak of the war, Dahlgren was busy testing rifled guns. An old idea which had been applied to small arms for centuries, rifling provided obvious advantages in the control and penetration of shells. But from the point of view of 1859, the increased pressures, especially with the fast-burning powders of the day, made guns built of any metal then available liable to burst on firing. Although aware of tests in Europe on the use of steel and wrought iron, Dahlgren relied on "the very superior character of our own cast iron," and hence supported smooth bores as against rifling.[12] Thus, the Civil War Navy fought with heavy ordnance which appears in retrospect highly transitional. Much strengthened by the built-up principle, the 9- and 11-inch Dahlgrens were nevertheless smooth-bore muzzle-loaders which threw solid shot as well as shells. The ordnance revolution was part way around in 1860, and the Navy, as personified in Dahlgren, was able to follow the turn but not to hasten its rate.

The major gains during the war itself in handling ordnance testing were, as is usually the case in this conflict, improvements in organization. With the reorganization of the Navy Department in 1862 by Secretary Gideon Welles, a separate Bureau of Ordnance was established, with Dahlgren as chief. At the same time he retained direct command of the "Department of Experiments" and as much of the Washington Navy Yard as he needed. By thus combining the freedom of action of a bureau chief with actual facilities directly under his eye, Dahlgren had what he had yearned for before the war — an adequate and independent administrative structure to deal exclusively with ordnance. Even so, routine tended to crowd out large-scale research. He complained that the 15-inch gun he developed during the war "as an experiment on a large scale" suffered because "circumstances impose the necessity of proceeding without full tests." [13]

Since the department never turned down an offer of a private firm to manufacture guns, the bureau was deeply involved in coöperating with private enterpreneurs in the production of ordnance. Dahlgren spent much time in furnishing specifications and standards, and in testing guns in experimental batteries. The complex nature of the relations between the bureau and private manufacturers is illustrated by Dahlgren's successful efforts to develop a supply of potassium nitrate, necessary for gunpowder, in place of the former complete dependence on British India.[14]

Dahlgren also had to deal with the swarms of inventors.[15] He recounts being called to the White House, where Lincoln had received a sample of gunpowder. The President went to the fire, "clapped the coal to the powder and away it went, he remarking, 'there is too much left there.' " In general, Dahlgren wished Lincoln and other high officers "would not meddle in such matters," for "projects for new cannon, new powder, and devices of all kinds are backed by the highest influence." [16] Most of the inventors he considered "unscrupulous scoundrels . . . whose sole aim is to rob the United States." [17]

Behind the interview with Lincoln lay a curious tale which illustrates the devious channels by which inventors approached the government. One Captain Isaac R. Diller of Illinois had both an idea for the manufacture of gunpowder and access to the President. Further, he lined up Charles M. Wetherill, a chemist in the government's agriculture establishment, to help him. Lincoln arranged for Wetherill to be released from his regular duties and also to receive his salary while preparing powder for testing. Dahlgren then got the job of evaluating the new explosive, for which Diller made extravagant claims. His real object was to keep the process secret until he could sell it to the government for $150,000, and Lincoln went so far as to recommend this expenditure if the powder passed the tests put to it by "the officers, or other skilled person or persons, I shall select." After a year's negotiation and testing the episode closed in December 1863, when the President declined to appoint a new board and gave Diller permission to patent his secret.[18] In addition to intervening in the domain of both the Bureau of Ordnance and the Department of Agriculture, the President inadvertently came dangerously close to opening up a large field for adventurers who might use government funds to develop secret processes and then sell the results to the government for a high price.

The Navy's efforts in steam engineering followed much the same pattern as ordnance. In the reorganization of 1862, B. F. Isherwood, who had grown up in the Navy with steam, became head of the new Bureau of Steam Engineering. By the close of the war, Gideon Welles reported that "nearly every variety of boiler and of expansive gear, of rate of expansion, and of saturated and superheated steam has been made the subject of accurate experiment, and it is believed that the files of the department contain the latest and most reliable information on these subjects." The quality of various types of coal and even of petroleum received extensive tests.[19] Isherwood published volumes of his *Experimental Researches in Steam Engineering* in 1863 and 1865, in which he worked out methods for determining the limit of expansion of the size of steam engines.[20] An important secondary effect of the war was the introduction of steam engineering into the curriculum of the Naval Academy and the general expectation that knowledge in this field was a necessary part of a professional officer's equipment.[21] The steam navy that blockaded the South was a make-shift affair, and its wholesale scrapping immediately after the war has obscured the effort and engineering skill required to launch it.

When the war broke out the problem of protecting the vital parts of wooden ships with iron plating forced itself on the Navy Department because of the recent heavy investments of the European nations in new ironclads and because of the activities of the Confederates. In the summer of 1861, Welles appointed a board made up of Commodores Joseph Smith and Hiram Paulding and Commander Charles Henry Davis to study the subject. While professing ignorance, they approved three of seventeen plans submitted to them, among them John Ericsson's *Monitor*, which was built in the New York area.[22] Large-scale construction of these vessels, whose most nearly novel feature was the revolving turret, was well under way even before the famous encounter with the *Virginia* in Hampton Roads. Instead of a bureau in Washington taking over this program, a sort of monitor office under Rear Admiral F. H. Gregory grew up in New York. This "draftsman's paradise" supervised all construction, with special emphasis on ironclads built by private concerns. Forty-five or fifty officers were under Gregory, stationed at the several contract shipyards.[23]

Thus the Navy in the Civil War came to terms with every important phase of the technological revolution that affected it. Under constant criticism from outside and riven by internal controversy, the department nevertheless managed to find officers well qualified to handle the new research technology and put them in administrative positions where they were able to act. In no important way did they further the naval revolution, but to keep pace with it was a major accomplishment which hinted at the government's potential ability to apply scientific procedures to technological problems.

The Army, while subject to some of the same forces as the Navy, especially in ordnance, made a far less articulate response. Eli Whitney's use of the principle of interchangeable parts while making firearms for the government early in the nineteenth century was a vague forerunner of a new technology in ordnance. A national foundry — an idea somewhat akin in administrative structure to a national observatory — was the subject of voluminous official correspondence from the 1830's onward, without any practical result.[24] The work of

Captain Thomas J. Rodman in determining the "curve of pressures" on a gun's barrel and in designing an iron gun cooled from the inside in casting paralleled Dahlgren's experiments in the Navy. The army officer went even further to try cakes of powder instead of grains to slow down the time of explosion. Like Professor Daniel Treadwell of Harvard, who was working in the same field, Rodman got no encouragement from the government in the 1840's. Only in 1859 did the Army accept his guns. Ironically, the outbreak of the war interrupted his research in powder.[25]

Another West Point graduate, Robert P. Parrott, who was connected with an important foundry in New York State, developed a wrought-iron rifled cannon built up in coiled cylinders.[26] Although manufactured in large quantities, these guns had an evil reputation in the service because of their habit of bursting.

In the case of small arms, the War Department's record is much worse than in heavy ordnance. Various breech-loading rifles and a good many repeaters using fixed ammunition were already on the market in 1861,[27] while inventors of all sorts came up with hundreds of new ideas. Yet, near the end of the war, the chief of ordnance, freely admitting the superiority of breech-loaders, was still plaintively calling for the appointment of a board of review to decide on standard weapons to be "exclusively adopted for the military service." [28] The abject failure of the Army to develop any system of testing makes the Navy Department's accomplishments more impressive.[29]

Military Aeronautics and Medicine

Of the many proposals for unconventional weapons — torpedoes, machine guns, rockets, liquid fire — none has been more often referred to as a harbinger of the total wars of the twentieth century than the observation balloon.[30] Actually used by the French revolutionary armies of the 1790's, the balloon had become familiar to Americans by exhibitions from the 1830's on, and a new vocation of "aeronaut" had developed by 1860. The use of balloons had been seriously proposed to Secretary of War Poinsett during the Seminole War.[31] In 1861 various experienced aeronauts with their equipment either volunteered for service or were sought out by some official.

Since ballooning clearly involved scientific principles in making and handling gas and in utilizing what little information the contemporary meteorology could give concerning wind currents in the upper air, Joseph Henry was early consulted. Because his opinion carried weight in the War Department, his reports did much to convince officials that ballooning was practical and militarily useful. Also, his endorsement of the work of Thaddeus S. C. Lowe gave that aeronaut a preferred position over his rivals for the headship of a balloon corps.[32] Technically ingenious and showing marked energy, Lowe conquered the problems of manufacturing hydrogen in the field, developed techniques of aerial observation, signaling, and telegraphy, and struggled to procure unusual supplies such as sulfuric acid amid the red tape of an inelastic Quartermaster Corps. In the summer of 1862, he had seven aeronauts in the field at one time.[33] Although unfamiliarity with observation technique and the difficulty of operating in the field made the balloons only mildly useful instead of a spectacular success, Lowe's main trouble came from administrative snarls. Neither he nor any of his aeronauts had military rank, and, as a civilian organization trying to coöperate tactically with the Army, the balloon corps could work efficiently only when an understanding officer such as General A. A. Humphreys gave it sufferance. Between 1861 and 1863, the corps was shunted from the Topographical Engineers to the Quartermaster Corps to the Corps of Engineers. On the eve of Gettysburg an effort to put it under the Signal Corps foundered when Colonel Albert J. Myer refused to accept it. The balloon train was disbanded at Washington with nearly two years of heavy fighting yet to go.[34] Aerial observation collapsed not because of any technological backwardness, but from the failure of the Army to incorporate a new procedure into its operations.

The Army's use of medicine, which had hardly yet attained the status of a science resting on the results of research, was undertaken on a huge scale.[35] The problems of disease in the mass armies and the thousands of battle casualties forced the moribund Army Medical Department to come to life. Completing the last full year of peace with the absurd total of 98 officers and an 1860 appropriation of only $90,000, the Medical Department by 1865 was spending $20,000,000 — more than the budget of the entire Army before the outbreak of the war.[36] An ambulance corps, field services, base hospitals, and mass medical examinations had to be created entire. The selection of doctors, systems of inspection, and supervision of sanitation necessitated an energetic central organization. Much of the credit for these administrative gains belongs to Surgeon General William A. Hammond, in spite of his dismissal late in 1863.

Because of the inadequacy of the Medical Department early in the war, the Sanitary Commission, appointed by Secretary of War Cameron, had widespread powers of inspection. Not the least of their accomplishments was the forcing of Hammond's appointment over the heads of many senior officers. Of the eight original members of the commission, three were primarily scientists. Bache was vice-president, while J. S. Newberry and Wolcott Gibbs, although doctors of medicine, had gained their reputations as geologist and chemist.[37] As the department became better organized, the commission withdrew from inspection and concentrated on relief — the prototype not of a research organization but of the Red Cross.

Army medicine only occasionally and incidentally undertook activities important to research. The large number of cases seldom encountered in peacetime practice stimulated an interest in records and statistics. The multivolume *Medical and Surgical History of the War of the Rebellion* is a monument to the emphasis on record-keeping in the Medical Department and was raw material for later research. The astronomer B. A. Gould turned his hand to statistics for the Sanitary Commission in his actuarial studies on the heights, ages, and peculiarities of soldiers.[38]

Besides reports, the Medical Department also collected specimens which became a nucleus for an Army Medical Museum. Even more impressive than the original idea is the fact that after the war the Army took over Ford's Theater in Washington to continue the museum permanently.[39] A striking example of the unpredictability of institutional development is the way in which the Civil War also gave an impetus to medical bibliography. In 1865 the Surgeon General's Library had only 2253 books. But a young medical officer and $80,000 of hospital funds left over from the war started the Army Medical Library on its way to becoming the unrivaled center of medical bibliography in the United States. John Shaw Billings made from a purely auxiliary collection of books an institution of prime importance not merely to the Army or even to the government but to the whole medical profession.[40]

In a few cases individual doctors in army service utilized their experience for research. S. Weir Mitchell, who had had a taste of French medical science during a year at Paris under Claude Bernard, took up neurology while an assistant surgeon at Turner's Lane Hospital at Philadelphia. His *Gunshot Wounds and Other Injuries of Nerves* (1864) was the beginning of a long and distinguished career in neurology.[41]

For medicine, as for ordnance and ship construction, the Civil War occurred just before revolutionary developments made the practice of its period almost immediately obsolete. The recent discovery of anesthesia had only partly solved the problems facing surgery. The germ theory of disease, already close to effective formulation in Europe, would soon make the medicine of the Civil War armies appear barbarous. Yet the government here for the first time dealt with medical and health problems on a large scale and could never again retreat to the complete indifference of the prewar period. The urge to collect information on the mass medical phenonema going on before the eyes of the army doctors led to the important permanent institutions of the Army Medical Museum and Library. Their collections continued to be of use after a new age had dawned.

In both technology and medicine the government groped for science almost unconsciously, driven by the needs of the hour. Research technology, now just beginning to enter the realm of the possible, was such a novel relation that only in a few fields, such as ordnance and steam engineering, did the government effectively practice it. Without previous experience, officials got good results where they succeeded in creating sufficient administrative framework to bring existing scientific knowledge and practice to bear. Their failures sprang from lack of organization. In no case did they attempt to create new knowledge, and they seldom attempted to work out new applications of known principles which had not yet been reduced to working models. They took both fundamental and applied science as it was.

The Services of Existing Agencies

A better measure of the relation of science to the war is found in the research agencies that already existed in the government. In the Smithsonian Institution, the Coast Survey, the Naval Observatory, and the Topographical Engineers, the problems of organization and personnel were more or less under control. The fact and form of their existence were already apparent. It is possible to ask of them what their science did for the war, and in turn what the war did to them.

The Smithsonian Institution, with its endowment fixed in size and committed in purpose, continued operations as nearly normally as possible in 1861, a difficult task not only because such old friends as Jefferson Davis were leaving the capital forever, but also because Secretary Joseph Henry himself felt that the Union was doomed and the South should go in peace.[42] Only his inflexible disdain for all things political led him to keep the Institution aloof from "inauspicious connexion" with the "exciting subjects of the day." After 1863 he would not even allow the lecture hall to be used by groups "over which the Institution has no immediate control," a rule reflecting the jeering he had received from an antislavery group when he insisted on an announcement of the Institution's neutrality.[43]

Nevertheless, Henry specifically recognized the connection of science and warfare, during which "truths are frequently developed . . . of much theoretical as well as of practical importance. The art of destroying life, as well as that of preserving it, calls for the application of scientific principles, and the institution of scientific experiments on a scale of magnitude which would never be attempted in time of peace." In this endeavor Henry claimed that the Institution "continually rendered active coöperation and assistance." [44] Instances of actual war work appear in the reports only occasionally, such as the manufacture of "disinfecting liquid" in the chemical laboratory.[45] The potential use of the Smithsonian building by the government at the outbreak of the war took the form of a request to quarter troops there. Henry, though keenly aware that this would put the Institution out of business, suggested that if the government wanted the building they should seize it, and that "it would be more in accordance with the spirit of the Institution to employ the building as an infirmary." [46]

In general, however, the Institution attempted to keep going as usual under difficulties. The museum continued to support expeditions where it could, even, in the case of Elliott Coues in New Mexico and Arizona, utilizing the Army in the old coöperative arrangement. Robert Kennicott was active in the Canadian northwest.[47] Collections came in steadily, and much effort went into the distribution of duplicate specimens, in an attempt to clear away the backlog sent over from the Patent Office. The meteorological work of the Smithsonian also continued with the publication in 1864 of a part of the observation of the period 1854–1859.[48]

But every phase showed curtailment by the war. The weather-reporting system was "very much diminished" both by the defection of the Southern states from the observing system and because public business crowded weather bulletins off the telegraph.[49] More serious, the war struck at the lifeblood of the Institution, its endowment. By bad luck some of the funds were invested in the stock of Southern states. As the war went on and inflation set in, "the increased price of printing and other articles" hampered the publication program, and with "the high premium on gold" all international purchases and the exchange service became more expensive.[50] Thus the Smithsonian, far from being able to expand its functions to include war work, found its position in the scientific life of the country relatively shrunk along with the real value of its endowment. As an Institution it responded to the call of the government in only one effective way — the lending of the services of its distinguished secretary.

Joseph Henry gave as much time as he could spare from his regular duties to the government. His reports on the balloon ascensions of Lowe show the influence of his advice. A personal acquaintance with Lincoln, which ripened during the war years, brought many requests from the White House on all sorts of subjects, even including the phenomena produced by a spiritualist medium.[51] While he continued active in his research for the Lighthouse Board, especially on navigation in Confederate waters, his interest in fog signals did not begin until 1865.[52] By far the largest amount of work done by Henry for the government came from his membership in the Navy's Permanent Commission, as will presently appear.[53]

The Coast Survey, at the height of its power in 1861 but vulnerable because of its theoretical lack of permanence and its dependence on service personnel, seemed in danger of being completely ruined by the war. Envious people "remarked to Mr. Bache, in a tone of condolence, but with a smile of satisfaction, that they supposed the coast survey would be stopped now." Such expectations did not consider Bache's "remarkable talent" for using his organization to coöperate with the "great movements of the day." He made special surveys, distributed maps, and sent out parties of his assistants to serve directly with forces in the field.[54] These activities extended to inland rivers as well as the coast.[55]

Bache was not content to let this coöperation take place on a low level or become involved in the ambiguities that wrecked the balloon corps. In May 1861 he proposed "a military commission, or advisory

council, to determine proceedings along the coast." [56] Composed of
Captain Samuel F. Du Pont, U.S.N., Bache, Major J. G. Barnard
of the Army Engineers, and Commander C. H. Davis as secretary,
this secret commission was not only "to condense all the information
in the archives of the Government . . . useful to the blockading
squadrons" but also to choose objectives for amphibious operations
along the coast.[57] Thus on Bache's initiative the Navy and the Army
both had ready access to the Survey's information and had an organization capable of translating these data into effective military decisions. The reports of the commission covered the whole coast, and
one of the fruits of its work was Du Pont's successful attack on Port
Royal.[58]

Like Henry, Bache took part in many war activities outside his
official position, for instance, the U. S. Sanitary Commission. But he
took his whole organization into the war effort far more than the
Smithsonian. Indeed, the Coast Survey's topographical activities are
one of the major applications of science to military use during the
war. Bache's military background, his political *savoir-faire*, and his
gifts of leadership largely account for this accomplishment, and the
major threat to the Coast Survey's position at the end of the war was
the superintendent's serious illness, destined to be fatal in 1867.

Inside the Navy the old-line organizations — the Naval Observatory and Hydrographical Office and the Nautical Almanac Office —
shaken up both by the abrupt departure of Maury and by the general
reorganization of the department, provided a chance for some re-
grouping to aid science. Charles Henry Davis, on duty in Washington
during the early days of the war, had the background from his
Nautical Almanac days necessary to appreciate the opportunity. Accordingly, in the reorganization of 1862 Davis became Head of the
new Bureau of Navigation, including the Naval Observatory, the
hydrographic functions which under Maury had been attached to it,
and the *Nautical Almanac*. Davis also wanted and tried to get the
Naval Academy, which he conceived at least potentially to be a scientific institution. Gilliss, after nearly twenty years' exile, came back to
head the Observatory and to restore astronomy as its major interest.[59]

With Davis as a leader the surveying-oriented science in the Navy
gained an administrative position much more favorable than it had
known in the days of its surreptitious development. Davis was now on
the same level as Dahlgren and Isherwood. In addition, his influence

was spread by his incidental duties, for example his membership of the ironclad commission and the Coast Survey's commission. Since Secretary Gideon Welles regarded Davis more highly as a scientific man than as a commander afloat, he spent most of the war in Washington. One of his early duties, as the most energetic member of the Office of Detail, which assigned officers to duty, he brought with him to the Bureau of Navigation. This accident gave both power and distracting concerns to the bureau, which soon found itself really in the officer personnel business. None of Davis's successors after the end of the war had much scientific interest; among the projects eventually allowed to die was his proposed scientific manual for officers.[60] Thus, for a brief moment only, from 1862 to 1865, science had an honored place in the Navy hierarchy and in Davis a recognized spokesman.

The Army's Topographical Engineers, heirs of surveying-oriented science and the flower of the old West Point education, had bad luck with the war. Their career tests very sensitively the extent to which the Civil War was a total war. On the one hand, their peacetime activities never completely ceased. The Great Lakes Survey, at the outset under the command of George Gordon Meade and later under other officers, continued its work, using military personnel during the entire course of the conflict.[61] Most of the officers were called in to Washington at the beginning of the war and reassigned to armies in the field.[62] Unlike the Coast Survey, the scene of their prewar topographical work had been in the West, removed from the battlegrounds on the soil of old states that had opposed federal surveys within their own jurisdiction. A report of General McClellan's at the end of 1862, while praising the Topographical Engineers highly, indicated the difficulty of their position. They often had to get information under fire, with the Army waiting for their knowledge before moving. They were so few in number that they had to depend on Coast Survey parties and "other gentlemen from civil life" for much aid. Furthermore, "it was impossible to draw a distinct line of demarkation between the duties of the two corps of engineers." They interchanged with the Corps of Engineer officers, mixing mapping and intelligence with construction work, and laying out defenses.[63] Many of the ablest officers, such as Meade, found promotions to high command faster in the general service and switched their commissions.

This loss of identity presaged the abolition of the Topographical Engineers as a separate entity in 1866, at about the same time West

Point ceased to be exclusively concerned with a surveying and engineering education. Thus one of the channels of science in the government which had been dominant before 1860 scarcely survived the war. The essential military contribution to exploration that had dominated the early period was on the way out. Strikingly, while army officers trained in science made great contributions to the war, no individual came to the fore to represent its interests on the same level with the triumvirate of Henry, Bache, and Davis.

The Genesis of the National Academy of Sciences

These three names — Henry, Bache, and Davis — gathered to themselves all the prestige that the old-line government science had developed. They were the men who, keeping their organizations together under trying circumstances, still had energy left over to devote themselves to the great questions raised by the war and to follow up the avenues it opened. Over the long view they were the culmination of the rise of the professional scientist in the government service. Each headed an important and integrated scientific organization within or close to the administrative framework. At the same time, they represented a gradual lowering of the level at which decisions respecting science were made. In Jefferson's time the President appreciated scientific matters and set the policy. The same could hardly be said of Van Buren, leaving Poinsett to make the real decisions at the cabinet level in the late 1830's. By the Civil War neither the President nor any cabinet member gave systematic attention to science. It rather fell, for the first time, into the hands of those professionals who had emerged at the level of bureau heads. Henry, Bache, and Davis were at the intersection of the two basic drifts — the rise of the professional and the drop of the level of effective authority in scientific matters. From this unusual historic position they were able to undertake what had always failed or missed the mark before — the coördination of the government's scientific policy.

Bache, Henry, and Davis had more in common than comparable official positions and long-standing friendship. They all moved in the shadow of that high-spirited group, the Scientific Lazzaroni. Bache as the chief had set forth the ideal of a scientific adviser to the government back in 1851, and in the meantime the group had attempted to control the scientific organizations of the country and to create new institutions more adequate to the position they thought

"Young America" should occupy. In early 1863 they were at the height of their power and activity. Bache had weathered the storms around the Coast Survey. The detested Maury had gone over to the rebels, which opened the way for Davis's Bureau of Navigation. Louis Agassiz was replacing a North Carolinian on the board of regents of the Smithsonian.

In Cambridge, where the research activities of the Coast Survey and the *Nautical Almanac* combined with the College to create a concentration of Lazzaroni members equal to that in Washington, Benjamin Peirce and Agassiz were deep in a fight to reform the university system of Massachusetts and remake the curriculum of Harvard in the interests of science. They claimed in the name of the Lazzaroni to control the president and corporation of the University and were able to force one of their number, the chemist Wolcott Gibbs, into a professorship over violent local opposition. Agassiz, Peirce, Wolcott Gibbs, and Benjamin Apthorp Gould were getting things done in Cambridge.[64] Bache and Davis had close ties with this group both officially and personally. Early in January 1863 Peirce wrote to Bache inquiring about the next meeting of the Lazzaroni.[65] Clearly the Cambridge and Washington members had much to talk over. By late January the meeting was set for about February 21, with Peirce planning to be in Washington from February 14 to February 23.[66]

Meanwhile the Washington Lazzaroni, Bache, Henry, and Davis, got together some time late in January to discuss the immediate possibility of setting up a "National Association under an act of Congress." Now that it came to a practical question Henry urged objections. First, he "did not think it possible that such an act could be passed with free discussion in the House — that it would be opposed as something at variance with our democratic institutions." His long experience with congressional whims gave this opinion weight. Second, he knew that "if adopted it would be a source of continual jealousy and bad feeling — an object of attack on the part of those who were left out." More than most of the Lazzaroni, Henry had friendships with scientists outside the chosen circle and knew both their power and their disposition. Third, "although it might be of some importance to the Government yet it would be impossible to obtain appropriations to defray the necessary expenses of the meetings and of the publications of the transactions." Fourth, "there would be great danger of its being perverted to the advancement of personal interest or to the support of the partizan politics." [67] The inveterate fear of politics and politicians which had shaped his Smithsonian policy and which a good many scientists shared appeared in these last two objections. Bache appeared to Henry to be convinced. At any rate he stopped talking about a national academy. The group determined instead to get as much of an advisory organization in the government as was possible with executive action. Davis claimed that the inspiration of a Select Commission in the British War Office led him to the idea,[68] but clearly Henry was the main driving force in this direction. The result was the Navy Department's Permanent Commission, destined to last the remainder of the war.

On February 11, 1862, within about two weeks of the trio's conversations, Gideon Welles sent a letter of appointment to Davis setting up a "permanent commission to which all subjects of a scientific character on which the Government may require information may be referred." Made up of Davis, Henry, and Bache, "this commission shall have authority to call in associates to aid in their investigations and inquiries." Neither members nor associates were to receive any compensation.[69] The commission, small, manageable, and without publicity, immediately went to work. For the next two months, according to Henry, it "occupied nearly all my time not devoted to the Institution and more than I could well spare." [70]

The Permanent Commission met frequently throughout the remainder of the war. They examined not only inventions, such as torpedoes and underwater guns, but also designs for warships. Often they could simply say of a proposal, "the plan is entirely crude and undigested, and does not require further notice." Sometimes they made tests and collected further information. Although not directly working for the Army, they soon found they needed the opinion of an officer skilled in land warfare, and General J. G. Barnard of the Corps of Engineers was added to the commission. By January 1864, they had more than 170 reports.[71] On occasion they used Benjamin Peirce as a consultant, and also J. E. Hilgard of the Coast Survey. The last report, dated September 21, 1865, was numbered 257. Thus the evaluation of inventions, at least for the Navy, became well organized in the hands of a predominantly civilian and scientifically capable committee. This was the nearest thing to a central war scientific agency achieved during the Civil War.[72] As a purely passive agency, however, dependent upon the chance suggestion, which never came, from uninformed and unorganized civilian inventors, the Permanent Commission never became a real research organization. It had neither the budget nor the personnel to select and attack problems independently. Only a few of the members, and they only fitfully, grasped the vision of applying science directly to the engines of war.

While the Permanent Commission got under way, however, Bache, Davis, and the Cambridge Lazzaroni had no real intention of heeding Henry's caution. What they needed was a friend in Congress, and Louis Agassiz produced one in Senator Henry Wilson of Massachusetts. He had evidently worked on the project for some time, for on February 5 Peirce reported that Agassiz had written to Wilson "to go ahead upon the National Academy of Science," and referred the Senator "to one . . . A. D. B. as our chief in all such matters, and as capable of furnishing him a complete plan fit to lay before Congress in 24 hours." [73] Thus the wheels were still turning, Wilson providing incidental assistance by getting Agassiz's expenses paid to Washington, ostensibly to accept his Smithsonian appointment, but miraculously including Saturday, February 21, the date of the Lazzaroni dinner. Agassiz had at first declined going to this meeting because "he had not the wherewithal." [74] But his decision to come and his tie with Wilson made certain that the days near February 21 would be interesting for the small coterie who desired a national academy. The impulse of the group was heightened by the war only in that it provided the occasion for their basically nationalistic enterprise. Their aim was to provide a young America, which they defiantly expected to survive its great trial, with a worthy counterpart of the Royal Society and the French Academy. Agassiz was fond of pointing to the founding of the University of Berlin in 1810 as an example of what a nation could do in a time of crisis,[75] but in spite of the glory of the enterprise an exquisite care was necessary when moving behind the back of Joseph Henry.

Agassiz reached Washington on the afternoon of Thursday, February 19, 1863.[76] Although Henry expected him at the Smithsonian, he "put up at Bache's." [77] That evening Senator Wilson called, and besides Bache and Agassiz, Peirce and Benjamin Apthorp Gould, forgathered for the Lazzaroni dinner on Saturday, were there. They perhaps had as a tangible basis for discussion a plan of an academy drawn up by Davis. The one known feature of this plan was a provision "creating the Academy with a dozen or twenty members, and allowing them to fill up the whole number by usual system of ballot." [78] However, the subsequent events suggest that counsels for speed and secrecy prevailed at the meeting. Perhaps Wilson pointed out that the close of the session of Congress was fast approaching. Any scheme that required consultation with even twenty men would delay passage, and before another session the opposition that Henry had so clearly foreseen both in Congress and among the scientists themselves would have had ample time to organize. When Wilson emerged from the conference he had with him the draft of a bill which named fifty scientists, incorporated them as individuals into a National Academy of Sciences, and gave them the power to perpetuate themselves by filling vacancies. Nothing in the law or in any public statement either at the time or later gave any clue to the identity of the authors of the list or their criteria of selection. But Davis's account points clearly to the meeting of the evening of February 19, thus putting the power entirely in the hands of Bache, Agassiz, Peirce, and Gould.

The Lazzaroni, of course, were on the list in full force, together with those so eminent in their fields that no one could possibly leave them off, such as, John Torrey, Asa Gray, Jeffries Wyman, and William Barton Rogers, who perhaps gained entry only because of the intercession of Senator Wilson. The government was very largely represented, not only by Bache, Henry, Davis, Dahlgren, and Gillis, but also by lesser employes and officers of both the Army and the Navy who had scientific interests. Of the omissions that might raise eyebrows, George P. Bond, director of the Harvard Observatory and far gone with tuberculosis, was anathema to the unforgiving Peirce.[79] Spencer Fullerton Baird, although close to Henry, was evidently too modest a zoologist for the lordly Agassiz.[80] And the chemist John W. Draper of New York was better known than many on the list.

Senator Wilson gave notice in the Senate on February 20 of his intention to introduce the bill, and on February 21 he did so.[81] It contained only three sections. The first listed the fifty incorporators. The second gave them power over their own rules and membership. The third provided for an annual meeting and stated that "the Academy shall, whenever called upon by any Department of the Government, investigate, examine, experiment, and report upon any subject of science or art, the actual expense of such investigations, examinations, experiments, and reports to be paid from appropriations which may be made for the purpose, but the academy shall receive no compensation whatever for any service to the Government of the United States." [82] This was clearly the advisory body envisioned in 1851 by Bache. Not mentioned, but implied positively by the limited number of places and negatively by the prohibition on compensation, was the honorary nature of membership. Appropriations could be had only in relation to a request from some department of government. A general statement of principles and many details vitally affecting the nature of the organization were left to the fifty incorporators.

Joseph Henry within a day or two, "on accidentally calling at the Coast Survey," found that "the whole matter was in the hands of Senator Wilson!" [83] He did "not approve of the method which was adopted in filling the list of members. It gave the choice to three or four persons who could not be otherwise than influenced by personal feelings at least in some degree; and who could not possibly escape the charge of being thus influenced." But he did not "make any very strenuous objections" to the plan, "because I did not believe it could possibly become a law; and indeed there are very few occasions when acts of this kind could be passed without comment or opposition." [84] This generally excellent judgment of the political scene reckoned without the possibility that Henry Wilson might find one of the few exceptional occasions.

Davis intimated that he and Agassiz went up to the Capitol for some lobbying on the bill,[85] but Wilson essentially depended on the pressure of adjournment. This was the lame-duck thirty-seventh Congress, elected back in 1860, and the session automatically ended at midnight, March 3, 1863. As the Senate rushed through matters large and small on that last day, Wilson followed a long line of private petitions by asking "to take up a bill which will consume no time, and to which I hope there will be no opposition . . . It will take but a moment, I think, and I should like to have it passed." When no objection arose, he suggested it was "unnecessary to read the first section of the bill, which merely contains a list of names of the corporators." A preoccupied Senate, after listening to the reading of sections two and three, passed the bill without a recorded vote and moved on to consider declaring a "Day of Prayer and Humiliation." [86] The House took up the Senate's bill considerably after seven o'clock in the evening and passed it without comment. Lincoln evidently signed it the same evening.[87]

In contrast to the lively and sometimes damaging interest of the Congresses of the 1840's in the National Institute and the Smithsonian, this demonstration of legislative pliability reveals remarkably little about the basic forces producing a National Academy. The Congress as an entity had not spoken in any positive sense. Still less had the nation spoken through its chosen representatives. That the great crisis of the day made the congressmen feel they were the agents of destiny is possible, but that they connected this particular legislation with their immortal reputations is highly unlikely. Wilson did prove he knew his way around in the last minute rush, and a little group of scientists, five at the most, proved they knew what they wanted. Indeed, at the time of the National Institute the politicians controlled these matters to the exclusion of the scientists. In 1863 the professionals only needed the politicians to put a legal rubber stamp on their arrangements. In his long career Henry Wilson showed little interest in or understanding of science.

The Attempt to Organize a Scientific Adviser to the Government

But the secrecy that had produced such remarkable results in Congress could not last. On March 5 Henry finally found out the bill had become a law.[88] Letters from Wilson to the fifty incorporators spread the news through the country, catching even some of the most eminent by surprise. It took little longer for the ones passed over to realize what had happened. On March 7, George C. Shaeffer, "a *savan* in one of the departments," walked into Davis's office and "flew out against the Academy in good, set terms." [89] John W. Draper wrote to Henry protesting his exclusion.[90] John Torrey reported to Asa Gray that "the whole matter was concocted by the party assembled at the Coast Survey." [91]

Gray found himself in a peculiarly delicate position concerning the new Academy. His position as the leading botanist in the country was unchallengeable, he was an intimate of Henry's, and his enthusiasm for the cause of the Union had made a distinct impression in scientific circles abroad if not at home. Yet in Cambridge he was heavily engaged against the Lazzaroni position both on the appointment of Gibbs and on educational policy generally. Further, as the leading expounder of Darwin's ideas in America, he had long been in conflict on scientific questions with Agassiz, who was ever more bitter against evolutionary theories.[92] As an officer of the American

Academy of Arts and Sciences, he also had his eye on the hierarchy of scientific institutions in America, fearing a national covering organization which would threaten the older societies.[93]

Benjamin Peirce, back in Cambridge, was sure that Gray "is trying to divide Henry from us . . . Ha. Ha. Ha. The botanist may find that [it] is possible to dig too deep for successful undermining." He was a bit alarmed, however, for he warned Bache "we must keep our eyes open," especially since Henry and Gray were both Presbyterians.[94] Here indeed was a crucial decision for Henry. With the plentiful fuel of dissatisfaction on every side in the scientific community, one word of disapproval from him would have set the match. But Henry was not the man to let the less than candid behavior of his Lazzaroni friends influence him. While making no secret of the course of events in February, and still placing "but little faith in appropriations of Congress," he concluded to attend the first meeting "and do what I can to give it a proper direction." [95]

The organization meeting at New York in April 1863, which 31 of the 50 attended, found the Lazzaroni in firm control. Although they made Henry presiding officer over the session, Bache became the first president. The four other general officers — J. D. Dana, Agassiz, Wolcott Gibbs, and Fairman Rogers — were all Lazzaroni. The class of mathematics and physics had Peirce and Gould, two of the original schemers, as its officers. Only the other class, natural history, went outside the select circle to honor the aged Benjamin Silliman, Sr., with the chairmanship.[96] This weakness emphasized that aside from Agassiz none of the ruling circle had any interest in natural history. The imbalance of the number of incorporators assigned to the two classes confirmed the dominance of mathematics and physics, which had thirty members to only fourteen for natural history; [97] six were unclassified.

Provisions for terms of membership and office had obvious implications for the urgent task of acclimatizing the National Academy, so exclusive and irresponsible in its beginnings, to a democracy. At first Benjamin Peirce had some idea of limiting membership to a ten-year term, with reëlection possible, so that the "whole *practical number* of the academy" would continue to be Lazzaroni.[98] But by the time of the New York meeting the constitution and rules, "elaborately prepared" by Bache, not only allowed membership without term but made the tenure of officers for life. None of Bache's henchmen on the committee of organization — B. A. Gould, Agassiz, Peirce, or Fraser — objected, leaving it to William Barton Rogers, the founder of Massachusetts Institute of Technology, to point out that life tenure of officers would "blast every hope of success." [99] As a result the term of officers was limited to six years.

An issue that generated even more heat at New York was the question of a loyalty oath. The Civil War was a struggle for Union, and the Academy, in the words of Senator Wilson, was an "element of power" to make the bleeding nation "one and indivisible." [100] What should they do about Southern scientists, who should in theory be included? But what of the actual state of war? Those who could not take an oath presented therefore a pressing issue, bringing into question the eighteenth-century notion, still officially announced as late as the instructions to the Wilkes expedition, that science was above the clamor of wars. Bache, Agassiz, Gould, and Fraser pressed for the oath, while Joseph Leidy threatened to resign. William B. Rogers opposed it, at the same time "appealing to his record as an old and consistent antislavery man." The lone reporter of this incident, the geologist J. P. Lesley, indicated that the discussion got even deeper into the realm of the relation of the Academy, and through it science, to the government. "Someone, I willingly forget who, argued that we would lose government patronage, unless we bid for it with the oath; I suspect it was only an unfortunate way of stating a higher truth, that we are the children of the government, and the academy is the creation of the government, and owes it an oath of allegiance as its first duty." [101] Those in favor of the oath not only pushed it into the constitution but prevented the recording of the vote. Some of the politic ones predicted that when the war was over the oath could be relaxed, which in fact occurred in 1872.

Although Bache emerged from the organization meeting in triumph, the opposition was far from dead. John A. Dahlgren sent in his resignation on May 14.[102] Whatever the motives, the withdrawal of the man in charge of one of the bureaus most closely concerned with technological change was a severe blow to the Academy's ambition actually to function as scientific adviser to the government. When the president tried to put W. B. Rogers on a committee, he countered by asserting that the rules "were to come up for final . . . action at the December meeting" and that he did not think it "expedient to enter upon the business of the Academy until that time." [103]

A clear test of the Lazzaroni interest came in the election of the president of the American Academy of Arts and Sciences in Boston late in May. Peirce lamented that, "to show their hatred of the National Academy, all its opponents combined to elect Gray as President and William B. Rogers as Rec[o]r[ding] Sec[retary]." [104] Gray did not actually take a public stand on the National Academy, concluding to "say and do nothing — at present, and see," but he considered it "strictly governed by Coast Survey and Agassiz clique." [105]

Even Benjamin Peirce began to waver by the fall of 1863, confessing to Bache that he feared "we have made a mistake in founding the Academy — and that it may not be so great a misfortune if its enemies were to succeed in overturning it." [106] Despairing of "the growing falseness of Washington," he felt that for "men seeking truth to meet there, is like a party of poets meeting in a cotton machine . . . of angels in the palace of Beelzebub or of imps in Abraham's buzzum." Thus the question had changed from whether a small group of American scientists could form and dominate a national academy to the more desperate one of whether all American scientists working together could make a corporate body survive in the harsh realities of 1863. "Amid the din of war," continued Peirce's lament, "the heat of party, the deviltries of politics, and the poisons of hypocrisy — science will be inaudible, incapable, incoherent and inanimate." [107]

Bache, less fastidious and much more experienced concerning the ways of the capital, did not delay his attempt to establish the Academy as the government's adviser. On his own authority he set up committees and had them ready to report at the meeting in Washington in January 1864. Most of the sixteen learned papers read there had nothing to do with the war and were given by members of the Lazzaroni,[108] but the committee reports were the heart of the advisory program. Joseph Henry concluded that they "exhibited a considerable amount of valuable labor of the kind much wanted by the government." He now felt that "if the members will all attend and resolve to do all in their power to support the establishment on just and unselfish principles the academy will do much good both in the way of advancing science and assisting the government." [109] An evaluation of the war service of the Academy depends on the record of the committees of 1863 and 1864.

Committee number one on weights, measures, and coinage was appointed May 4, 1863, at the request of the secretary of the treasury.

Since Bache had long been in charge of the office for the Treasury Department, the appointment of this committee is directly attributable to him. Under the chairmanship of Joseph Henry, the group studied the problem, which was of course as old as the republic, until 1866, when a standing committee succeeded them. In spite of the conservative views of Henry, the committee's report, rendered in 1866, played a part in the passage of the bill legalizing the metric system.[110] Thus the Academy here served a useful purpose, although not during the war or on a subject directly related to it.

Committee number two, appointed May 9, 1863, was to consider means for protecting the bottoms of iron ships from injury by salt water. Significantly, the request from the Navy Department came by way of Admiral Davis. The committee reported at the January 1864 meeting that it could recommend no scheme, and while it suggested experiments to be undertaken, the Academy had no money and the Navy did not proffer any. The committee disbanded without making any contribution to the war effort.[111]

Committee number three, to study the magnetic deviation of the compass in iron ships, also came through Davis's influence. Actually, before the Academy was formed the secretary of the Navy had already set up a commission on the subject with Bache as chairman. The Navy simply allowed the Academy to honor itself by appointing an already existing committee. This group, while developing nothing new in the way of theory and calling in an "expert" to do the testing, did correct compasses on 27 vessels.[112] The group did no more work after September 1863, and presumably the Navy thereafter handled correction routinely in the compass stations established at the various yards. The role of the Academy was one of supererogation.

Committees number four and five were the result of requests directly from Bache and Davis acting in their official capacities. Number four dealt with the evaluation of a hydrometer invented by Joseph Saxton, a member of the Academy and an old employee of Bache's. The committee recommended in favor of the adoption of Saxton's hydrometer for revenue purposes, but without any effect.[113]

Davis asked the Academy to tell him whether to continue the current charts and sailing directions of Maury, always an enemy of the Lazzaroni and now doubly damned as a rebel. In the light of Maury's enduring reputation, the conclusion formally adopted by the Academy, that his charts "embrace much which is unsound in philosophy, and little that is practically useful," casts something of a reflection on the committee's ability to keep its personal and scientific judgments separate. The full report was both extensive and careful, however, developing the same opinion of Maury's work which had timidly appeared in the *American Journal of Science* during the 1850's. A second resolution urged that "such information should continue to be collected." [114]

Taken together, the committees make it clear that Bache and Davis personally inspired all the requests that came to the Academy in 1863. The total results were modest enough, and when the Maury question and the metric system are ruled out as unrelated to the war, the remainder of service is negligible. The four committees of 1864, while not emanating directly from Bache and Davis, made no better record. The best opportunities came on a joint group of three each from the Academy, the Franklin Institute, and the Navy Department, to experiment on the expansion of steam. Isherwood as well as Davis served for the Navy. But after some progress reports in 1865 and 1866 the committee became inactive. In 1880 the investigations were reported not yet completed, and after that they dropped from sight.[115] No wartime committees date later than May 2, 1864, when Grant was just beginning his campaign with the Army of the Potomac and nearly a year of fighting lay ahead. When compared with the 257 reports made by the Permanent Commission, the advice rendered by the National Academy appeared slight, and the efforts of Bache and Davis to channel government business to it only rarely succeeded.

The National Academy's Fight for Survival

The torpidity that overcame the Academy in the spring of 1864 coincided with the serious illness of its architect, Bache.[116] With their "darling chief" incapable of further duty, the Lazzaroni maintained their quarrelsomeness without the energy and direction that the greatest government scientist of the whole period had imparted to them. Agassiz came close to precipitating a mass resignation of naturalists by fighting the election of Baird in the summer of 1864. But the failure of this effort meant that the little group which had controlled decisions so easily in New York in 1863 had met an open rebuff. Henry's support helped put his modest assistant secretary into the select fifty, and he warned Agassiz that in "this Democratic country we must do what we can, when we cannot do what we would. We must expect to be thwarted in many of our plans." [117] Any rosy hopes of congressional appropriations were now dead, and the few publications that the Academy managed had to be paid for out of the pockets of members.[118]

Through the years 1865 to 1867 attendance at meetings trailed off. With Bache ill, James D. Dana resigned the vice-presidency, leaving Henry as a kind of receiver for the organization. At the meeting of January 1867, only seventeen attended, reading seven papers.[119] A desultory but questioning discussion among scientists had begun again concerning the nature and functions of the Academy.[120] But Bache, even in his death, gave his creation the impetus to survival. By leaving it the bulk of his estate and by making Henry feel an obligation to continue it, he gave the support and the leader necessary to weather the crisis. Henry wrote, "I very reluctantly accepted the office of President and I was principally induced to do so at the earnest solicitation of Mrs. Bache, who since her husband was the first president, and because his fortune after her death will be under the care of the Academy, is exceedingly anxious that it should be perpetuated." Henry was "far from desiring that it should expire in my arms; but how to preserve its life and render it useful is a different problem." [121]

Between 1867 and 1872 Henry developed a three-point program for the salvation of the Academy. First, he made the prime consideration for membership in the Academy "original research," with no one "elected into it who had not earned the distinction by actual discoveries enlarging the field of human knowledge." This shifted the emphasis in the purpose of the Academy from practical service to the government to the recognition of "abstract science." [122] The second point of the program was to reduce the number of meetings to one a year and hold them in Washington. The third point was to increase the membership beyond fifty.

Wolcott Gibbs and B. A. Gould, both old Lazzaroni "who have thus far considered themselves the essential elements of the society," [123] put up some opposition, and Benjamin Peirce eventually resigned, but the Academy was no longer a fighting issue. An act of Congress removed the ceiling of 50 on membership, opening the way for the admission of 25 new members at once in 1872. By taking in several who earlier had been passed over, the Academy killed much of the appeal of the rival organizations that had sprung up,[124] and younger scholars selected because of contributions to abstract research gave some hope for the future. By Henry's policies the life and dignity of the Academy were assured, while the gifts of Bache and others provided it sufficient money to operate as an honorary society if not as a research establishment for the government. No longer the chosen instrument of a small and spirited group, the Academy was kept alive because leaders such as Henry, who had no great positive program for it, realized that to allow such an organization to die would do severe damage to the prestige of science in America.

The relations of the government and science between 1861 and 1865 show no overwhelming stimulus of research by a war that was only beginning to resemble the conflicts of the twentieth century. A general shake-up of administrative arrangements is evident, with both positive and negative results. The old army-civilian exploring expedition barely survived the war and never regained its dominant position. The possibility of a new relation of science to technology appeared within the services, but little of the experience carried through the severe retrenchments which began even before the fighting stopped. The National Academy was a monument to Bache's dream of putting science in the service of the government, not to any actual wartime accomplishment, and its main potentiality was to serve as a rallying standard when a new crisis should come.

Nevertheless the America that emerged from the Civil War found itself in a different era which demanded and achieved new scientific institutions, both inside and outside the government. These important changes sprang not from the military necessities of the struggle but from the dynamic forces that were transforming all of American culture.

VIII

THE EVOLUTION OF
RESEARCH IN AGRICULTURE
1 8 6 2 – 1 9 1 6

IN sharp contrast to the government's hesitant stimulus to research technology and to the National Academy, the outbreak of the Civil War had a profound effect on the position of agricultural investigations. With the heirs of John C. Calhoun gone from Congress and the agricultural states of the Northwest in the ascendancy, increased support for information to farmers was almost a foregone conclusion. The United States Agricultural Society with its long agitation for a department found that the core of its opposition had seceded. Justin S. Morrill was as ready as ever with his land-grant college act.[1] Unlike the National Academy, these measures were the creations of politicians, not of scientists. Indeed, Bache in his 1851 address proposing a scientific organization to aid the government did not even mention agriculture as a field of possible action. Nor did anyone connected with the Patent Office or with agricultural studies appear on the list of the Academy's incorporators. Journalists, professional consultants, gentlemen farmers, enthusiasts for manual-labor farm schools, and a few chemists under the spell of Liebig were the core of the pressure groups agitating for the agricultural measures.[2]

The Legislation of 1862

The Republican platform of 1860, emphasizing the alliance between the industrial East and the agricultural West, set the stage for early action on a more effective substitute for the impotent and unpopular efforts of the Patent Office.[3] Lincoln backed a proposal of his secretary of the interior, Caleb Smith, to create a bureau of agricultural statistics, while the commissioner of patents in his report for 1861 suggested an industry department with a mechanical, an agricultural, and a commercial bureau.[4] A bill for an independent agricultural and statistical bureau passed the House of Representatives while several versions failed in the Senate before a compromise emerged which drew the opposition of only seven votes in the House and thirteen in the Senate.[5]

The act itself gave ample attention to science as an aid to agriculture, putting most of the emphasis on the programs already begun by the Patent Office. The new department was to "acquire and diffuse . . . useful information on subjects connected with agriculture in the most general and comprehensive sense of that word, and to procure, propagate, and distribute among the people new and valuable seeds and plants." Although lacking a place in the cabinet, the commissioner as the head of an independent department was to gather information — either in books, by correspondence, by collecting statistics, or by "practical and scientific experiments (accurate records of which experiments shall be kept in his office)." The law provided for a chief clerk, and mentioned the service of "chemists, botanists, entomologists, and other persons skilled in the natural sciences pertaining to agriculture." [6]

Five days after Lincoln signed the organic act of the Department of Agriculture, the Homestead Act became law, opening a new chapter in the management of the public domain and the expansion of American farming into the West. As consolation for the states no longer heavily endowed with public lands, a land-grant college bill now found favor in Congress. An "act donating public lands to the several States and Territories which may provide colleges for the benefit of agriculture and the mechanic arts" [7] passed both houses in June 1862 and became law on July 2. The culmination of a long-standing pressure which had made itself felt in the 1850's and even before, the land-grant college idea received its final form from Representative Morrill of Vermont, who had been its leading advocate since 1856.[8] The newer Western states, which opposed the act because it gave potentially equal endowment of land to each state, east or west, found themselves in a minority without the Southern allies who had aided them in the 1850's.[9]

To what extent the land-grant colleges were supposed to be scientific institutions is questionable. Morrill's original idea seems to have emphasized direct instruction to the agricultural and industrial laboring classes. The variety of the schemes put forward resembles the babel that greeted the Smithsonian bequest. By 1867, Morrill was quoted as saying that the institutions envisioned in the act were not agricultural schools but "*colleges,* in which science and not the classics should be the leading idea." [10] The states, which had the responsibility of administering the funds from the sale of the federal lands, responded with several institutional patterns — endowing existing colleges, setting up new schools, dividing the funds between agricultural and mechanical purposes. No colleges set up under the act had any effect during the war period, and before 1887 only a few institutions had a real research program. The scarcity of secondary schools among farming people, the lack of trained teachers and of adequate textbooks, and the absence of incentive to study agriculture in college hampered the early attempts. In 1872–1873, nearly half of the 26 colleges had fewer than 200 students each, and only 4 institutions enrolled over 400. The agricultural and mechanical departments were even smaller. [11] Thus, the land-grant colleges with their federal endowment were important only as a framework for future development. [12]

In spite of their failure to provide solutions for the many problems facing Northern agriculture during the Civil War, the acts of 1862 mark a genuine turning point for science in the government. Up to this time scientific institutions had had a questionable constitutional status because they were tied to other much-disputed internal improvements. All the great accomplishments of the prewar years — the Smithsonian, the Coast Survey, the Naval Observatory, and agricultural research in Patent Office — had evaded the constitutional issue. In establishing the Department of Agriculture and granting public lands for colleges the Congress proceeded on the unspoken but definite assumption that its power "to lay and collect taxes . . . for the common defense and general welfare of the United States" obviously warranted federal sponsorship of scientific research. Although opponents could and did invoke states' rights against federal scientific activity, the outbreak of the Civil War had ruled in favor of Alexander Hamilton's interpretation of the general-welfare clause as clearly as it presaged the triumph of Hamilton's vision of an industrial nation. From this time on, Congress proved itself at least occasionally willing to establish permanent bureaus with ample grants of power explicitly stated in organic acts. [13] With the Constitution no longer a stumbling block, the era of bureau-building had begun.

The Early Years of the Department of Agriculture

The name of the first commissioner of agriculture should have given his department a scientific bent. But Isaac Newton was a Pennsylvania dairy farmer who had made deliveries at the White House and other prominent residences in Washington, to the incidental advantage of his political ambition. Working up through the United States Agricultural Society, he served as commissioner of agriculture in the Patent Office under Lincoln until the act for the independent department had passed. Inheriting the old rooms in the Patent Office building, the commissioner's new agency did little more at first than change its name.[14]

In his first annual report, Newton announced an expanded program, which called for the collection of information generally, including statistics, the distribution of seeds and cuttings, and the answering of farmers' queries. For more strictly scientific work he proposed "a chemical laboratory" for analysis of "various soils, grains, fruits, plants, vegetables, and manures" and professorships of botany and entomology.[15] Clearly, although the commissioner wished to use science and believed that its cultivation benefited agriculture, he had no specific ideas as to what problems could be solved or how they should be attacked. Hence his proposal amounted to the pursuit of the disciplines of chemistry, botany, and entomology along academic lines, with the hope that the farmer would gain increased awareness of his surroundings and indirectly obtain higher production. Nevertheless, an organization thus devoted to research thinly spread over whole scientific disciplines had to justify its existence by specific discoveries of value to the farmer, or it would be forced to modify its approach. Perhaps it could become a great agency for basic research in the life sciences, but to aim at that alone, as if it were the National Museum, ran the danger of losing essential support.

Newton's scientific appointments followed the pattern of academic disciplines. The first chemist, C. M. Wetherill, who had a first-rate German education, began analyzing the composition of sugars, syrups, and other products, publishing a report on the *Chemical Analysis of Grapes* in 1862. But his preoccupation with Captain Diller's gunpowder led to a quarrel with Newton and his leaving the department.[16] Whatever glory the early Division of Chemistry deserves stem from the work of Wetherill's successors, Henri Erni and

Thomas Antisell. As early as 1864 Erni, finding that the only chemist in the government was likely to be kept busy regardless of his research interests, complained that "a great number of letters of inquiry regarding scientific problems in agricultural and general industry have been answered by me, frequently involving extended chemical and microscopic examination." [17] Antisell, a veteran of the western surveys before the Civil War, looked toward mineral and metallurgical investigations in addition to agriculture as a proper sphere of activity.[18]

In entomology Townend Glover, who was carried over from the Patent Office, continued his long tenure under the department, industriously describing insects until his death in 1878. William Saunders, a Scotsman with experience as a nurseryman, became superintendent of the propagating garden, where he centered his experiments on plant introductions. Some of this work touched such economically important plants as eucalyptus and navel orange trees. Beyond his nursery, Saunders' interests ranged widely. He converted Newton's ill-fated experimental farm into a kind of arboretum surrounding the new department building and was prominent in various public landscaping projects. He was also one of the founders of the Grange — the Patrons of Husbandry — and served as its master for the first six years after 1867.[19] But his work, hampered by lack of facilities and by the political necessity of distributing free seeds in vast quantities, never went very far into any question. Saunders remained at his post until 1900.

A statistics bureau, library, and museum rounded out the activities of the department under Newton. But these surface organizational accomplishments cannot hide a most unpromising start for the new department. By 1865 the agricultural journals and societies were calling for a change, and after Newton became ill in 1866, President Andrew Johnson tried to remove him. But the Senate, spiting the President in all things great and small, refused to confirm a successor, leaving Newton in office until his death in 1867.[20]

The commissioners of agriculture after 1867, although some were abler than Newton, had political backgrounds and short terms of office. By 1870 the divisions of the department were chemistry, horticulture, entomology, statistics, seeds, and botany. The employees of these divisions devoted themselves to whatever agricultural problem came their way within their discipline. This organization by fields of learning was further emphasized by the creation of a division of microscopy.[21]

When an urgent problem arose, the department seldom had the resources to cope with it. For instance, the spread of the railroads in the West after the Civil War and the opening of the trails suddenly converted the so-called Texas fever of cattle from a local nuisance into a national and international danger. Cattle brought from the southern states, though remaining healthy themselves, infected northern animals in large numbers with a disease that often proved fatal. Commissioner Horace Capron had to look beyond his own staff to find men for the work that Congress almost forced upon him by appropriating $15,000 for the investigation of animal diseases.[22] The leading researcher, Dr. John Gamgee, a British veterinarian, obtained most of his assistance from doctors borrowed from the Army. Only the report on the statistics of the progress of the disease came from a regular department man. Gamgee knew that practical cattlemen often blamed ticks as the carriers of the disease, but he concluded that "a little thought should have satisfied any one of the absurdity of the idea." [23] In the light of later discoveries, this classic bad guess shows the young department to very poor advantage.

More significant perhaps than the conclusion was the general procedure of the entire project. Most of the studies started with the assumption that a fungus caused the disease. Indeed, it is clear that Army Surgeon John Shaw Billings was called in because of his interest in cryptogamic botany rather than as a doctor of medicine. The older systematic studies were far from providing the solution of a practical problem in the microscopic realm.[24]

The two main causes of the ineffectual showing on Texas fever were, first, the inability to organize a long-term intensive research effort, and second, the lack of adequate concepts of disease transmission. The latter was the fault of the department only because it supinely assumed that all concepts must be delivered ready-made from abroad. In 1869 Louis Pasteur and the others still had much to do to make the germ theory a workable tool, and in waiting for them the department was behaving much like the rest of American science. Yet Pasteur in these years was working on problems just as earthy as those facing the department — diseases of silkworms, production of wine and beer, anthrax, chicken cholera, cattle pleuropneumonia. Out of his studies, so practical from the point of view of the French farmer, came revolutionary scientific theories. Pasteur's great dictum that there were not two forms of science — pure and applied — but only science and the application of science, described a situation with which the department had to come to terms.[25]

The tensions that beset the department in finding appropriate scientific means to accomplish its mission to the American farmer are well illustrated by the vicissitudes of the division of botany in the early years. When Capron became commissioner, in 1867, Joseph Henry saw a chance to further his favorite policy of self-denial by transferring to the Department of Agriculture the huge mass of unmounted specimens of plants from the government surveys that had piled up at the Smithsonian. Herbarium work in classification and nomenclature, basic to all forms of plant investigation and then the dominant interest of American botanists, was of great potential importance to agricultural studies. Yet the gap between the systematic botany of the time and an applied science of agriculture was very broad. Occasioned partly because the botanical knowledge of any one species was too thin, the rift proved even greater because the great crop plants — wheat, corn, and the rest — were precisely those about which the botanists knew least and which their tools, designed to study plants in nature, were least able to penetrate.

With the blessing of Asa Gray and John Torrey, the arbiters of botany in America, C. C. Parry became the botanist of the Department of Agriculture and, taking over the mass of specimens that Henry self-righteously renounced, began to make of them a national herbarium in much the same way that Baird was creating a national zoological museum within the Smithsonian. Parry was a competent collector who had learned both botany and marked frontier individualism in the hard graduate school of the Western surveys.[26] Things went well enough under Capron, but when Frederick Watts, a Pennsylvania lawyer-farmer, became commissioner in 1871, Parry began to get into trouble. His habit of corresponding directly with outside botanists such as Gray irritated the commissioner and the chief clerk, who "not only dictates as to how I shall cary [sic] on the details of my division, but is constantly making foolish and unnecessary corrections of my official letters requiring frequent recopying." [27] After several scenes the commissioner curtly wrote Parry that his "services as botanist of this department will not be required after this date." [28]

The botanists of America promptly declared war on the department. Using Joseph Henry himself as courier of their messages, they badgered Watts into justifying himself. Besides attacking Parry as "wanting in perspicuity," he claimed that the business of the department was to "render the developments and deductions of science directly available to practice," and that the botanist should study "vegetable physiology, their relations to climate, soils, and the diseases of plants, which are principally of fungoid origin." Not only had Parry failed to do this, but the "routine operations of a mere herbarium botanist are practically unimportant." [29] The breakdown in understanding between the botanists and the department was almost complete. Watts's brave words about the mission of botany would have rung truer if Parry's successor, George Vasey, had not continued the old policy for another twenty years. The National Herbarium gradually became a major institution in its field, but in 1896 it was retransferred to the Smithsonian Institution,[30] its logical and proper keeper.

The Parry fiasco pointed up a number of weaknesses within the department. The political position of the commissioner, the power of the chief clerk over the scientists, their helplessness in the face of capricious dismissal, and a disdain for the counsels of the scientific community outside the government, came starkly to light. It is true that the botanists tended to concentrate on what the department could do for their science instead of what it could do for the farmers. Yet such episodes explain why Asa Gray had the fixed opinion that only a monarchial government could effectively support science. "Neither our Congress nor our executive department can be depended on for attending to any such thing wisely or honestly." [31]

Total appropriations for the department, although fluctuating year by year, scarcely progressed for about twenty years. The $199,770 in 1864 compared to $199,500 in 1880 [32] is some indication of a stable level of activity at about one-half the plane of the prewar Coast Survey, the only gain coming from the fluctuation in the value of money.

The newly established journal *Science* used the amount spent for 1881 ($256,129.68) as an occasion for the bitter comment that "the results obtained by this class of expenditures have hitherto been, out of all proportion, small." The anonymous author claimed that the department "cannot be accounted competent to carry on continuous scientific researches." Far from suggesting that it go into works which "must necessarily run on consecutively from year to year," he urged the commissioner to accept his fate, since rather "than try to grasp the unattainable, it will assuredly be wiser to study special finite questions as they present themselves." To do this "the best means is the employment of special scientific men of approved competency, each one to grapple with his own particular question in such place and manner as he may deem fit." [33]

The Development of Bureaus and the Problem Approach

This invitation to farm out the department's business to private hands and to give up the attempt to build a federal scientific organization was both the counsel of despair and the revelation of a widespread desire among American scientists to restrict the government to as small a sphere as possible. The scandals of the Grant administration and the tremendous burst of economic power, which overshadowed the political squabbles of the period, helped foster this extreme antipathy toward all sorts of government enterprise, for which the *laissez-faire* philosophy of William Graham Sumner was already furnishing a rationale. About this time the nation was also recoiling from a gigantic program of intervention, the reconstruction of the Southern states.

Within the scientific community itself rapid changes made a restrictive attitude toward the government more plausible. Graduate education was beginning to take hold in a few universities and with it rose academic research. German ideals of seminar and specialization were beginning to find favor at the new Johns Hopkins University. Products of the Sheffield Scientific School at Yale and of Agassiz's teaching at the Lawrence Scientific School at Harvard were spreading the gospel of research across the country. The traffic of young Americans to European centers increased steadily and these new scientists, dedicated and specialized, were finding positions back home that allowed them to use their advanced training. Beside this stimulating scene the politics-ridden atmosphere of the capital seemed bleak indeed. Give up Washington, give up the government, and let Yale, Harvard, and Cornell take over the agricultural research of the country when and if they pleased.

Yet other voices spoke another chorus. By 1880 the transformation of the United States from an agricultural to an industrial nation was far advanced. Mechanization and commercialization of the farm, the building of the railroad network, and the creation of a single international market for agricultural products were making a new world for the farmer which he did not understand. The westward migration was invading the Great Plains and the Rocky Mountain area where the physical environment made the old rules of thumb for agriculture worse than useless. The Homestead Act created 160-acre farms in areas where the realistic size of a unit should run to thousands of acres. Although the dirt farmer himself often despised book learning and certainly did not look to science as his salvation, possibilities of research producing relevant answers to the urgent problems confronting agriculture were apparent to the pressure groups responding to the farmers' plight. The men who had this vision could reach their state governments and through Congress the federal government and the Department of Agriculture. They could not so easily place their needs before Harvard University, whose agricultural adjunct, the Bussey Institution, was withering away in the face of opportunity.

Between 1880 and 1897 the Department of Agriculture answered the call. The evolution of its organization was somewhat halting and painful, and no Bache emerged as a driving leader, but scientific excellence, thriving in unexpected places, became increasingly more prominent. As its ability to furnish answers increased, the department gradually evolved an adequate social and political mechanism, the government bureau. With its roots far back in the pre-Civil War period,[34] when the Coast Survey, for instance, achieved some of its attributes, this entity began to mature after 1875.

Although no single organization actually realized all its facets, the ideal new scientific bureau had clearly defined characteristics. In the first place, the center of its interest was a problem, not a scientific discipline. Instead of a chemist who tested both fruits and fertilizers, the problems of growing particular crops or improving animals became central, and the bureau mobilized teams of experts from various disciplines to attack each one. Such an approach required on the one hand stability to concentrate on a given line of investigation over a period of years, and on the other hand the flexibility to shift resources as the problem changed.

Thus the ideal bureau chief sought continuity by means of a grant of power in an organic act of Congress that stated the mission of the agency comprehensively and then stood unchanged over a period of many years as a kind of little constitution. The Department of Agriculture's organic act was a model that was to reappear at the bureau level. To gain flexibility, the bureau chief fought against appropriations for specific purposes and favored lump sums which could be shifted when the problems changed their shape. This emphasis on mobility arose because the new chief not only concentrated on problems but also took an active part in seeking them out. Not content to await congressional action or public protest, he strove to find problems and answers in anticipation of need.

In the second place, the ideal bureau aimed at a stable corps of scientific personnel which was not only competent but also loyal to the bureau and confident that its work was important to the country. As will appear much later, the Forest Service and the Public Health Service carried this to the length of putting their personnel in uniform.

In the third place, the ideal bureau established as harmonious relations as possible with many groups outside itself. Congress naturally was of great importance, but a house of Congress as a single deliberative body discussing scientific policy in full debate seldom existed at all in these years. Hence bureau policy had to aim at keeping a small number of senators and congressmen who sat on the right committees informed about the nature and progress of its work. For many years Representative James W. Wadsworth of New York kept abreast of the developing agencies in the Department of Agriculture. The legislators thus 'trained would then protect the bureau's interests, often getting substantive action by resorting to riders on appropriation bills.

The ideal bureau looked carefully to its relations to other agencies. Sometimes it was a child of an older group, drawing on its parents' tradition and personnel, as the Fish Commission grew out of the Smithsonian. Sometimes it had children of its own, as the Bureau of Animal Industry begot a Dairy Division which later broke away completely. Sometimes of course, one bureau and its growing tradition crossed the path of another.

To be safe from the vagaries of politics, the ideal bureau needed a group interested in its problems, but outside the government. Conservation would later spawn the most active of these associations, and the Coast Survey had used the AAAS from its founding. These groups often blended economic and scientific interests, and to the extent that they deserved the opprobrious title of lobbyist they were often politically useful.

In the same way, the ideal bureau had to have a working understanding with at least one university — the source of trained personnel and, even in 1880, of considerable research service. Sometimes the link was with a long-established institution, as the Geological Survey's alliance with Yale. But agriculture, thanks to the "cow colleges" set up under the Morrill Land Grant Act, was in the process of shaping a special university system for its own purposes.

Land-grant colleges meant dealing with states, which, although no longer able to claim sovereignty over scientific enterprise, still had formidable rights to press against the scientific bureau's natural interest in centralization. John Stuart Mill's dictum that "Power may be localized, but knowledge, to be most useful, must be centralized," [35] had its greatest relevance to agencies whose very business was knowledge. Yet the different terrain, weather, and crops of the various states might provide an opportunity to experiment and collect facts in the diversified localities where the problems lurked. The ideal bureau, while jealously guarding the lines of communication by which local information flowed into a central clearinghouse, could by using the states reach out farther from Washington and get down closer to local problems.

Even though it might begin as a research institution, the ideal bureau found that success in science inevitably bred responsibilities in two directions. As soon as it had proved itself, the bureau had to furnish routine services related to its problem, as the Naval Observatory had for long been the keeper of the nation's time. Much more significantly, success in research, implying ability to control a problem, involved the ideal bureau in regulation. The understanding of an animal disease led to the drawing of lines of infection and the definition of danger areas. These led to quarantines, and the bureau found itself with the sanction of law to enforce its scientific theories.

The advent of regulation based on scientific investigation had profound results. For the bureau as a whole it meant larger appropriations although not necessarily more funds for research. To carry out inspection and enforcement meant large staffs often consisting of people with an outlook different from scientists. By bringing the results of science directly home to the people, regulation increased public awareness of the bureaus more than any amount of scholarly publication. In a larger way, regulation meant the yoking of science to the fundamental operations of the government. Indeed, it opened the way for the belief that the rational ways of science could solve the problems of the nation more effectively and impartially than the chaotic clash of political interests.

Entomology, Animal Industry, Plant Industry

The first rumblings of a new approach came in the field of entomology.[36] With Townend Glover's health failing, the Department of Agriculture had so small a reputation in the 1870's that, when an unusual flight of locusts invaded the eastern Great Plains, Congress turned to another agency entirely. C. V. Riley, the state entomologist of Missouri, had written that the Department of Agriculture "might employ the large sums it now fritters away in the gratuitous distribution of seeds, to better advantage in . . . sending out . . . a commission; but the people have lost all hope of getting much good out of that institution." Others considered federal action ridiculous, the *Nation* asserting that "the Agricultural Commissioner will scatter the seed broadcast over the land, while the national entomologist will follow closely on his trail and exterminate the bugs that may attack ripening grain. We only want now another Commissioner to harvest the crops." Riley tried in vain to get respectability by having the National Academy consulted on appointments.[37] Congress finally provided a commission in the Interior Department, where the Western surveys were vigorous.

The United States Entomological Commission, with Riley as chairman, got out into the field to study the locust, publishing several bulletins. Despite its lack of practical results, the commission was important. In the first place, its research revolved around a problem. A national calamity called for a scientific answer, and the federal government responded, "a real encouragement to scientific workers as well as to farmers." [38] In the second place, the commission, instead of disbanding at the end of its appropriation as had so many other piecemeal responses, merged into the Department of Agriculture and perpetuated itself and its approach as the Division of Entomology. Riley succeeded Glover in 1878, and, although his political activity led to exile for a time, he remained until 1894.[39] The commission itself was transferred to Agriculture in 1880 because of the reorganization of the surveys in Interior.[40]

Between 1881 and 1885, congressional appropriations to the Division of Entomology increased from $7000 to $42,900, a plane which with minor variations held until 1906.[41] Although the commissioner of agriculture did the testifying before Congress until the 1890's, Riley was an industrious politician who sometimes did more harm than good in his efforts to wheedle appropriations.[42] Yet "economic entomology" was the new order of the day. The division investigated scale insects on orange and fruit trees, and pests affecting other staples, for instance, cranberries.[43] Assistants went to the scene of any important insect outbreak. Such agents played a part in the first really dramatic victory over an insect pest, the introduction of an Australian ladybird that was the natural enemy of a ruinous scale on citrus fruit trees.[44] The establishment of the state experiment stations greatly aided this field work. Older activities of course continued. Identification and classification, which had once been the main preoccupation of the entomologists, remained as important background work not only for the division but for many other branches of the government.[45] Thus taxonomy became a routine responsibility.

The ancient dream of a silk industry in America occupied much of the time and funds of the division in these years. In part an effort to diversify the agriculture of the South, it was definitely pressed upon the division from outside. By 1908 scientists had proved what they had believed in the first place, that silk culture was uneconomical in the United States without a protective tariff.[46] The external relations of the division gained strength with the establishment of the Association of Economic Entomologists in 1889, which aided in getting favorable legislation.[47]

By 1894, when Riley gave way as chief to his assistant, Leland O. Howard, the Division of Entomology had moved toward becoming one of the new scientific bureaus. Yet it still had a long way to go. It tended to let problems come to it instead of seeking them out. When in 1889 the gypsy moth first appeared in Massachusetts, the state began a program of extermination, which it abandoned on the brink of success in 1901. In 1905, with the pest out of control, the fight had to begin again with the help of the federal government. "Knowing what we do now," wrote Howard, "it would seem that the Federal Bureau of Entomology might fairly be blamed for lack of foresight in not warning Congress and the other States of the great danger and in not appealing to Congress for funds with which to prosecute radical work. As I look back, the idea seems never to have occurred to us." [48]

The development of the division between 1894 and 1915 rounded out the structure of a new scientific bureau. The increasing introduction of insect pests from abroad in the same period that large-scale production of single crops made American agriculture more vulnerable created problems of national importance. The San José scale in Eastern orchards, the boll weevil in the cotton belt, and the mosquito's prominence as a carrier of disease were the most spectacular challenges, which the division met with increasing promptness and vigor. Beginning in 1895, Howard carried on an active campaign for quarantine legislation that would allow him to check the accidental introduction of foreign pests. The opposition of nurserymen and other importers was so strong that several years and intensive efforts at educating individual congressmen were necessary to get a Federal Quarantine Act passed in 1912.[49] Although a separate Horticultural Board at first had control of actual quarantine, the Bureau of Entomology had much to do with it.[50]

After 1906, when the Division of Entomology became a Bureau in name, its appropriation quickly mounted, from $84,470 in 1906 to $262,110 in 1907 and $829,900 in 1916.[51] This increase reflected a general rise in activity, more regulation, and especially intensive work on the boll weevil. Following its policy of carrying "the laboratory to the problem rather than the problem to the laboratory," the bureau set up headquarters and experimental fields in infested districts. This led to the danger of logrolling on the pattern of river and harbor improvements. In spite of pressure from congressmen, one of whom told the chief "with his voice trembling, that his reelection depended on his success in retaining a certain station," [52] the bureau maintained its mobility in shifting its activities to follow its scientific problems. The boll weevil yielded to no single simple remedy, but out of the information came procedures by which cotton could be produced in spite of the pest. By 1916 the metamorphosis of the Bureau of Entomology into a new scientific agency was virtually complete, and it was proving its worth so regularly that its position in the government was not only secure but taken for granted.

The impulse toward problem research followed closely in the field of animal diseases. By the 1870's the rationalization of the cattle industry caused by railroad building to the West had continued into international trade, producing an export business of live cattle to Europe. The concern of the British government may have been in part economic, but the presence of disease coming from the United States gave a perfect reason to impose drastic restrictions. The federal government, not the states, felt the full weight of international pressure. In spite of the earlier investigations of Texas fever and pleuropneumonia under the commissioner of agriculture, the secretary of the treasury was first to react to British restrictions. In 1878 and 1879 Secretary John Sherman issued a series of orders which provided in turn for inspecting cattle, prohibiting export entirely, and a ninety-day quarantine. Amid all the clamor from cattle exporters Sherman pointed out that it "is hardly just . . . while no efficient measures are adopted in the United States to even systematically ascertain the extent of cattle disease in this country, and much less to adopt efficient measures for its suppression, to complain that the British Government adopts effective measures." [53] The secretary of the treasury had scanty legal power to set up an inspection and none at all to undertake the research on which to base an effective policy. Confusion was worse because the British were condemning Western cattle for fear of pleuropneumonia, which existed only in the East. [54]

While Congress drifted, a host of other animal-disease problems stacked up on the government's doorstep. Texas fever still had no solution. The beginning of shipments of American meat abroad in refrigerator ships emphasized the danger from hog cholera and trichinosis. Continental Europe joined in the game of restricting American meat products. The Department of Agriculture created a Veterinary Division in 1883 under D. E. Salmon, but could do little without new power and more money. Commissioner George Loring reached for outside help by calling a convention of livestock breeders in Chicago, which appointed a committee to work for national legislation. [55]

In 1884 the Congress in the organic act for a Bureau of Animal Industry created whole at a single stroke a new type of scientific agency. There were many who opposed it, claiming that it was unconstitutional to give special aid to one industry, that it would saddle the taxpayers, that disease was a myth, and that state regulation was sufficient. [56] These objectors had a less rocklike constitutional position than had the John C. Calhouns of an earlier day, and like them did not prevail on a direct vote. The organic act put most of the weight to setting up a regulatory system to control pleuropneumonia, involving the right to prevent the export of affected livestock and by coöperation with the states to regulate the disease domestically. It even provided that two agents who would investigate diseases be "practical stockraisers or experienced business men familiar with questions pertaining to commercial transactions in livestock." But the chief had to be a "competent veterinary surgeon" whose duty was "to investigate and report upon the condition of domestic animals in the United States, their protection and use, and also inquire into and report the causes of contagious, infectious, and communicable diseases among them, and the means for the prevention and cure." [57] The Bureau of Animal Industry thus had most of the attributes of the new scientific agency at its birth — an organic act, a set of problems, outside groups pressing for its interests, and extensive regulatory powers.

Regulation paid off handsomely on pleuropneumonia. By rigid inspection and destruction of affected cattle the disease had been checked by 1890. In 1892 the secretary of agriculture officially declared the United States free from it.[58] But Texas fever was more stubborn because shrouded in mystery which could be dispelled only by an increase in basic scientific knowledge.

The investigators of the 1880's had a much more favorable scientific atmosphere for an attack on Texas fever than had Gamgee back in 1868. Pervading all was the immense quickening of the biological sciences in the wake of Charles Darwin's publication of the *Origin of Species* in 1859. More concretely, the work of Pasteur, Robert Koch, and others had by this time done much to establish the germ theory with animal diseases as the subject of the most telling experiments. The discovery by C. L. A. Laveran of malaria parasites within the red corpuscles of the blood had suggested a new place to look for microörganisms that might cause a disease. With bacteriology entering its greatest period of expansion, adequate tools were now at hand for the attack on Texas fever, still protected by unexpected quirks in its life history.

The Bureau of Animal Industry brought to this problem the insight of several men of varying background working as a team. Salmon, the chief of the bureau, had studied the fever for a long time and had accurately established the border of the permanently infected district, healthy cattle from which would infect Northern stock. This was not only the basis for a crude quarantine applied in the fever season, but also an important clue to the nature of the disease.[59] In 1889 a team of young Cornell graduates started to work systematically. Theobald Smith, a doctor of medicine inspired by the new techniques of bacteriology, did the major microscopy, while Cooper Curtice gathered laboratory material from animals dead of the disease. F. L. Kilborne looked after the cattle in the experimental fields near Washington.[60]

Almost immediately a microörganism in the red corpuscles was singled out as the probable cause, and at the same time a tick of a particular species whose range coincided with the permanently infected district came under the same suspicion by the investigators that had long been held by cattlemen.[61] But, as Smith later wrote, "after it had been shown that the disease failed without ticks, everything was still to be done." [62] A series of decisive and well-designed experiments in the pens near Washington not only proved the agency of the secondary host but caught the fact that the disease was passed from one generation of the ticks to the other. Smith worked out the nature and course of the disease, while Curtice did a most intensive study of the life history of the tick.

When the final report on Texas fever appeared in 1893, not only did the American stockgrowers potentially benefit, but government science had made a first-rate contribution to basic knowledge. Not perhaps on a par with the work of Pasteur and Koch, the discovery of the role of a secondary host in transmitting disease nevertheless had that quality of opening up great possibilities for the future which is the hallmark of a basic discovery.

Although Theobald Smith soon went on to a Harvard professorship, the Bureau of Animal Industry still had to struggle with Texas fever. After fourteen years of experimenting on chemicals to destroy ticks, the bureau began in 1905 a campaign to eradicate the disease with state and local coöperation. A Field Inspection Division, set up in 1912, became a Division of Tick Eradication in 1917,[63] a permanent activity with no end in sight. This routine responsibility, which in itself involved an elaborate research program, is as much a part of the story of government science as was the superb original discovery itself.

The Bureau of Animal Industry, precocious with its organic act and regulatory functions, led the department into the twentieth century both in its size and in its structure. An appropriation of $812,000 in 1895, over two and one-half times that of the Division of Entomology, became $4,427,860 in 1910 and $7,880,026 in 1915,[64] over 25 per cent of the whole department total. In 1897 the breakup into Inspection, Dairy, Pathological, Biochemic, and Miscellaneous Divisions marked a widening sphere of interest and increased specialization. In 1906 meat inspection, as a result of pressure generated by Upton Sinclair's book *The Jungle*, brought the bureau increased appropriations and underscored its permanent position as the largest and most fully developed line agency in the department.[65] Meanwhile, the quality of research did not suffer from prosperity, as, for instance, the work of Marion Dorset on hog cholera demonstrates.[66]

The economic and scientific pressure that revolutionized the department's work on domestic animals operated as strongly if less dramatically on the efforts to help crop plants. Yet the very strength of the department's earlier interest in this field meant that reorganization had to start at the bottom instead of the top. The gardener William Saunders and the botanist George Vasey were in firm control, and the munificent appropriations for free seeds were under the thumb of congressmen who cut this particular melon with the same zest they applied to river improvements and postmasterships. Hence the problem approach and the new sciences applied to plants had to creep up from below.

In spite of the existence of the Division of Microscopy, which was supposed to do analyses for the whole department, little knots of scientists whose work was built around the use of the microscope in pathology and bacteriology began to appear within the Division of Botany. A section of mycology, the study of the fungi, first brought Frank Lamson-Scribner into the service. By 1888 it was called the Section of Vegetable Pathology, with Beverly T. Galloway beginning a distinguished career as its head.[67] Under him a group of young men applied the same technique to plant diseases that was yielding spectacular results in the Bureau of Animal Industry. Erwin Frink Smith, who was there from the beginning, became one of the great pioneers in plant pathology. His studies of peach yellows and crown gall not only had great value to farmers but opened up new avenues in the bacteriology of plant diseases and in the pathology of cells, a study leading directly to cancer research.[68]

In 1895 this group became the Division of Vegetable Physiology and Pathology at the same time Lamson-Scribner came back to head a Division of Agrostology, which specialized in grasses and had added responsibilities as the Great Plains became an important agricultural area.[69] At the same time the Division of Microscopy was abolished as "an absurdity," since "the microscope has come into daily, almost hourly, use in nearly all scientific laboratories." [70] The National Herbarium went back to the Smithsonian, where it could become a tool of basic science more easily while still providing the department with identifications when needed.

Thus by 1900, in addition to the Division of Botany, Divisions of Pomology, Vegetable Physiology and Pathology, Agrostology, Gardens and Grounds, and Seeds, operated parallel to one another. Together they had appropriations of $268,400,[71] intermediate between Entomology and Animal Industry. The secretary of agriculture then used his power to group these into a Bureau of Plant Industry, which Congress recognized in the next appropriation act.

The problem approach had won out, although regulatory activity did not come until 1917.[72] The change in results which accompanied this shift showed up clearly in that old scandal, seed distribution. Although no part of the government's scientific program had been so notorious for so long, congressional interest kept up free distribution until 1923. Yet the basic idea of gathering plants abroad with which to enrich and diversify American agriculture — the old dream of Henry Perrine — was not in itself the root of the trouble. In 1898 the department first got authority to test seeds bought on the open market and to obtain rare plants and seeds from other countries, an idea that old Commissioner Watts had scoffed at a quarter of a century before.[73] Under this authority grew up a whole new breed of explorer, not without popular appeal.[74] Some climbed the Himalayas or drowned in the Yangtse pursuing new plants. From their work came introductions of real importance — durum wheat, Sudan grass, Smyrna figs, soybeans. In most of these cases the plants were not unknown earlier. For example, durum wheat had been introduced by Russian immigrants in the Dakotas long before Mark A. Carleton brought it in for the department. What he did was to study carefully the Russian soils and climates where it flourished and then to select test localities for trials extending over several years. It was no accident that he was a plant pathologist in his own right.[75] Plant introduction and its offshoot, the distribution of seeds, were worthwhile activities only when accompanied by long-range testing and careful control. As the department increasingly applied research, an odorous political practice gradually disappeared. In this case more science meant less politics.

The Bureau of Plant Industry, with the Bureau of Entomology and the Bureau of Animal Industry, became the core of the department's research establishment. Their contemporary, the Bureau of Chemistry, followed other lines of development,[76] and the growth of the Weather Bureau, the Forest Service, and the Biological Survey belongs to stories to come later. In 1880 the forerunners of the three line agencies had convinced nobody. Yet even then they were beginning to become new scientific bureaus. By 1897, although receiving only modest support, they had begun to show results that were paying off consistently to the farmers of the country and on occasion were achieving the level of basic scientific discoveries. By 1916 the bureaus commanded large funds, and at the same time by regulation and inspection had become real powers in the government.

Coöperation with the States

From its beginning the department found itself forced to consider the States.[77] Farmers in the 1870's were seeking relief from their economic problems more insistently in the state capitals than in Washington. Besides the Granger laws, this agitation produced its share of state boards of agriculture, some with an interest in research. More important, the land-grant colleges provided a link between the federal and state governments. Although these were hard years for the colleges as free land and overproduction made a college education for a practical farmer seem an absurdity, the devoted professors and their few but equally devoted students began to realize that the way out was new knowledge. By the middle 1870's states began to set up experiment stations, which aimed to provide the land-grant colleges both outdoor and indoor facilities for research.[78] Although the great farmers' organizations such as the Grange sometimes looked askance at book learning, each land-grant college became a nucleus of the advocates of science in agriculture.

To these scattered friends of research, two major factors pointed toward participation of the federal government in the experiment-station program. First, the precedent of the Morrill Act suggested the availability of a share of the national domain which was then being liquidated so rapidly. More fundamentally, the principle of the centralization of knowledge suggested that only systematic interchange of information among the state experiment stations and the Department of Agriculture could produce efficient results. In 1872 Commissioner Watts called a convention of two delegates from each agricultural college, state agricultural society, and board of agriculture, which assented to Senator Morrill's motion to request additional donations of land. From this time on the Senator from Vermont introduced a long series of bills for land grants, often compromising with the advocates of aid to common schools. Finally in 1890 he secured the "Second Morrill Act," which appropriated for each land-grant college a sum gradually increasing to $25,000 annually. In 1900 an act permitting this to come from the treasury if the sale of lands failed [79] meant that the support had become permanent.

Meanwhile the advocates of the experiment stations continued to hold conventions and introduce bills to Congress. In 1883 the president of Purdue University, admitting that "the results actually accomplished by the national grant in the two decades now closing are not satisfactory," claimed that the colleges had done most "in the direction of scientific training and investigation. The founding of the national schools has caused the study of science to assume new importance in all higher institutions." [80] From such opinions came the support for the repeated attempts to establish "national" experiment stations supported by money drawn from the treasury. The colleges thought the control of the department too ominous in these versions, but in 1887 a bill was introduced by Senator William H. Hatch of Missouri which made the stations purely state institutions aided by federal land grants. With all the interested groups mollified, this version became law.

The passage of the Hatch Act changed the Department of Agriculture from a single central agency into a nexus of a system of semi-autonomous research institutions permanently established in every state. No other scientific activity of the government had attained such a spread, which, because of the political fecundity of the American people in spawning states, more than adequately covered every physiographic region. Since each station was attached to a land-grant college, the department gained an in-law kinship with an equally numerous group of institutions which were gradually earning the right to be called seats of higher learning.

Among the more immediate results of the Hatch Act was the formation of the Association of American Agricultural Colleges and

Experiment Stations, an outgrowth of the earlier conventions. While much of its work concerned the improvement of agricultural education, the Association allowed the department official representation. As an external group watching the interests of agricultural research within the central government and at the same time having roots in every state, this organization gave the external support without which no scientific agency could long endure.[81]

Yet coöperation and good will could not hide the fact that state institutions financed by federal money were in "an anomalous partnership" [82] if measured by the conventional standards of legal and constitutional theory. The department almost immediately reacted by establishing an Office of Experiment Stations, whise director, W. O. Atwater, had been director of the Connecticut station and a leader in the movement behind the Hatch Act.[83] While the office concentrated on establishing amicable relations with the stations and with foreign institutions engaged in similar research, a hint of coördination as well as coöperation entered. "It is the duty of this Office," wrote the director, "to indicate lines of inquiry, furnish . . . advice and assistance, and to 'compare, edit, and publish such results' of their work as may be deemed necessary." [84] By linking its line agencies to the experiment stations, the department began a long evolution toward a centralized research center. The commissioner of agriculture foresaw this when he predicted in 1888 that such a center could relieve the state stations "of much costly and laborious scientific work and enable them to devote their energy the more completely to the things that are of practical interest to the farmer," and serve as a model of "what an experiment station is and how its work is most successfully accomplished." [85] This idea was the embryo of the great Beltsville Research Center of the twentieth century.

Also implicit was the possibility of trouble over the role of basic research.[86] Even before the passage of the Hatch Act, the magazine *Science* beat the drum for an experiment station as "primarily a research institution, intended to promote the science of agriculture, and capable of the highest and most permanent usefulness, only when it fulfills this intention." [87] E. W. Hilgard, the great soil man at the University of California, answered that the stations could survive only by rendering "to the agricultural population the scientific aid which they so sorely need when brought face to face with new and untried conditions and factors in a new country." He objected to any separation of the practical and more fundamental projects in the administration of the stations.[88] The magazine's final retort foreshadowed the chronic dilemma of the stations. "There appears to us comparatively little danger that the work . . . will be too rigidly scientific, and too far removed from the apprehension of the farmers. There is a constant pressure upon a station for immediately useful results, and any station refusing reasonable conformity to it will not enjoy a long life." [89]

Financially the federal government gradually felt bound to step into the experiment stations to safeguard its investment. Unqualified officers, superficial work, and diversion of funds to other purposes were in the background when in 1894 Congress gave the secretary of agriculture power to get from the stations an annual statement of expenditures.[90] The Adams Act of 1906 strengthened both financial support and control by the federal government, restricting funds to original research.[91] Yet the states, undeterred by these few strings, put an ever larger investment into the stations. In 1887–1888 the federal government furnished 82 per cent of the support of.the stations, while the states contributed only 10.6 per cent. By 1906 the federal government furnished less than half,[92] while in 1920 the figure was down to 18.9 per cent.[93] The total in all categories climbed steadily.

Cabinet Status and a More Elaborate Department

Neither the Hatch Act in 1887 nor the elevation of the department to cabinet status in 1889 effected an overnight change. Nevertheless, by the mid-1890's the old department of the commissioners was rapidly disappearing. In 1890 the whole organization was comparable in size to the Coast Survey or the Geological Survey alone. The secretary of agriculture had a valid if exaggerated point when he said that "this department, though representing the greatest interest in our domestic affairs, is the one of our national departments endowed with the smallest appropriation and receiving the least consideration." [94] Within five years the department entered into a new era.

One factor in the new department was the built-in source of scientific personnel — the land-grant colleges — which, aided by the experiment stations, were now turning out young men devoted to the ideal of a scientific agriculture. Largely isolated from the rest of the university movement and defensive in their attitude toward the lack of tradition and atmosphere in their own colleges, these new men replaced the varied assortment of foreign-trained or self-trained older scientists. A homogeneous and self-consciously separate group began to set the tone of the department.[95] Most of them not only were educated at land-grant colleges but often interspersed teaching and research on the campuses with tours of duty in the department. Thus the central organization itself tended to become a kind of university. Significantly, in the 1890's the dream of a hundred years earlier, a national university, hovered lightly over Washington and received much of its impetus from agriculture.[96]

Closely related to the rise of the land-grant scientists was the development of a modicum of central control within the department. When the commissioner was elevated to secretary, room appeared for an assistant secretary, who undertook to coördinate the scientific work. The first appointee, Edwin Willits, although a lawyer and politician by background, was president of Michigan Agricultural College.[97] Nevertheless, even a little control displeased the older division chiefs, such as C. V. Riley.[98] The second holder of the office, Charles W. Dabney, was the president of the University of Tennessee and a German-trained chemist with extensive experience in the land-grant college system,[99] which from this time on became the major source of assistant secretaries.[100]

Besides solidifying the land-grant tradition in the higher appointments of the department, Dabney concerned himself especially with applying the civil-service merit system to all personnel, including the heads of scientific bureaus. The reformers' favorite method for improving the government in the late nineteenth century, a competitive classified service, had made a beginning with the Pendleton Act in 1883, and had in general caught on first for the more routine positions.

Scientists greeted civil-service reform somewhat gingerly. They lost heavily by frequent rotation in office and by the appointment of party hacks to places requiring scientific training. After Cleveland's election in 1884, the shadow of the "tall, gaunt Southerner in a white linen duster, with the corners of his mouth stained with tobacco juice" that fell across the desk of the assistant entomologist was as disturbing to the morale of scientists as were requests for campaign contributions. Insecurity was a major emotion in all the scientific offices. Yet even under the old conditions science did not always suffer. L. O. Howard admitted that during the fifty-three years after

1878 he did not know of "a single case in which a scientific man lost his position for any reason other than incompetency." [101]

The difficulty with the merit system was that scientific ability was notably hard to measure by formal examination. Freedom to appoint a politically important incompetent could also be used to appoint the brilliant but perhaps unconventional scientist. As Howard put it, "my personal acquaintance with entomologists and with teachers of entomology was very great, and I always felt that I could pick my assistants much better than any Commission with its series of examinations." [102] At the beginning of Cleveland's second administration in 1893, only 698 of 2497 employees of the department were covered by the classified service. Three and one-half years later the only employees above the rank of laborer not under classifications were the secretary, the assistant secretary, and the chief of the Weather Bureau.[103]

Dabney, fully recognizing the difficulty of applying the examination system to scientific positions, began the long search for a workable solution. He concentrated on developing a list of eligibles from among the recent graduates of the land-grant colleges and, since little graduate training in the country applied directly to agriculture's problems, emphasized that the department itself "must become more and more a training ground for scientific experts." [104] Such an in-service program he conceived of as a kind of national university within the civil service.[105]

In practice the department largely relied on recruitment of young men by examination, completion of their qualifications while on the job, and promotion through the various classifications up to bureau chief.[106] The coming of the classified service thus accomplished the substantial objectives of giving the department stable leadership from qualified people at the bureau chief's level. At the same time the system accentuated the corporate nature of the department, setting off as a separate society the scientific personnel already markedly conditioned by a land-grant college background.

Formal civil-service requirements did little to change the deepening problem of holding the best people. In 1897 Dabney saw that the "bright young men who are making reputations rapidly leave the Department to go to the colleges, universities, experiment stations, and even to manufacturing and other industrial corporations." [107] In 1913 the secretary of agriculture, pointing to the low maximum salaries, struck the same note when he asserted that because of "the great demand for such men in this country and abroad, the department is constantly losing men it ought to keep, and it is unable to find an adequate supply of just the right type of men to replace them." [108]

Implicit in the problem of holding personnel, which doubtless lowered the long-run effectiveness of the department, was the enrichment of American science as a whole by the men who departed. Theobald Smith, the obvious symbol of the exodus, left the department in 1895 at the age of 36 for a distinguished research career at Harvard, the Massachusetts State Board of Health, and the Rockefeller Institute.[109] A more typical case taken at random, which reveals clearly the role of the department as a contributor to science, is E. D. Merrill. Against all advice he took a $1200 job as assistant agrostologist in 1899. He "always looked on my two and one-half years work in Washington as my post-graduate course in taxonomy, even though to a very large degree I was placed on my own resources." After several years in other government service in the Philippines, he returned to important posts at the University of California, New York Botanical Garden, and Harvard.[110] After graduate training in the universities grew to large proportions, the department's unpublicized service became harder to isolate as a separate influence on American science. Yet it has continued to enrich the personnel of a number of sciences that touch agriculture.

When William McKinley became President in 1897 he appointed as his secretary of agriculture the head of the experiment station at the land-grant college of that most proudly agricultural of states, Iowa. Tama Jim Wilson, an ex-congressman, stock farmer, journalist, and professor, was an admirable representative of the outside forces that were shaping the department.[111] Taking over the direction of scientific work himself, Wilson reigned through the administrations of McKinley, Theodore Roosevelt, and William Howard Taft — an unprecedented 16 years.[112]

The inauguration of McKinley was in addition significant to those broad economic and social forces which shaped the larger environment of the department. The farmers had tried political action through the Populist Party and Bryan's campaign, where they staked all on a free-silver monetary policy. After their failure, they had to come to terms with a predominantly urban and industrial United States. Free land on the frontier no longer existed as a symbol of independent husbandmen. In a society expanding in population and industry, but not in geographical extent, agriculture had to become efficient to a degree unparalleled earlier. The department had shown between 1880 and 1897 enough of the possibilities of science to become increasingly the focus of this effort.

The Bureau of Chemistry

In the new atmosphere of industrial America the Bureau of Chemistry's development diverged somewhat from the other line agencies. Under the old department it was the most important division and hence tended to cling to coverage of the whole discipline of chemistry. When it did shift to the problem approach, it found itself dominated by the peculiar nature of the particular investigations it chose.

In the very early years, as the only conspicuous and relatively well-equipped laboratory in Washington, the Division of Chemistry inherited from other departments of the government all sorts of routine testing jobs entirely unconnected with agriculture.[113] These interruptions used up much staff time and disrupted long-range research on agricultural problems without providing any alternative focus. But since a laboratory as a social organization was a new and rare thing,[114] the division had to suffer these intrusions as long as it contained the government's only example of this powerful scientific innovation.

A little later, the division, based on a discipline instead of a problem, suffered the same disintegrating tendencies that disfigured the Division of Botany and destroyed the Division of Microscopy. As money became available in appropriations for chemical equipment, it went to the people who had problems to solve instead of to the Division of Chemistry. The Bureau of Animal Industry, its Dairy Division, and the Division of Vegetable Physiology and Pathology had laboratories and chemists of their own before 1900.[115] The Division of Chemistry could hold on to its position as general chemical experimenter for the whole government only by tremendous expansion. Harvey W. Wiley, the indomitable chief who ruled the division and bureau from 1883 to 1912, made a try at keeping this position by setting up a contracts laboratory in 1903,[116] an attempt described by one of his colleagues as "about as reasonable as to confine all typewriting to a single bureau." [117] Chemistry was too large a subject and its uses too ubiquitous to permit great centralization. Long before 1903 the survival of the Division of Chemistry depended on its quest for problems to solve.

The first major project that commanded attention from the department's early chemists was not a fortunate one from the point of view of American farmers as a whole. The dream of producing sugar in the temperate regions of the United States was as old as the dream of raising silk. Sorghum as a source of sugar had beguiled the department from the Civil War days. When Wiley took over in 1883 he extended sugar research to the pilot-plant stage. After sorghum as a sugar producer (though not as a source of grain and forage) proved a pipe dream, Wiley vigorously pushed sugar beets and determined the belt where maximum results in raising them could be expected.[118] The outcome of these years of effort was a highly specialized industry which could exist only under tariff protection.

The analysis of soils as one of the classic and most useful applications of chemistry to agriculture never seems to have caught Wiley's imagination. Hints of the magnitude of soil problems drifted into the department with the Weather Bureau after 1890, and the Geological Survey's studies attracted some attention. When Dabney, as assistant secretary, became alarmed by the soil-erosion problem in the South, he stirred up a joint study by the Weather Bureau and the Divisions of Chemistry, Forestry, and Botany. As a result of this activity the Weather Bureau soil work became independent and then in 1897 merged into a Division of Soils under Milton Whitney.[119] In 1901, both Chemistry and Soils became separate and independent bureaus.[120] Although destined for further administrative vicissitudes in the twentieth century, soils as a problem unit for research in the department had achieved coherence, but outside Wiley's domain.

Meanwhile Wiley found a problem that gradually engrossed both him and his bureau. As food production and distribution became large-scale, the consumer found himself at the mercy of an impersonal system in which his ability to test the purity and healthfulness of food products went down while opportunities for adulteration and spoilage increased. As agriculture became more a part of an industrial civilization, the purity of foods and drugs became a grave concern. Its very nature suggested the application of science to establish standards for regulation, which in turn required enforcement. Unlike earlier programs of regulation such as those concerning cattle diseases, a pure-food-and-drug policy was essentially a service to the consumer and might very well cost the agricultural producer money and trouble.[121] In pushing into this field Wiley thus not only rigorously coupled research to regulation but also tended to change the definition of the group served by the Department of Agriculture. With food and drugs under its control, it could not be simply a farmers' department.

As early as 1880 the problem of pure foods and drugs began to press the government, which, except for some legislation regulating the importation of drugs, had left the problem to the states. Inquiries often sought out the Division of Chemistry as the laboratory of all work, making the commissioner complain that all he could do was inform his "correspondents that under the present standard of commercial morality, nothing is safe from adulteration . . . that the power of the government ceases at the custom-house; and that the general regulation . . . can only be done effectually by a rigid system of inspections." [122]

When Wiley took over the Division of Chemistry in 1883, he established a section on food adulteration and also began studies of preservatives.[123] After 1889 he had a specific appropriation for these investigations, and the Division of Chemistry became the recognized center of the study within the government. Had Wiley loved soil analysis, and had some treasury department official rather than he been a chemist and publicist with imagination, the administrative history of pure-food-and-drug research might have been quite different. The division's *Bulletin 13* on food and food adulterants appeared serially between 1887 and 1902, laying the scientific groundwork for a political campaign for legislation. In 1903 Wiley set up a drug laboratory to establish standards of purity with which to attack patent medicines and remedies.[124] Thus Wiley gradually converted his Bureau of Chemistry into an efficient experimental research laboratory to support inspection and regulation of foods and drugs.

Every session of Congress after about 1880 saw several food-and-drug bills introduced, some placing control in the Department of Agriculture and some in an independent commission. Wiley, getting his chance in 1900 when a Senate committee appointed him as a scientific expert,[125] proved not only an effective witness but also a popular promoter of his cause. In 1902 he got legislation to make experiments with his "poison squad," volunteers who lived on a diet containing controlled amounts of food preservatives.[126] When the muck-raking atmosphere of Theodore Roosevelt's administration proved congenial to legislation, Wiley not only testified before committees and furnished congressmen with arguments but also aided in drafting the bill that became a law in 1906.

The impact of the regulatory function on a scientific bureau is evident in the Bureau of Chemistry. In 1888 the personnel for all purposes consisted of one chemist, two assistant chemists, seven laboratory assistants, and three others.[127] In 1897 the number had risen to only 20,[128] and in 1906 to 110. One year later the bureau had 250 people, and 425 in 1908. Appropriations went up: [129] 1906, $130,920; 1907, $395,920; 1908, $697,920.

With the Bureau of Chemistry making the examinations of foods and drugs under the act, Wiley became a center of political pressures entirely out of the ordinary for the head of a scientific bureau. He dramatized himself as the watchdog of the kitchen and the incorruptible enemy of the "whiskey rectifiers" and the "patent medicine brethren." [130] Despite the fame, however, he did not see eye to eye with either the secretary of agriculture or the President. He described Tama Jim Wilson as having "the greatest capacity of any person I ever knew to take the wrong side of public questions, especially those relating to health through diet," and he felt that Roosevelt had received undue credit for the passage of the act.[131] Thus, although the bureau made steady progress in enforcing pure-food-and-drug standards, it became a correspondingly serious problem to the responsible political chiefs to be sure in the face of pressure that Wiley's facts were straight. With the complexities of the industrial age, science in solving problems for the government inevitably raised the question of how expert but possibly fallible scientists could be kept in control by inexpert and undoubtedly fallible but responsible officials.

Tama Jim tried to soften the clash between his crusading bureau chief and angry manufacturers by appointing a Board of Food and Drug Inspection, consisting of Wiley, the solicitor of the department, and a young chemist named F. L. Dunlap. Wiley, considering the board a "usurpation of authority," claimed that the other two by voting him down made it "impossible to bring any cases against certain classes of offenders." [132] He never reconciled himself to the board arrangement nor to his fellow members, whom he regarded as tools of the interests.

Wiley further had the bad luck to rub President Roosevelt the wrong way. Since he opposed additions of any kind to foods, some cases involving benzoate of soda and saccharin got up to the President, who, because his doctor gave him saccharin every day, reportedly declared that anybody who says it "is injurious to health is an idiot." [133] More soberly, Roosevelt felt that the "trouble with Dr. Wiley is that to my personal knowledge, he has been guilty of such grave errors of judgment . . . as to make it quite impossible to accept his say-so in a matter without a very uneasy feeling that I may be doing far-reaching harm to worse than no purpose." Nevertheless, "I have such confidence in his integrity and zeal that I am anxious to back him up to the limit of my power wherever I can be sure that doing so won't do damage instead of good." [134] To solve this classic dilemma of the administrator faced with doubts about his technical experts, Roosevelt turned to other experts. He appointed a committee of chemists to advise the secretary of agriculture on pure-food-and-drug questions. With Ira Remsen, president of Johns Hopkins, as chairman, the so-called Referee Board was made up entirely of outstanding chemists. It would consider only questions disputed "among eminent authorities." While the members were to devote only part of their time to government work, they were to be "paid liberally for the time employed." They could make "all experiments necessary to reach a decision on the questions submitted." Although their function was purely advisory, Roosevelt hoped that "when the Referee Board speaks, it will be the final word on the subject as far as the United States is concerned." [135]

The Referee Board promptly backed a relaxation of the prohibition on benzoate of soda and thereby earned Wiley's undying enmity. But Remsen found the duty distressing,[136] and neither the department's board nor the Referee Board continued active for long after Wiley left the bureau in 1912. Although he himself never enjoyed the luxury, his principle that the chief of bureau not only conduct the research but make the decisions won out in the end.

In spite of the acrimony that accompanied regulation, constructive lines of research also emerged from the pure-food-and-drug program. For example, the more general problem of the preservation of foods under refrigeration increasingly underlay spoilage and contamination. As early as 1900, department people became interested

in the transportation of perishables, and in 1901 continuous study
began in the Bureau of Plant Industry. In 1908 the Bureau of Chemistry set up a Food Research Laboratory which contributed many improvements to applying refrigeration to food handling both by freez-
ing and by cold storage. Thus the department provided the essential
research for a new industry which was emerging as a result of economic changes.[137]

Spreading the Results of Research

The coming of pure-food-and-drug legislation, so indicative
of the impact of the industrial and urban changes that were transforming American society, was one of the government's outstanding accomplishments in the Progressive Era. Yet Wiley led his
bureau so far beyond the objective of service to the American farmer
that the transfer of pure-food-and-drug functions to some other department was implicit almost from the beginning, actually taking
place in 1940.

Meanwhile the department, growing constantly as a research and
regulatory agency, gradually faced up to the fact that it had to become an institution for popular education as well. To make a real impression on American agriculture, research results had to get into the
hands of the farmers, who were often closely wedded to habitual ways
and contemptuous of the faith in science that radiated from the land-
grant college graduate. The state experiment stations and colleges,
along with the agricultural journals, had borne the brunt of extension
work for many years in a disorganized way. Seaman A. Knapp, who
had helped push through the Hatch Act and had distinguished himself by introducing rice on the Gulf Coast, was one of the first to
grasp the large dimensions of the problem of educating the mass of
farmers. He found that by inducing Iowans versed in good farming
practice to settle in Louisiana he could teach the natives by actual
demonstration. In the fight against the boll weevil, which resisted
dramatic and easy extermination, he found that the practices of seed
selection, deep plowing, wide spacing, and rotation would enable the
farmer to grow a crop despite insects. To get this across he set up a
demonstration farm which convinced as well as taught. Significantly,
part of the early support of the program that stemmed from this
work came from the Rockefeller Foundation.[138] In 1914 the Smith-

Lever Act put the Extension Service on a separate and permanent basis.[139] One feature of this law was the "50-50" plan by which each federal dollar was matched by one from the states.[140]

By the eve of World War I the Department of Agriculture had reached maturity. In 1862 it had been merely a hope and a promise. In 1880 it had convinced few that it was worth the money expended. In 1897 it had proved extremely useful but the whole department was still only a little larger than a good-sized bureau of that day. In 1913 it was a $24,000,000 business[141] and had 14,478 employees.

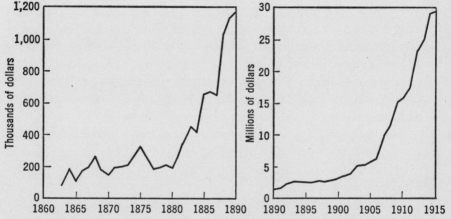

Fig. 1. Total appropriations of the Department of Agriculture, 1862–1915. [Source: Chief of the Division of Accounts and Disbursements, "Report," *Annual Reports of the Department of Agriculture for the Year Ended June 30, 1915* (Washington, 1916), 251.]

These were classified as: 1812 in scientific investigation and research; 1323 in demonstration and extension; 687 administrative and advisory; 6021 in regulatory work; 4635 clerks and lower.[142]

In general the department was weak and the bureaus strong. Secretary of Agriculture D. F. Houston, finally recognizing the need for reorganization, pointed out in 1914 that "the three leading lines of departmental work — the regulatory, the research, and the educational — had become in a measure intermingled in the various bureaus, so that no satisfactorily clear view could be had of them in their entirety either in any bureau or in the department as a whole." Admitting some lost motion and overlapping, he hoped to organize the three functions so that "each would reinforce and foster the other."

To do this he rejected the idea of separating them into different divisions and suggested instead "the definite outline . . . in each bureau of these three groups of activities." [143]

The permanence and the solidity of the department's administrative structure were unquestioned. It was an integral part of the machinery of government. Both in the realm of disease and in the control of pure-food-and-drug standards the department had proved itself able to administer programs that made a vital difference to large numbers of people. More important, it had harnessed research to these regulatory activities in such a way that both gained. Research workers had data and experience gathered by regulatory activities that would have become static and oppressive without information flowing in the other direction as well.

The results of research from the department's stations and laboratories emerged in an ever-swelling stream. By 1916 no other great economic interest in the United States could boast such a research establishment for the application of science either in or out of the government. With farming divided into such small individual units it is difficult to visualize this establishment growing to any stature in private hands. Indeed, no comparable agricultural research organization existed anywhere else in the world. On a few problems, such as Texas fever, the department's research had been brilliant. On many more it had proved useful. The demonstrably ineffective programs such as seed distributions had steadily lost ground as the greatest innovation of the period, the new scientific bureau, had gained.

The land-grant colleges were both a source of supply for the department's personnel and its representatives in the states. Through the extension service it reached individual farmers throughout the land. With the line bureaus at the center, the department was a self-contained system which approached the ideal of centralizing knowledge while partly localizing its administration.

IX

THE DECLINE OF SCIENCE IN THE
MILITARY SERVICES

1865–1890

THE exploration-centered activities of the military, which with the Coast Survey dominated pre-Civil War government science, emerged from the conflict badly disorganized but hopeful. The Corps of Topographical Engineers was gone. West Point was no longer exclusively an engineering school. Specialized officers had gone over into general service. Yet many expected that when the Army and Navy had shaken out their volunteers they would resume operations as before. The eventual disappointment of this expectation and the decline of the scientific function in the military is one of the basic drifts of the last third of the nineteenth century.

The Navy's Observatory and Its Overseas Exploration

Some of the more permanent of the old organizations maintained and even raised standards of excellence within their own spheres. The Naval Observatory, hampered by Maury's lack of emphasis on astronomy and by his sudden departure, began under Gilliss a new program which gained speed only haltingly because of the need for new equipment.[1] In 1866, with a new transit circle and with the work of the Hydrographic Office formally detached, the Observatory began a program of fundamental research in astronomy.

Some who contributed mightily to astronomy and physics worked for the Navy in this period, but they did not stay long. E. S. Holden, a West Pointer, began as an assistant at the Naval Observatory in 1873 but resigned in 1881 for a career in private institutions that carried him to the directorship of the Lick Observatory of the University of California.[2] A. A. Michelson, a graduate of the Naval Academy, developed his lifelong interest in the velocity of light while teaching there. But he left Annapolis in 1879 at the age of twenty-seven for European study and a distinguished career at several universities.[3]

The scientific luster of the Observatory came from the work of Simon Newcomb, who held the rank of professor of mathematics and later was head of the *Nautical Almanac*. By any reasonable standard Newcomb was one of the great scientists in America and one who commanded high respect abroad. He might well have repeated the familiar pattern of the period by leaving the government with his career largely before him. In 1875 President C. W. Eliot of Harvard, bent on creating a real university, offered Newcomb the directorship of the Harvard Observatory. On this occasion, Grant's secretary of the Navy, G. M. Robeson, gave his estimate of the Navy as a scientific outfit by telling Newcomb: "By all means accept the place; don't remain in the government service a day longer than you have to. A scientific man here has no future before him, and the quicker he can get away the better." Nevertheless, Newcomb, whose wife was a granddaughter of Hassler, chose to stay with the government, both because he felt the opportunities greater for the particular problems in mathematical astronomy he wished to pursue, and because he "did not believe that, with the growth of intelligence in our country, an absence of touch between the scientific and literary classes on the one side, and 'politics' on the other, could continue." So he resisted the academic lure, giving the Naval Observatory and the *Nautical Almanac* a golden age.[4]

Another distinguished astronomer worked for him on the *Nautical Almanac* in these years. G. W. Hill performed mathematical prodigies on the theory of the motions of Jupiter and Saturn and developed a new method of calculating the motion of the moon.[5] Yet Newcomb fought in vain to get Hill's salary raised from $1200 to $1400, concluding that it would be hard "to find a more impressive example than that afforded by [his] career, of the difficulty of getting the public to form and act upon sane judgments in such cases as his." [6]

Military control was periodically reaffirmed by the appointment of line officers to command the Observatory, a policy that drew increasingly severe criticism in scientific circles.[7] Indeed, in 1885 a committee of the National Academy suggested that it be called, "as the present observatory was styled originally, the 'National Observatory of the United States,' and that it . . . be under civil administration." For the Navy's use they suggested an enlarged establishment at Annapolis, saving the one at Washington for the control of "those who have made astronomy their life-work." [8] But the Navy held on, and in many ways advanced astronomical science in this period.

In 1873 Alvan Clark and Sons, the great telescope builders, completed an instrument for the Naval Observatory which was the best of its kind for its day.[9] In 1893 new buildings and a new location in northwest Washington provided more favorable surroundings.[10] By this time the Naval Observatory had achieved a permanent place in the government's scientific establishment, but with the retirement of Newcomb from the *Nautical Almanac* in 1897 its great days had passed. Its activities ran strongly to routine, such as the time service, and appropriations remained small and static.[11] The future of government science lay in newer and stronger agencies, while the future of astronomy in America lay at Yerkes and Mount Wilson.

In overseas explorations also the Navy showed a still lively but gradually weakening interest. In 1866 Admiral Davis set the stage for a concerted attack on the Isthmian Canal problem when he stated in a report requested by Congress that "there does not exist in the libraries of the world the means of determining, even approximately, the most practicable route for a ship canal." [12] In 1869, on orders from President Grant, the Bureau of Navigation sent out a whole series of simultaneous expeditions to examine in detail every possible route. To judge and correlate the findings, the President appointed an Interoceanic Commission made up of the chief of the Bureau of Navigation of the Navy, the chief of engineers of the Army, and the superintendent of the Coast Survey. The naval parties, under the command of such veterans of the pre-Civil War effort as R. W. Shufeldt, completed their work in five years. For the first time the United States had sufficient reliable data on which to base a canal policy. According to a recent student, their measurements are substantially unchallenged today.[13] As with their predecessors, the railroad surveys, the really tough part was drawing the conclusions. Most of the individual commanders fell in love with their own routes, underemphasizing the difficulties, while the commission as a whole decided for Nicaragua.[14]

The varied factors that dominated the actual course of canal effort from this point on — diplomatic, technological, and economic — are sufficient reason why this was the last large-scale exploring effort on the old pattern by the Navy. Gross geographic facts were no longer in serious doubt, nor were they sufficient to answer the questions a new age was asking.

Surveying and chart-making in the Navy changed from a general scientific activity aimed at an over-all view of little-known regions to a routine activity of a small and specialized office. The Hydrographic Office, cut off from the Naval Observatory in 1866, lost most of Maury's wide-ranging interest in oceanography and concentrated on making charts of specific areas outside the continental United States. In 1872 the office received an appropriation of $50,000 for "surveying the Pacific," but no more money was forthcoming.[15] The effects of the resulting abandonment of a comprehensive plan were still discernible during World War II.

Meteorology Under the Army

In the Army, the pattern of immediate postwar activity and gradual disenchantment was even more marked than in the Navy. The administration of meteorology is a clear example. This orphan had never found its proper niche. Neither the Army Medical Corps nor the Naval Observatory, the Smithsonian, nor the Patent Office had been able to develop a real bureau to study the weather. Yet the government was the obvious agent to undertake this work, the science had just entered a period of rapid advance, and the need for forecasts was pressing more economic groups all the time.

The Smithsonian network having succumbed to the war, Joseph Henry, whose interest in a volunteer system had never extended to spending much money, [16] was foremost among those urging a permanent service within the government. As usual, he was anxious to relieve the Smithsonian of anything anyone else was willing to do.[17] At the same time pressure for a comprehensive forecasting service became particularly strong. Increase A. Lapham, a local savant of the Milwaukee area who had long participated in meteorological observations, was especially desirous of providing storm warnings for the Great Lakes. He was fortunate in getting his petitions in the hands of a congressman, H. E. Paine, who happened to have studied with Elias Loomis, one of the great authorities on weather of an earlier generation. This coincidence opened up a legislative avenue. At about the same time, Cleveland Abbe of the Cincinnati Observatory had actually begun to collect data and issue forecasts in 1869, although he conceived his effort as one to popularize his observatory, where his main interest up to that time was astronomy.[18] These purely civilian efforts brought forward strong arguments for the value of meteorology to commerce and agriculture.

The question where a weather bureau should fit in the government had no obvious answer in 1870, before any new scientific bureau had clearly emerged, and when the very concept was still nebulous. But the military had a strong claim because of its services to science in the past, a stable personnel under discipline, and the obvious economy, so often appealed to before, that the soldiers were already being paid anyway. More specifically, the Army had an officer who was already trying desperately to set up a weather service. Colonel Albert J. Myer was a former Army surgeon who had done much to get a Signal Service established during the Civil War. At the end of hostilities, the cutback was so drastic that Myer found himself the chief signal officer with two enlisted servants, two clerks, and a budget of $5536. Actual duties had to be performed by a few officers and men borrowed from the Corps of Engineers.[19] The Army clearly had no desire to stay in the telegraph business, which was the main concern of signaling at the time. Myer, badly needing a function, put together the interest in meteorology he had developed while in the Medical Corps with the appropriateness of using a telegraph network for weather data in an unheeded suggestion to the secretary of war early in 1869.[20] When Lapham's memorials seemed to be striking sympathetic ears, Myer went to Congressman Paine "greatly excited and expressed a most intense desire that the execution of the law might be intrusted to him." [21] The result was a joint resolution, approved in February 1870, "for taking meteorological observations at the military stations in the interior of the continent and at other points in the States and Territories of the United States, and for giving notice on the lakes and at the seacoast, by magnetic telegraph and marine signals, of the approach and force of storms." [22]

Although not mentioned in the resolution, the Signal Service had clearly won the prize. Two basic considerations dominated the form of the new agency. First, the Army was not interested in the results of meteorology for its own use but merely to provide occupation for one of its branches. Second, a strong humanitarian and economic appeal stressed the need for immediate and practical results. The strength of this last urge is indicated by appropriations, which went up from $15,000 to $250,000 between 1870 and 1873, a period in which War Department expenditures as a whole went down sharply. By 1885 the appropriation broke a million dollars.[23]

The year 1870 was still too early for Myer to have the pattern of a new scientific bureau before him. In terms of the ideal agency such as later developed in the Department of Agriculture, the Signal Service as a weather bureau had an organic act and theoretically the only stable personnel system then known in the government — the military. Yet the need for immediate results so dominated it that from the beginning its routine operations of gathering data and forecasting swallowed up both funds and energy. With the theoretical working concepts of the science just beginning to emerge,[24] either the basic discoveries in meteorology so necessary for improved forecasting had to come from outside the service or a research program had to be started.[25] Myer proved himself an able organizer and hired Cleveland Abbe, who had a Coast Survey–Naval Observatory background, as a civilian meteorologist as early as 1871.[26] Yet it remained for his successor, General William B. Hazen, to face squarely the need for basic research.

In 1881 the new chief signal officer set up a division in Washington called the "study room." In addition to Abbe, William Ferrel came over from the Coast Survey and several other civilians with the title of junior professor began to undertake research in meteorology independently of the daily grind of recording observations.[27] Although some of the work, such as the study of standard instruments and preparation of tables for reducing observations, was background for operations, problems of atmospheric electricity and solar radiation and the preparation of general works such as Ferrel's *Recent Advances in Meteorology* were more basic.[28]

Meanwhile at Fort Myer near Washington, Hazen tried to transform his military post into a meteorological school both for his commissioned officers and for enlisted observers. Abbe, Ferrel, and others came over to lecture.[29] Some of the enlisted men were college graduates who had responded to special inducements to join the Army to learn to be weather observers.[30] Pointing out the utility of weather data in agriculture, Hazen also encouraged states to set up weather services. By 1884 thirteen had responded.[31]

Hazen's program, intelligent and forward looking, actually increased the difficulty of the Signal Service's position by creating tension between civilian and military personnel. At the same time it aggravated misunderstandings by stressing basic research in a supposedly practical operation. Even worse, the oft-cited advantages of military discipline and a career service for officers were nullified by absurdities of Signal Service organization. No intermediate rank existed between the chief signal officer, a brigadier general, and second lieutenant, making it impossible for an ambitious young man to get promotion without transferring to some other branch of the Army.[32] The college-graduate privates formally complained that an officer swore at them during drill, which led to their being court-martialed and the officer reprimanded. The magazine *Science*, caustically suggesting that the punishments should have been reversed, used the episode to damn military control of the Weather Service.[33]

Every president from Grant to Harrison recommended a reorganization. As early as 1881 the secretary of war ordered an investigation which led an assistant inspector-general to conclude that the weather observations had "no natural connection with the military service." [34] General John A. Logan introduced in the Senate a bill for transfer to the Interior Department which never emerged from committee. Nevertheless, surface incidents accumulated, contributing to a feeling of uneasiness that set off one of the fullest congressional investigations of the nineteenth century. The so-called Allison Commission of 1884 to 1886, although ranging to far broader topics than weather,[35] listed the Signal Service first among its objects of interest and tackled head-on the question of civilian versus military control.[36]

Hazen found himself defending the military and the study room at the same time. Concerning the role of basic science he testified that the "Navy have a corps of professors distinct from the working part of the Navy. So in the Coast Survey and in other bureaus they have a corps of theoretical workers who work to find out facts, and it seems to have been found necessary all over the world." [37] Under the sympathetic questioning of Congressman Theodore Lyman of Massachusetts, Ferrel was able to point out that some thought had to be given to keeping up with other governments and that it was doubtful whether private universities could do work "in such a manner as to suit the service precisely." Lyman, thinking of the Coast Survey, voluntarily concluded that "this is not a new thing in the scientific administration of the Government, that there should be at the Signal

Service certain professors, part of whose time is occupied on some of the more abstruse questions." [38] But disbelief also showed itself. When Abbe tried to show the connection between "abstract science" and the daily work of the Service, Hilary Herbert of Alabama satirically remarked, "The expression 'abstract science' is not used now as it was a few years ago. The scientists claim that it is all practical, do they not?" [39]

However loyally Hazen stood by his civilian scientists, he adamantly insisted on military control of the service. Citing fourteen years of experience and the expenditure of millions, he claimed that the service would be "wholly unsuited to any plan but a military one, and to attempt to transfer or change it would be the loss of system and plan and a long step backward." He even doubted whether a civilian agency could succeed "without the obligations of regular enlistment and military discipline." [40] He felt that those who wanted a transfer were "some members of the National Academy," his own "personal enemies," and those who "have a grudge against the service." [41] Certainly many scientists, whose views found voice not only in the National Academy's report to the commission but also in *Science*, were strongly opposed to the military. They were the same who urged civilian control of the Naval Observatory.

Abbe supported Hazen, even though he felt that civilians had "executed nine-tenths of the advances and improvements in the work of the service." [42] Most of the Army personnel under the chief signal officer (except of course the college-bred privates) naturally followed their chief, but his superiors did not. Secretary of War Robert Todd Lincoln considered the service too large an expense to his otherwise impoverished establishment. Nearly half its employees were civilian, and their work had "no relation to the Army whatever." Indirectly he stressed the decline of the unique role of the military in frontier regions when he pointed out that the Army now owned only 2800 miles of telegraph lines.[43] Philip Sheridan, the commanding general, went further, attacking even the signaling functions of the service and recommending that Fort Myer be abolished. He thought everybody "who is intelligent enough" was taught signaling "at the different military posts throughout the whole country," and that reports of this activity to the chief signal officer were "more a matter of courtesy than anything else." [44] The Army wanted science no more than science wanted the Army. Whatever internal dissensions brought it about, Hazen, the enlightened reformer, stood almost alone for a lost cause.

Three members of the Allison Commission wished to abolish the Signal Service and set up a civilian weather bureau under the secretary of war. But since three others opposed any change, no legislation grew directly out of the investigation. But by the end of 1886 Hazen became ill and the secretary of war abolished the study room.[45] A. W. Greely, the new chief signal officer, seemed to favor military discipline but took a compromise position. His increased emphasis on agricultural meteorology gave an indication where the final pressure was coming from.[46]

Meteorologists, led by H. Helm Clayton of the Blue Hills Observatory, joined with agricultural interests to develop a campaign among the land-grant colleges, experiment stations, state boards of agriculture, and the Grange to capture the weather bureau. With no opposition except a few grumbles from Greely, a law of 1890 set up both the Weather Bureau in the Department of Agriculture and a permanent Signal Corps in the Army.[47] As Greely's appropriation dropped to a mere $31,000 [48] in 1892, he started up the long road of developing a true Signal Corps for the Army almost from the bottom. Meanwhile in the same year the Weather Bureau received an appropriation of $889,753.50,[49] nearly 40 per cent of the Department of Agriculture's total budget.

Although the head of the Weather Bureau alone of the chiefs in the Department of Agriculture was appointed directly by the President, the first appointee, Mark W. Harrington, was a civilian scientist of standing. President Harrison thus set a precedent that resulted in long tenure and selection with the advice of scientists.[50] Under the Department of Agriculture the Weather Bureau achieved stability disturbed only when the rise of aviation dictated a shift to the Department of Commerce in 1940.[51] Cleveland Abbe stayed on in various capacities until his death in 1916 — a service of 45 years. As late as 1895 he still pined for Hazen's study room and its atmosphere of basic research.[52]

Polar Explorations

One offshoot of the Signal Service period of meteorology was a change in the nature of the government's participation in polar exploration. In direct line from Elisha Kent Kane, the expeditions of

I. I. Hayes, Charles F. Hall, and G. W. De Long continued the tradition of combined government-private support and the dominance of the geographic problems dramatized by the loss of Sir John Franklin. But in 1879 an International Polar Conference at Hamburg chose 1882–1883 as a year for a concerted attack on Arctic problems by means of simultaneous observations.

Since meteorology moved to the forefront in place of gross geography, the Army Signal Service manned the two stations for which the United States was responsible. The station at Point Barrow, Alaska, performed its duties without incident, but the expedition to Ellesmere Island stumbled into adventure and dramatic tragedy.[53] Led by A. W. Greely, then a lieutenant, the party combined its meteorological, magnetic, tidal, and pendulum observations with dashes northward toward the pole. Although dissension appeared early, the enterprise was turned into a desperate scrape by the failure of the supporting expeditions of 1882 and 1883. When relief finally reached the party in 1884, only seven remained alive. Neither Secretary of War Robert Todd Lincoln nor the Navy relief commander showed to advantage in the tragedy, which was a sensation in its day and an important conditioning factor in the Signal Service reorganization discussions. Yet in a more profound way it emphasized the government's stake in polar exploration and laid the groundwork for a really scientific interest in Arctic and Antarctic problems — a highly specialized field in which the military continues to figure, and in which the "International Year" as a technique of coöperation is still important.

The Corps of Engineers

While the loss of the weather service neatly summarizes the declining role of the military as a reservoir of general scientific resource for the government, the really crucial battle of the postwar year was over whether the Corps of Engineers could hold their earlier place as the explorers of the West. In one sense the Engineers were in a strong position in 1866. With all doubts about the constitutionality of internal improvements removed, the appropriation for river and harbor works jumped several fold and steadily remained high.[54] But the very opulence tended to absorb the Corps in routine civil engineering. Much work was done under contract, and no general plan emerged because the Corps considered itself simply the executing agent of Congress, refusing even to suggest a project without a specific order.[55] Its position in scientific research depended entirely on whether it could maintain its preëminence in exploring and surveying the great West, especially without a distinct Corps of Topographical Engineers.

This story deserves special attention, because it is not simply the decline of science in a military service but rather the metamorphosis of the old general-purpose military survey into a civilian agency which became a new scientific bureau much like those then developing in the Department of Agriculture.

X

THE GEOLOGICAL SURVEY

1867–1894

AFTER the Civil War the American people, whose dominant passion became the exploitation of their national resources, began to look on the great trans-Mississippi West with different eyes. This subcontinent, almost wholly the property of the federal government, was rapidly changing from the preserve of soldiers and mountain men to a potential source of wealth for many groups of citizens. In the middle 1850's the transcontinental-railroad problem had been in the hands of military-scientific exploring expeditions. By the end of the war a railroad was actually under construction; daring entrepreneurs supported by land grants and federal money drove the golden spike in 1869. With the completion of the first line and the frenzied projection of others, the buffalo and the Indian had lost control of the region, and as they declined the Army of the frontier gradually found its reason for existence vanishing.

Three Predecessors

As everyone expected from its early preëminence, the postwar Army took up surveying the West with renewed vigor as soon as it had again become a regular outfit, despite the disappearance of the Corps of Topographical Engineers. Yet its first major effort — the Geological Survey of the Fortieth Parallel, begun in 1867 — profoundly reflected the new forces at work in the West. The area — a strip about one hundred miles wide running east to west from Colorado to California — was essentially the route of the Central Pacific. Since the road itself was already surveyed and nearing completion, the object of the expedition was to determine the natural resources along the way, with particular emphasis on mining.[1]

Instead of a West Point graduate with military rank, a geologist commanded the party. Clarence King, product of Sheffield Scientific School, had learned both the West and the technique of field geology with J. D. Whitney's California state survey during the war years. His extraordinary personal charm and his friendship with the proper literary people assured him a reputation as a kind of Greek god and Renaissance man rolled into one. The main author of the fortieth parallel idea, he was largely responsible for its presentation and passage through Congress. While the Corps of Engineers took over the administration, King had a free hand in planning his survey and hiring his all-civilian staff.

Although he preserved general natural history as his scope, publishing on the botany and ornithology of the region, King concentrated on geology and adopted unprecedentedly high standards. The topographic map, necessary to place accurately the geologic data, was on a scale of four miles to the inch with slopes indicated by contour lines rather than the customary and usually vague system of hachures, or shadings. Extensive and systematic use of the microscope in the study of rocks introduced a new technique into Western geology. From the work a comprehensive view of the geologic history of the mountain region of the West emerged. Although he paid attention to the mining industry, King's real aim and real triumph was in pure geology.

Starting from the California end, he worked his way eastward in 1868 and 1869 to the Wasatch Mountains in Utah and then set up headquarters in New Haven to work up the collections. In the summer of 1870 he suddenly received orders from the chief of engineers to take the field again as Congress had continued the appropriation without any prompting.[2] This led to a leisurely study of the volcanoes of the Pacific Coast until he could continue eastward progress in 1871. By the end of 1872 he had reached eastern Wyoming where the survey could tie into other studies, thus connecting the continent in a single geologic profile. With the job done, King left the field and spent most of the rest of the decade publishing the results. When the seven volumes stood complete in 1880, not a trace of his administrative organization remained. The work had ended, complete, at a total cost to the government of $386,711.[3]

More orthodox than its role in sponsoring the King survey indicated, the Corps of Engineers began its own program in 1869 when Lieutenant George M. Wheeler began a military reconnaissance in Nevada and Utah.[4] By 1872 he had congressional authorization to make "military and geographical surveys west of the one hundredth meridian." He was interested in quickly producing maps of the whole area that would be sufficiently accurate for military use. His scale, as much as eight miles to the inch, rendered his topography, without contour lines, often not detailed enough for the geologist or engineer.[5] In accordance with the traditional union between surveying and natural history, he recognized the secondary aim of collecting "all the information necessary before the settlement of the country, concerning the branches of mineralogy and mining, geology, paleontology, zoology, botany, archaeology, ethnology, philology, and ruins." [6] Wheeler used civilian scientists as well as officers drawn from several branches of the Army. For instance, in 1876 he had seven lieutenants, and two assistant surgeons who doubled as zoologist and botanist. An astronomer, three geologists, a mineralogist, and an ethnologist were civilians.[7]

As it gradually embraced the whole West, the Wheeler survey began to look permanent. In time, it yielded somewhat to the changing problems wrought by settlement by including in its aims the classification of the public domain into "arable and arid portions, the former divisible into those sections which are susceptible of cultivation, those in which irrigation can be had, and into mining, timber, and grazing sections." [8] The ten years after 1869 cost $499,316, with an appropriation for 1879 of $50,000.[9] Pay of the military and rations not being included, the actual outlay now was a good deal more.[10] The Wheeler survey had all the characteristics and traditions, the advantages and drawbacks of the pre-Civil War military exploring enterprise in the West.

At the same time the advance of settlement and land legislation such as the Homestead Act were carrying the Department of the Interior ever farther into the Army's Western preserve. Long immersed in the mechanics of parceling out the public domain, the General Land Office had only occasionally shown any tendencies toward science.[11] J. D. Whitney, who as director of the California state survey had shown what science could do with the management of Western mineral lands, claimed in 1865 that "I have accumulated many facts which demonstrate, in the most conclusive manner, that the most profound ignorance of everything connected with the subject exists in the General Land Office and the Census Bureau." He felt that "the Government is likely to be greatly misled and may do a great injury to the country, if it allows itself to be guided by recommendations emanating from the Department of the Interior." [12] But the General Land Office did spawn a scientific survey even if its own politics-ridden organization could do very little directly. Starting with an appropriation of $5000, Ferdinand V. Hayden began a geological survey of Nebraska under its direction. Adopting the old title, "U. S. Geologist," Hayden gradually reached out into other parts of the West. By 1873 his group was called the "Geological and Geographical Survey of the Territories." [13]

Hayden, a man of great energy, had had an education at Oberlin College and at medical school in Albany. But he was a self-made scientist, brought up in the hard graduate school of the Western surveys of the 1850's. A professor at the University of Pennsylvania when he began his work, he became a full-time administrator as his ability to obtain political support for the organization grew. From Nebraska he moved into Wyoming and then Colorado and New Mexico. In 1871 he was in Montana, his work laying the basis for the creation of the first National Park, Yellowstone.[14] By geologizing in the Great Salt Lake Valley in Utah he invaded the heart of Clarence King's territory. Indeed, the possibilities of duplicating the work of others, made so likely by his own and Wheeler's geographical spread, were further increased by his ranging widely not only through geology but through all of natural history.

One of the sources of Hayden's rise to eminence was his network of friendships with leading scientists. He could call on an imposing array of collaborators in special fields,[15] and his lavish publication policy encouraged them. A notable example of this method of creating good will was the Western tour of two intimate friends and collaborators of Darwin — Sir Joseph Hooker and Asa Gray. The first botanist of Great Britain accompanied by the first botanist of the United States toured the whole West in the comfort of the new railroads.[16] At their leisure they prepared a comprehensive discussion of the vegetation of the Rocky Mountains to appear as a publication of Hayden's.[17] Because of this journey both Gray and Hooker were always ready to lend Hayden support and write testimonials for him, while their names on an article in his *Bulletin* further enhanced his own reputation. But a long series of such publications did not add up to a planned survey.

Concentrating less on maps than on general geology and natural history, the Hayden surveys stacked up a great deal of data, some of it important, without getting very close either to the mining industry or to the more pertinent problem of classifying the public domain before it passed into private hands. The connection with the General Land Office and the Department of the Interior was thus nominal, which did not prevent Hayden from getting his appropriations. Soaring one year to $115,000,[18] they usually came to $75,000. By 1878 he had received a total of $615,000, nearly twice as much as King and more in actual cash than Wheeler.[19] The Hayden survey was thus on the surface the most imposing and respectable outfit in the field. It was the civilian side of the old military-civilian team expanded to new proportions.

The Powell Survey

Three surveys were, however, not enough to exhaust the scientific possibilities opening up in the changing West. A man of imagination and insight could make a name for himself and get money out of a Congress that seemed anxious to have still more surveys. John Wesley Powell brought few formal qualifications into the already crowded arena except a fertile imagination and an insight that grew with experience.[20]

A schoolteacher who chose an interest in science rather than a conventional college education (one term at Oberlin College, Hayden's alma mater, was plenty for him), Powell rose to the rank of major during the Civil War and served as an artillery officer for Grant even after losing an arm at Shiloh. Afterwards, as a professor of geology at the Illinois State Normal University, he secured state funds for the museum of the Illinois Natural History Society with which he took a group of students to the Colorado Rockies in 1867. Before starting West, however, "the Major" went to Washington to get some help from the Army. Through his old commander, Grant, now secretary of war, he got authority to draw rations for his party from Army posts. This minimal subsidization was important largely in introducing Powell to the frenzied postwar atmosphere of Washington.

In April of the next year, 1868, Powell was back at the capital. This time he wanted to explore the Grand River — a headwater of the Colorado — with a purpose more definite than an outing for students. In a letter to Grant he pointed out, first, that the Colorado's canyon would be the best geologic section of the continent and, second, that a survey of the Indians of the Colorado plateau would be of value to the War Department. Already Powell had discovered two problems in the West that others were doing little to solve. The Grand Canyon had seldom been seen by trained observers, much less explored.[21] It was the last completely blank area on the country's map, and hence the last of the type of problem that had challenged Lewis and Clark. But Powell saw beyond the vacant spot on the map to the great possibilities for geology in the unknown region. His reference to the Indians indicated that he saw in them, who had figured so largely in all previous explorations as a hindrance, a field for study, an opportunity soon to vanish.

Although Grant was again ready to allow the issuance of rations, the commissary general ruled it illegal because Powell was neither a civilian employee of the government nor in the Army. The only way around was to get a law passed by Congress. Illinois members such as Senator Lyman Trumbull could help, but most of Powell's assistance came from Joseph Henry. Although the Smithsonian, as usual, could put up no money, a letter from the secretary was a valuable asset. Endorsing the project and stating that Powell would study the hydrology of the river in relation to agriculture and irrigation, Henry incidentally emphasized the Indians, an old interest at the Smithsonian. Although Congress grumbled about setting a precedent, Powell's bill finally passed. He set off for the Grand River with some instruments from the Smithsonian, authority to draw rations from the Army, and the rest of his backing from educational institutions in Illinois.[22]

The next year, 1869, Powell got the same bit of support from the government, this time to tackle the Colorado River itself. His passage in boats through the Grand Canyon was one of the last epic American explorations of completely unknown geography, and the major emerged a famous man. Not the least remarkable aspect of his feat was the way in which he used an old-style daring adventure to establish himself as the head of a scientific survey of the region he had discovered. The Colorado basin more than fulfilled his expectations as a great outdoor museum of geology and anthropology.

In Washington in the spring of 1870 Powell was able to get Representative James A. Garfield, a member of the Smithsonian board of regents, to sponsor a bill which, when passed, gave him $10,000 for a "Geographical and Topographical Survey of the Colorado River of the West." It was placed at first in the Interior Department. But when Powell told a member of the appropriations committee that his collections went to the Smithsonian Institution, the congressman "accidentally sent my whole work there." [23] Still holding on to his professorship at Illinois Normal, he drew no government salary and continued to choose as his assistants members of his family and immediate circle of friends. Yet he projected a comprehensive geologic study of a most interesting region and made himself over into a scientist competent to deal with it. Besides the dazzling view of geologic history afforded by the Grand Canyon, Powell grasped the dynamics of the forces that produced such results, introducing, for instance, the concept of base leveling.[24]

In 1873 Powell moved a step further into the orbit of the government by cutting his Illinois connections and moving to Washington. Significantly, both he and Hayden found the opportunities of surveying the West so great that they shed their academic posts within a year of each other.[25] In a larger way the move is an ironical indication that, amid all the corruption of the Gilded Age, Washington was beginning to be in fact the cultural and scientific center that the old Columbian Institute had hoped to create.[26] Although Powell had a room at the Smithsonian, where he frequently consulted Henry and Baird, he usually worked at home.

A new action in 1874 extended the role of what was now known as the Powell survey to the "Geographical and Geological Survey of the Rocky Mountain Region" and returned it to the Department of the Interior from the Smithsonian.[27] But the major still had a free hand, and with only slightly increased appropriations began to push several lines of research. Since he could not follow them all himself he increasingly delegated not only the execution of the work but the planning and even the selection of problems. His brother-in-law, Almon Thompson, an able man in spite of his nepotistic appointment, took over topography. But in geology the staff gained two top-notch professionals. G. K. Gilbert, hired away from Wheeler, undertook what proved to be a classic study of the Henry Mountains in Utah. Clarence E. Dutton, borrowed from Army Ordnance, worked out the geology of the Grand Canyon itself and the surrounding plateaus.[28]

Meanwhile Powell became increasingly interested in the Indians, dispensing with a military escort and collecting information on their total culture. Through them he began to see the influence of the land on its occupants. By observing the people who already had extensive experience living in the arid West — both Indians and the Mormons — he formed a connected series of propositions which related the geologic and geographic base to the settlers who were already coming in. The dominant fact of the environment of the continent west of the one hundredth meridian was aridity. Much of the surface was unfit for cultivation and the remainder required irrigation. For agriculture to exist in this region, legal and social arrangements had to take account of these basic conditions. The Homestead Act, based on the experience of the humid East, was unrealistic, as were laws governing water rights. The aim of science in the West should be to prepare the way for settlements by classifying the land according to the environmental possibilities, and the land laws should make this possible. This concept not only gave science a definite use but made it the key to the settlement of the new territory.

In 1874 Powell struck this theme when he said that there "is now left within . . . the United States no great unexplored region, and exploring expeditions are no longer needed for general purposes." Instead "it is of the most immediate and pressing importance that a general survey should be made for the purpose of determining the several areas which can . . . be redeemed by irrigation." [29] Increasingly he turned his parties toward stream measurement and irrigation surveys. By 1878 he had enough data to publish his *Report on the Lands of the Arid Region of the United States, with a More Detailed Account of the Lands of Utah*.[30] Here he sketched in an incomplete but powerful way his entire concept of the needs of the West. The Powell survey thus harnessed topography, geology, and anthropology into a program of far-reaching and even revolutionary consequences to the method of distributing the public domain to settlers.

The Struggle for Supremacy

With all these surveys in the West, it is easy to overlook another outfit which was senior to them all. Upon the death of Bache in 1867 Benjamin Peirce himself took over the directorship of the Coast Survey. Although without administrative experience, this famous mathematician did give the activities of his highly competent agency a new turn. He began the triangulation of a transcontinental arc along the thirty-ninth parallel to connect the Atlantic and Pacific systems, now nearing completion.[31] Besides staving off the old threat that the Coast Survey should fold up when it had finished mapping the coast, this extension carried into the West all the traditions of accuracy and the aim of studying the physics of the earth that had developed under Hassler and Bache. If maps had to be made there, why not let this demonstrably efficient and scientific agency do it? In 1878 the name became the Coast and Geodetic Survey.[32] Expenditures had crept up gradually to $857,100 in 1876.[33]

This varied surveying activity in the West, marking on its positive side the desire of the government to turn science to account in the rapid development of an empire, soon degenerated into an interlocking series of quarrels. When in the summer of 1874 Hayden fell in with one of Wheeler's parties in Colorado, an argument over jurisdiction brought the scandal out in the open.[34] The American people became dimly conscious that science in the government did not organize itself automatically, that some kind of over-all policy was urgently needed. Scandal, corruption, waste, and duplication that were making headlines by 1874 in the wake of the excesses of the Grant administration seemed a priori as likely in the surveys as elsewhere.

Simple duplication, doing the same thing in the same way in two parts of the government, is usually easy to spot and to correct by administrative action. When the study of science is a prominent activity, an argument can even be made that the normal process of seeking to learn something new renders duplication unimportant if not impossible.[35] As President Grant blandly remarked in his answer to the congressional investigation of the surveys in 1874, "where the object is to complete the map of the country, to determine the geographical, astronomical, geodetic, topographic, hydrographic, meteorological features . . . it seems to me a matter of no importance as to which Department of the Government should have control of the work." [36] Whatever agency could perform most economically was all right, but in his eyes the Army and the "scientific gentlemen" of the Corps of Engineers naturally fulfilled these requirements. Yet the conflict waxed bitter precisely because the surveys were all different — following their own systems, pursuing their own aims, and even emphasizing entirely different branches of science. To divide the trans-Mississippi region into five sectors would not produce a single uniform map. To consolidate the surveys meant a choice of one among

several approaches. Thus the government faced in the surveys a problem far more serious than duplication and had to make decisions that
affected the very nature of its relations with science.

The 1874 investigation revolved largely around Wheeler and
Hayden, who personified the issue of military versus civilian control.
Both men charged bad faith and made statements weighted with personal rancor. Wheeler, while insisting that he could not speak for the
War Department, made it clear that he felt his survey should remain
with the Corps of Engineers. Hayden called for science not only by
qualified specialists but under the control of scientists acting on civilian
authority. Admitting that he used the Army for escorts in the field,
he claimed that civilians could do the scientific work more economically and resisted consolidation under military control. Powell entered
the fray to demonstrate that Wheeler's maps were "so inaccurate as
not to be available for geological purposes," and to emphasize that
military escorts only provoked fights with the Indians. He strongly
favored unification under one department, and expressed his willingness to enter the Interior Department in a position subordinate to
Hayden.[37]

The investigation, producing no change, only led to a cutthroat
competition for money. The Powell survey, the newest and smallest,
had to fight for its life, especially since few publications had appeared.
Although the major had sided with Hayden in 1874, the two increasingly clashed. After an argument about a mining report, Powell suggested to Secretary of the Interior Carl Schurz a division of authority
giving Hayden natural history, leaving only ethnology for himself.[38]
This willingness to subordinate his personal ambitions and the proposal of a side field into which to retreat strengthened Powell's position in the coming crisis. By this time he was ready to propose a
broader program than his own advancement, a noteworthy distinction in Washington in that era.

In March 1878, a resolution calling for a report on the possible
consolidation of the surveys touched off the debate anew. Within a
month Powell had published his *Report on the Lands of the Arid Region* as a theoretical base for reorganization. But in May another event
of great importance to the history of American science put the affair
in a new light. Joseph Henry died, full of years and honors, leaving
the presidency of the National Academy vacant. His cautious policies
had saved that organization and allowed new blood to enter it at

the cost of its position as active adviser to the government. Among the younger members were many geologists and paleontologists, including Othniel C. Marsh, the vice-president who temporarily fell heir to Henry's place.[39] This Yale professor, who was making epochal discoveries among the fossils of the West at his own expense and who had unseated one of Grant's secretaries of the interior by uncovering graft on the Indian reservations, suggested by his presence that the Academy might take a more active role in the survey controversy than his patriarchal predecessor would have allowed.

The possibility came to fruition in June 1878, when Representative Abram S. Hewitt of New York called for a review by the National Academy of the whole subject of the surveys. Hewitt, as a member of the appropriation committee, had to incorporate the request in the sundry civil bill. He adopted this clumsy procedure because his interest came not from a pertinent committee assignment but from his outstanding qualifications to grapple with the problem. As a manufacturer of iron who had aided in the early fumblings of research technology during the Civil War and who had some appreciation of the relation of science to both industry and mining, he was one of the few congressmen genuinely alive to the dimensions of the problem.[40] Moreover he was a friend not only of Marsh but also of Clarence King, who possibly first hatched the idea of resorting to the National Academy.[41] This stroke was clearly inspired by scientists and was a disappointment to the Army. It not only moved the investigation from a public committee room to the eternally closed doors of an exclusive society but it gave the community of professionals a definite advisory voice in shaping legislation. The Army Engineers were shut out, and, because Marsh and Hayden had had trouble in the past, only Powell could consider the procedure enthusiastically.

Marsh attempted to appoint a committee both impartial and familiar with the science involved. Chairman James D. Dana's experience went back to the Wilkes expedition. J. S. Newberry, who had been on several prewar surveys, was a onetime student of Hayden's who now spoke most harshly of him. W. P. Trowbridge of Columbia was a West Point graduate who had served on the Coast Survey. Simon Newcomb's credentials were self-evident. William B. Rogers, the old anti-Lazzaroni who had now rejoined the Academy, and Alexander Agassiz, son of the great Louis, completed the committee.[42]

The letters submitted to the Academy by the heads of the interested surveys are significant largely in revealing the divergent points of view. The acting chief of engineers reaffirmed the Corps's traditional role as mappers of the West and appealed to the British ordnance survey as a precedent. The land commissioner, describing his work purely as running lines, largely by contract, was nevertheless willing to take over geologic and geographic surveying — a reference to the existing loose connection of the Hayden survey to the Land Office. Hayden himself rather sourly defended the *status quo*, asking, "Is the country ready for a department or bureau of science?" [43]

Powell gave the committee a full statement of his ideas on the nature of government science. Branding the "prosecution of the work by a number of autonomous organizations . . . illogical, unscientific, and in violation of the fundamental law of political economy," he demanded, first, one general management, and, second, "that the division of labor should have a scientific base." Geology should be one department and mensuration another. "If ethnology, botany, and zoology are to be embraced in a general scientific survey, each subject should have but a single organization, with a single head subordinated to the general plan." [44] After a slashing attack on the General Land Office, "a gigantic illustration of the evils of badly directed scientific work," [45] he stressed land classification as the true aim of government science in the West.

He was aware, as some have since forgotten, that even before the Civil War "the statesmen of America who compose and have composed our National Legislature have been not averse to the endowment of scientific research when such research is properly related to the industries of the people." [46] To bolster this contention he analyzed the scientific budget of the government for several years past. The Engineers and the Coast Survey of course dominated. His curt dismissal of the Department of Agriculture, "where scientific investigations are pursued to some extent," indicates not only its lack of prestige in 1878 but the relative independence of two great evolutions within the government. Yet Powell's proposal was essentially the same as that which soon changed the scientific bureaus in agriculture — the problem approach. *"A geographical and geological survey, to be permanent, vigorous, and efficient, should include the survey of the public lands and be subsidiary thereto."* [47]

Should, then, government science support zoology and botany? Using the rule that "the endowment of science by governments should be very limited and scrupulously confined to those objects of research which . . . could not be undertaken by individuals," Powell said no. A survey should stay close to "utilitarian demands" and occupy a position "analogous to . . . the Coast Survey, Signal Service, Naval Observatory, Agricultural Department, and Light-House establishment." Special inquiries, such as that into "the ravages of locusts," were all right, but "no one will urge that it is the duty of the government to add half a dozen workers to the great army of independent investigators" already working at botany and zoology. Nor should the government assist in adding to the clutter of publications on natural history.[48] These theories, of course, had barbs which pricked the salient features of the Hayden survey.

Ethnology, however, was different. Not only was the opportunity to study the Indians "in their primitive condition" fast disappearing, but the Indian problem occasioned by the coming of the white man to every important valley in the United States "*must* be solved, wisely or unwisely." The government needed knowledge of Indian laws and customs in order to shape a policy that might avoid "the blunders we have made and the wrongs we have inflicted." [49] Since Powell's interests were running strongly toward ethnology, it was his own personal escape hatch from the controversies over mapping and geology.

The report of the National Academy's committee followed Powell's system of ideas. His private secretary remarked that "it sounds wonderfully like something I have read — and perhaps written — before." [50] The work was divided into surveys of mensuration and those of "geology and economic resources of the soil." For the first purpose, mapping and topography, the Coast Survey should be transferred to the Interior Department from the Treasury and become the Coast and Interior Survey. Also under the same department a United States Geological Survey would study "the geological structure and economical resources of the public domain." The two new organizations would replace not only the old Wheeler, Hayden, and Powell surveys, but also land parceling by the Land Office. A commission should then be set up to reconsider and codify the "present laws relating to the survey and disposition of the public domain," where "the system of homestead preëmption and sale in accordance with existing laws is both impracticable and undesirable." [51] The Academy was thus attacking a cherished American legal and social institution head-on, suggesting that science could produce something new and better.

The Creation of the Geological Survey

In November 1878, the Academy, after discussing its committee's report for three hours, adopted it by a vote of 31 to 1, the lone dissenter being Edward Drinker Cope, Marsh's paleontological rival who worked for the Hayden Survey. Yet the Engineers took it hardest. General A. A. Humphreys, who had already protested the absence of a government official on the deliberations, committed what Simon Newcomb called "a sort of hari-kari" by resigning from the Academy.[52] Undismayed, Marsh lined up imposing support from the President, the secretary of the interior, the secretary of the treasury, General W. T. Sherman, and the superintendent of the Coast Survey. Confident that the plan would go through Congress, he saw in it a victory that "will help the Academy very much."

In the House of Representatives James A. Garfield and Abram Hewitt emerged as the champions. The Republican from Ohio prefaced his oration by defending the theory of private enterprise in science and admitting the government's right to enter only when research was necessary for official functions, and impossible on private resources alone. Hence he recommended that to keep from competing with citizens the surveying work should be restricted "plainly and narrowly" and consolidated "under one responsible head." [53]

The Democrat Hewitt, speaking from the "peculiar experience which I happen to have had with reference to the growth of industry in this country," took a broader ground. Admitting that he had favored military control until Powell's arguments had converted him, he defended the Academy as "the only body who could form a proper judgment and render a wise decision." Brushing aside "the petty dispute" among the surveys as "belittling the real question involved in this magnificent conception," he urged the House to "place this work of national development and the elements of future prosperity on a firm and enduring basis of truth and knowledge." Hewitt claimed with some authority that the measure "commends itself to the judgment of the men who have been most energetic and successful in the development of our resources, the 'captains of industry' of our time." [54] The consolidation of the surveys thus appeared in Congress justified in terms of private enterprise and blessed by those who were reorganizing American economic life on an unprecedented scale.

But any measure that touched the hearts and pocketbooks of actual and potential Westerners as intimately as one recommending fundamental changes in land laws was bound to meet an opposition indifferent to the merits of scientific reorganization. In addition, the government was split between the parties with the House controlled by the Democrats, and most legislation originating there in this session failed.[55] As the debate developed, the Westerners extended their attack to the Academy and its "visionary scientists." A representative from Colorado appears in the *Congressional Record* as saying: "This Academy has never published but one work, and that was a very thin volume of memoirs of its departed members. [Laughter.] And if they are to continue to engage in practical legislation, it would have been well for the country if that volume had been much thicker. [Laughter.]" [56]

Under the onslaught the House passed a weakened version by 98 to 79, and the Senate dropped the whole idea by voting to abolish all the surveys but Hayden's.[57] Hewitt then resorted to his strategem of authorizing money in the appropriation bill for objects not mentioned in other legislation. By providing a salary for a director of a geological survey, Hewitt stuck a whole organic act into a subordinate clause. The new officer was to direct "the classification of the public lands and examination of the geological structure, mineral resources and products of the national domain." Although the old Wheeler, Powell, and Hayden surveys were specifically discontinued, and the Geological Survey placed in the Department of the Interior, the Coast and Geodetic Survey remained in the Treasury Department and any real revision of the land laws had disappeared except for a study commission.[58] A large appropriation bill had a good chance in the closing hours of a session, and this one became law March 3, 1879, the last day of the Forty-fifth Congress.

Immediately the struggle between the former rival surveys shifted to the executive, revolving about the appointment of a director. The victory of the principle of civilian control eliminated Wheeler, leaving as the most prominent candidate Hayden, who actively sought the post and busily spied on Powell.[59] But the major, who by this time had amply demonstrated his abilities for political strategy, was too clever to be trapped into an exhausting struggle. In the same appropriations bill with the surreptitious organic act of the Geological Survey appeared an innocuous item providing for "completing and preparing for publication the contributions to North American ethnology, under the Smithsonian Institution, twenty thousand dollars." [60] Serious about retiring from geology and genuinely fascinated by the Indians, Powell retreated to his old haunt, the Smithsonian. It was good strategy, but more than that, for Powell had without fanfare established a small but significant scientific bureau which he would never desert. Meanwhile Clarence King belatedly emerged as a candidate for the Geological Survey and with Powell's strong backing managed to get to President Hayes through Secretary of the Interior Carl Schurz. Since King's own survey was almost complete, he had escaped without scars from the great struggles of 1878 and 1879.

King, who appeared to his friends as the "best and brightest man of his generation," [61] began the new bureau by dividing his organization into four sections to correspond with geographical divisions of the West and by laying special stress on the mining industry, about which almost nothing was known, "either technically, as regards the progress and development . . . in methods, or statistically, as regards the sources, amounts, and valuations of the various productions." [62] For instance, he undertook a study of the Comstock Lode.

King found himself distracted almost from the beginning. Ill for a time, he increasingly allowed his attention to wander to gold mines in Mexico. Even more troublesome was the ambiguity of the organic act. Did the "national domain" mean the whole country or just the public lands? King adopted the principle that he could operate only on public lands, at the same time vainly urging Congress to change the law, making clear the right of the Survey to operate all over the nation.[63] Without entry into state and private lands, comprehensive mapping, the part of the National Academy program that Congress had not specifically enacted, was impossible. Nor could the Survey follow geological formations in their entirety, an important part of King's mining studies. Congress did nothing; King went to Mexico. When Garfield became President in March 1881, King submitted his resignation with relief.[64]

Powell's Formative Policies

Before any wind rose from Hayden's quarter, Garfield appointed Powell as director of the survey which his ideas had already shaped. Without a thought of dropping his new Bureau of Ethnology, Powell moved it in with the Geological Survey and began an uninterrupted reign of more than a decade as the government's leading scientific administrator.

The major dealt with all his problems creatively. By the time he had been in office three months he had cut through a complicated and delicate problem to formulate and publish an American system of geologic nomenclature, anticipating an international conference on the subject.[65] Pulling in King's regional offices to Washington, Powell reaffirmed his belief in the maturity of the capital as a scientific center which had figured in the founding of the Cosmos Club in his parlor.[66] Estimating that he needed half a million dollars a year instead of King's $156,000, he essentially reached the goal in three years.[67]

As Powell gained momentum his path increasingly conformed to that of the head of the ideal new scientific bureau. Having selected his objectives in anticipation of needs rather than in response to pressure, he actively sought authorization for what he intended to do. For instance, wanting to unscramble the confusion in the classification of Indian tribes, he secured a request from the director of the census to prepare "a classification of Indians by their linguistic affinities." This gave him an official cover under which to launch a large and arduous bibliographic study.[68]

In a day when patronage was important and often used for corrupt purposes, he played the game as skillfully as any spoilsman, doling out minor jobs to the friends and relatives of congressmen. At the same time he kept meticulous books and allowed no hint of personal interest or laxness to compromise him. The results were even more spectacular than the rise in appropriations indicates. He was able to get his annual portion as a lump sum, allowing complete discretion in selecting his problems and in shifting from one to another.

Perhaps his greatest stroke was in securing authority to nationalize the Survey. After a bill to extend it into the states had met the fate of King's effort, Powell asked his friends on the appropriations committee of the House, who had done all the legislating for the Survey anyway, to allow a minor change in the wording of the section of the sundry civil bill that served as an organic act. The phrase added was "and to continue the preparation of a geological map of the United States." [69] To make a geologic map, he had to make a topographic map first. The little phrase gave him all he needed to embark on the program the National Academy had envisioned for the Coast and Interior Survey which Congress had killed in 1879. The advocates of states' rights were caught off guard while *Science* magazine applauded a truly national geological agency, pointing out that T. C. Chamberlin had traced a terminal moraine through thirteen states and one territory for 3000 miles.[70]

The same personnel system with which Powell connived at congressional patronage served him well to build up a scientific staff not only competent but high in morale and initiative. Besides Gilbert and Dutton, carried over from the old Powell survey, he retained some of the best of Hayden's men, who remained loyal to the major even when under pressure from their old faction. For instance, Henry Gannett directed all the mapping. Whenever Powell could, he also picked up young college men like Bailey Willis.[71] Even a name that might seem to reek of nepotism — Arthur Powell Davis — in reality represented a thoroughly competent scientist.

Most revealing of the way in which Powell turned the lax rules of the day to the account of freedom for research was Lester Ward's presence on the Survey staff. Ostensibly a paleobotanist, he spent much of his time — during office hours as well as after — writing his *Dynamic Sociology*. Powell respected Ward and thought his work worth while, so he let him go his own way. That a social philosophy which stressed planning for human progress grew up in the Geological Survey has at least a symbolic significance.[72]

To expand his field operations even further than his permanent staff would allow, Powell hired college professors as part-time workers, to serve during summer vacations. N. S. Shaler and W. M. Davis of Harvard, and J. S. Newberry of Columbia, among the most prominent academic geologists in the country, were on the list, immeasurably strengthening the position of the Survey in the universities.[73]

Powell capped his system of alliances with the scientific community of the country by a characteristic arrangement. He hired O. C. Marsh as division chief of paleontology, gave him a budget of $15,000 a year, including a salary of $4000, and let him continue to live and keep his collections in New Haven, Connecticut.[74] Wealthy, and politically influential, the president of the National Academy thus worked for the Survey. Since Marsh spent much of his own money on fossils, the main difficulty was the disposition of the type specimens on which Marsh based new species. Would they go to Yale or to the National Museum, where Survey material was assigned by law? Powell and Spencer Baird of the Smithsonian, working together, evidently made a gentlemen's agreement with Marsh, allowing him to retain government property, labeled as such, in New Haven.[75] These princely terms gave Marsh freedom to project a monumental series of monographs on vertebrate paleontology, while Powell had a rock on which to found his political position. Incidentally, Marsh's employment meant that his bitter rival Cope, Hayden's paleontologist, could not work into the Survey.

The emphasis on paleontology is evidence that Powell, while insisting on practical ends for government science, did not take a short view of the usefulness of basic research. Indeed, his ideas strung together in such great skeins that he naturally included basic and background problems, seeking them out as the ones most necessary to solve. In the early years, the Survey concentrated its geologic work on areas of theoretical as well as economic interest and devoted both time and energy to basic research. In addition to Marsh's establishment in New Haven, the director set up a chemical laboratory in Washington and repeatedly stressed its potential importance.[76]

But before much could be accomplished Powell needed his detailed map of the country to show geologic structure, and hence also a topographic survey. Using his smuggled authority of 1882, he not only put many of his parties on map-making but sought to extend his range by coöperative arrangements with the states. Three years before the Hatch Act for agricultural experiment stations and without specific authority of his own, Powell made an agreement with Massachusetts to share the cost of a map to follow the style of the national Survey.[77]

He hoped to make his series so complete that the maps would serve not only for geology but for land classification and land use, and indeed for the whole government and the people for every purpose. Using up to a third of his appropriation for maps, Powell thought he could finish the job in a quarter of a century for about $18,000,000 — enough to jolt even those congressmen who thought he couldn't do it in a century for a hundred million.[78] The major's very success made him a conspicuous figure in the government. Both his methods of administration and his active attitude toward authority raised eyebrows. While in agriculture science crept up to a larger role in affairs almost unnoticed, Powell saw the new relation of knowledge to the affairs of the nation and announced it articulately. As the Geological Survey developed into a new scientific bureau, Powell became the personification of the new bureau chief.

With the firm establishment of the Survey, a scientific function of the government which had begun with Lewis and Clark completed its metamorphosis from a series of *ad hoc* military expeditions to a permanent civilian agency. Just as the West Pointers and the frontier collectors were well adapted to the wilderness era, the new Survey was in tune with the rapid settlement of the public domain.

In one respect the Geological Survey was incomplete in 1885. It was purely an information agency, with no legal tie to its proper problem. However clearly Powell saw land classification in the arid West as the true end of the scientific work of his bureau, he had no power to enforce decisions. But his awareness of its potential role as a land-classifying agency meant that he would move in this direction at the first opportunity. Although the Survey was not a conservation agency that applied science to the use of natural resources, Powell's desire to do so made it the fountainhead of a new approach to the public domain.

XI

THE ALLISON COMMISSION AND THE

DEPARTMENT OF SCIENCE

1884–1886

BY 1884, the strain of the forces released by the new scientific bureaus was so great that Congress set up a Joint Commission "to consider the present organization of the Signal Service, Geological Survey, Coast and Geodetic Survey, and the Hydrographic Office of the Navy Department, with a view to secure greater efficiency and economy of administration of the public service." [1] The place of meteorology in the Signal Service was, of course, an issue. [2] Duplication in mapping seemed to dictate the choice of the other agencies. Powell's personnel and fiscal management were marked for inquiry, especially by his and O. C. Marsh's scientific enemies. But beneath all these issues lay the more fundamental one of where and how science should fit into the government's structure. [3]

Proposals for a Department of Science

Made up of three members each from the House and Senate, the Joint Commission popularly took the name of its chairman, Senator W. B. Allison, but the best informed member was Representative Theodore Lyman of Massachusetts. One of the great rarities of American history, a trained scientist as a congressman, Lyman had studied with Louis Agassiz, who was also his father-in-law, and had participated with him in the administration of the Museum of Comparative Zoology at Harvard. [4] Elected as a civil-service-reform independent, he had little chance for reëlection, but he used the closing months of his single term to launch the commission for which he had such preëminent qualifications.

Lyman's first step was to write to Marsh as president of the

National Academy, asking him to appoint a committee to study the organization of the surveys of the chief countries of Europe, and to recommend methods of coördinating the scientific branches of the government.[5] Marsh appointed a broadly representative committee. Although it suffered an annoying loss when neither Simon Newcomb nor Colonel Cyrus Comstock was allowed by the Navy and War Departments to serve because of possible conflict with their official duties,[6] the remainder brought forward a program.

Their main practical recommendation affected the Coast Survey. Admitting that it would soon be "confined principally to the interior, and then the policy of consolidating its hydrography with the work of the navy hydrographic office will be open for consideration," [7] the National Academy committee still stood up for the civilian agency against military control and urged a reorganization along the lines of the unrealized proposal in 1878 for a Coast and Interior Survey.

But their real interest lay in establishing a general plan. The government should not undertake what "can be equally well done by the enterprise of individual investigators." It should coöperate with universities but not compete with them, and should also confine itself "to the increase and systematization of knowledge tending 'to promote the general welfare' of the country." Management of a scientific bureau they considered more difficult than "that of a purely business department," for it requires "a combination of scientific knowledge with administrative ability, which is more difficult to command than either of these qualities separately." Under the existing system they saw each bureau head "absolutely independent . . . controlled only by Congress itself, acting only through its annual appropriation bills. We conceive that this state of things calls for measures of reform." [8]

The scheme they proposed was worthy of its predecessors — the national university, the National Institute, the Smithsonian Institution, or Bache's National Academy. The committee claimed to be "stating only the general sentiment and wish of men of science, when it says that its members believe the time is near when the country will demand the institution of a branch of the executive . . . devoted especially to the direction and control of all the purely scientific work of the government." A Department of Science! Had not photography and telegraphy, "the electric light, the electric railway" created capital of hundreds of millions? "None who have ever lived with open eyes during the development of these results of purely scientific investigations doubt that the cultivation of science 'promotes the general welfare.' " [9]

Even a committee of nonpolitical scientists could see that their proposal stood no chance in Congress in a period when legislation normally got through only by stealthily clinging to appropriation bills. Hence, without even waiting for a rebuff they suggested a substitute. "Should such a Department be now impracticable, should public opinion not now be ready for it," they proposed the reorganization of scientific work into four bureaus: (1) Coast and Interior Survey, (2) Geological Survey, (3) meteorological bureau, and (4) a physical observatory to study "the laws of solar and terrestrial radiation . . . and . . . other investigations in exact science." The last organization was to take over weights and measures from the Coast Survey and to set up electrical standards. All four bureaus would be placed in one already existing department. Because the secretary "would probably find it impracticable to enter into . . . all details," proper coördination would come from a permanent commission "to prescribe a general policy for each of these bureaus." On this body would be the president of the National Academy, the secretary of the Smithsonian, two civilians of high scientific reputation, an officer of the Corps of Engineers, a professor of mathematics of the Navy, the superintendent of the Coast and Geodetic Survey, the director of the Geological Survey, and the officer in charge of the meteorological service. Approval of the commission would be necessary on estimates for appropriations.[10]

The discussion that stemmed from the National Academy committee's report probed both the Department of Science and the commission idea, the testimony making echoes in the press. No one was better qualified or more ready to speak than Powell, the new-style bureau chief. Eliminating all the agencies that used science in construction work — public buildings, rivers and harbors, lighthouses — he concentrated on those which conducted original investigations. With the bureaus being investigated he grouped the National Observatory, the Fish Commission, and the National Museum. As usual, he made no mention of agriculture. Drawing on his own experience, he laid down general considerations for the organization of the agencies that must discover new facts and principles. First, since all the investigations are interrelated and interdependent, they "should be placed under one general management." Second, science "must be controlled by the facts discovered from year to year, and from month to month, and from day to day." Operations must be led by the men who are actually performing the work, involving constant consultation and changes of plan. The director largely selects men "who have a genius for research" and lets the plans come up from them. "It will thus be seen that it is impossible to directly restrict or control these scientific operations by law. The general purpose of the work may be formulated in the statutes, and the operations may be limited by the appropriations." A statute could go no further because "if the operations themselves could be formulated by law, the facts would already be known and the investigation would be unnecessary." Hence the bureau "should be left free to prosecute research in all its details without dictation from superior authority in respect to the methods to be used." [11]

While agreeing with the National Academy committee that the Department of Science proposal recognized the principle of "unified administration," Powell attacked the commission idea. Civil and military heads of bureaus would not mix, because the "military officer plans and commands; the civil officer hears, weighs, and decides." Also the commission would come in conflict with the department heads in just the way Newcomb and Comstock had been forced off the National Academy's committee because "the military Secretaries did not desire to have their subordinates deliberate on questions of policy affecting the conduct of the Secretaries themselves." [12]

In place of the commission, Powell had a simple solution — turn everything over to the Smithsonian Institution. Implying that the public officers on the board of regents were a sufficient link to the government, he cited as historical evidence the performance of Secretaries Henry and Baird. In discussing what bureaus should go to the Smithsonian, he laid the groundwork for an interbureau feud by recommending that the hydrographic work of the Coast Survey be given to the Navy in exchange for the Naval Observatory. [13] The major gave his views to the Allison Commission believing their actions would "ultimately affect the deepest interests of all the people," and would promote or retard "scientific research itself, which is the chief agency of civilization." [14]

The Navy's first scientist, Simon Newcomb, although barred from the Academy's committee, also got a chance to testify, his basic ideas paralleling those of Powell except for the Smithsonian afterthought.

He felt that "the evils in the scientific Bureaus" came from "the want of adequate administrative supervision." The heads of departments "cannot master the details of scientific work, and the law affords them no real assistance." As a result, the bureaus are neither "managed on sound business principles" nor have "the proper scientific criticism and control." The one adequate remedy in Newcomb's eyes was "to place all the scientific work of the Government . . . under a single administrative head, to be selected by the President as he selects the heads of other Departments." This officer would oversee the same group of bureaus Powell had outlined and possibly "some of the numerous chemical laboratories which the Government now supports." He was unwilling to let the military take over the Coast Survey because "control and criticism require system of administration quite foreign to that of the Navy." [15]

In distinct contrast, Secretary of the Navy William E. Chandler, who was trying to create a modern fighting force afloat, testified that all scientific or art work for the government "should be conducted within and under the direction of that Department which needs the scientific assistance . . . and that it would be a most anomalous proceeding to erect as a governmental department a department of science or of art." It would "set the scientists to grasping Government," to searching through the various political departments, removing the agencies that use applied science and turning them over "to a department the head of which should be only a scientist." Thus aware, perhaps more than the National Academy, of the penetration of science into the federal structure, the secretary also dimly recognized that the government may properly choose, "within certain limits, to go beyond the use of science and art in the actual conduct of governmental affairs and undertake to foster science or art, or agriculture." Such an activity should go into "that department with which it has the most natural relation." [16] Thus his interest was that of the cabinet officer attempting to preserve a clear chain of authority.

The discussion of a Department of Science and its alternatives extended beyond the Allison Commission itself and recapitulated the hopes of a whole century since 1787. The younger R. W. Shufeldt, an army officer stationed at Fort Wingate, New Mexico, saw in a Department of Science the salvation of those military men who wished to specialize in science without being interrupted by general duty. He would unify the government's scientific work under a secre-

tary of science chosen by the President from the National Academy
with the heads of the eleven bureaus actually selected by that body.[17]
Officers whose interest was pure research could get long terms of
duty in the new department, thus preserving that general interest in
natural history and natural philosophy which was part of the officer's
code of another day.

The magazine *Science*, which supported the National Academy
committee and especially the proposed Department of Science, regularly tilted at the alternatives. Powell's solution was not satisfactory.
The Smithsonian could hardly take on the direction of all the government's scientific work under the terms of Smithson's will. Besides,
management by scientists would not be achieved, since only one was
among the twelve regents. Henry's and Baird's policy had always
been centrifugal, establishing a line of research and maintaining it
only until they could get someone else to take it. "Now it is proposed
to reverse the process and send separate institutions back home!" [18]
Shufeldt's scheme gave too much power to Academy members, who
as specialists "are the last who should control a department." [19]

Even the national university reappeared. Cleveland's secretary of
the interior, Lucius Quintus Cincinnatus Lamar, symbol of return of
the South to national politics, claimed that if the scientific bureaus
"could be combined as integral parts of one scientific institution"
and "should a university be erected thereon with a superstructure
commensurate with the foundations, it would be without a rival in
any country." [20] But *Science* felt that, without tenure of office in the
civil service, a national university would never work, and free education in the hands of congressmen would be "so much injurious
patronage." [21]

Alexander Agassiz on Government Science and the Theory of Laissez Faire

A broader view of the problem came from Alexander Agassiz, who
commanded great respect. Besides the luster of his name, he was at
the height of his powers as a scientist and head of Harvard's Museum
of Comparative Zoology, a distinguished if atypical research institution. At the same time, he was making a vast fortune as owner and
manager of copper mines in upper Michigan. His own person thus
intimately combined private capital and science. Incidentally, his
family had had fruitful associations with the Coast Survey for forty

years. He raised the general question of the relation of the central government to science. The idea of a national university was "very dazzling; but is it wise for the Government to enter directly as a competitor into the field of higher education?" Recognizing that "the friends of a paternal government would like to see the science of the country centralized, and the work of the bureaus gradually absorbing all the best available men . . . making Washington a great scientific centre," he looked abroad to the Old World, where he saw the contrast between France and Germany. The latter nation, which only a few years before had been a maze of petty states, had had completely decentralized institutions. At the same time, it was the scientific leader of the world in Agassiz's view. France, on the other hand, had always had tightly organized institutions centering in the French Academy. "It would be a great disaster should Washington ever become the Paris of the United States." The history "of our scientific bureaus has been such as to suggest nothing but disaster from the centralization of science at the capital." The bureaus were already bringing close the time "when no man of science, and no university even, could hope to attack many of the problems naturally within their scope, without at once seeing the same work undertaken by the directors of official science."

What then should the government do? Agassiz favored efficient administration and a single cabinet head to present claims to Congress. Beyond that, "moderate centralization, allowing of great competition, is the ideal of scientific activity, and the government should limit its support of science to such work as is within neither the province nor the capacity of the individual or of the universities, or of associations and scientific societies." [22] This classic rule, which Henry had invoked for the Smithsonian and which Powell had used to advantage against Hayden in 1878, had never before had an exposition so clearly related to *laissez-faire* political theory. Although Agassiz had not opposed a Department of Science, he had by fundamentally questioning the government's sphere attempted to restrict the role of science even below its contemporary level. Some of his words about bureaus which by "capacity . . . for indefinite expansion" are "constantly encroaching upon the field of individual activity" seemed to convey a personal as well as a theoretical bitterness. To understand the bearings of Agassiz's outbreak, it is necessary to look at the events of the full year of the Allison Commission, 1885.

On March 4, 1885, Grover Cleveland became President — the first Democrat since 1861. In the change of Congress, Theodore Lyman had lost his seat in the House and on the Allison Commission. Scientists generally were apprehensive of Cleveland, and pressure for jobs for Democrats outran the available places. As a result, a general hunt for corruption and official wrongdoing began, with the scientific bureaus getting a very different kind of scrutiny from any they had had in the early days of the Allison Commission.[23]

Hilary Herbert, congressman from Alabama, who had said little before March 4, now came forward to flay Powell and the Geological Survey with data fed him by Hayden and Cope, who still nursed old wounds.[24] But the major had seen well to his bookkeeping, and he weathered every storm.[25] At the same time, he had had to defend himself against critics in the Coast Survey, who, questioning the accuracy of his maps, felt themselves pushed by the Navy afloat and not yet established in the interior.

The once mighty Coast Survey proved to be the weakest link in the scientific establishment. J. E. Hilgard, although he had actually run the outfit from the later days of Bache, became superintendent only in 1881 when he was old and ill, inheriting simple financial customs reminiscent of Hassler. A committee from the Treasury Department, headed by a political friend of Cleveland's, F. M. Thorn, thought it had found an easy mark. Its evidence against the Survey, Thorn later admitted, was "mainly *ex parte* affidavits, some true, some false, some mistaken, some since retracted, and more or less wild gossip since disproved." The Survey people "were not confronted with the witnesses, and did not cross-examine them nor appear by counsel." [26] But the poison did its work. Hilgard had to resign, and the career service became so demoralized that *Science* feared "a year or two more such as the last will leave nothing worth preserving of an organization which was once the pride of American applied science." [27]

In general, the scientific community of the country was aghast. The AAAS passed a resolution approving the "high character" of the Coast Survey's work and demanding the appointment of a superintendent "of the highest possible standing among scientific men." [28] Alexander Agassiz, emerging the foremost defender of the beleaguered agency, pointed to Hilgard's methods as those of Bache and Peirce, and he also protested that the investigation was "somewhat autocratic." To the charge that the Survey had distributed "scientific salt in the shape of gifts . . . and has also done its full share in the way of personal favors to Congressmen," he answered that "there never has been a 'political scientist' at its head." Here he implied that the new and unconventional activities of Powell as head of the Geological Survey were the real sources of the staid Coast Survey's grief. "It is time that the system of indiscriminate scientific assistance, given by heads of bureaus to institutions and individuals, and never contemplated or sanctioned by Congress, should be discontinued. It has brought nothing but discredit upon the official science of the country." Liberal appropriations were all right, "but let the requisitions be so complete and detailed as to invite a fair and open criticism." [29]

President Cleveland actually offered the superintendency to Agassiz, who declined because of ill-health and because "four men can be named, two already in the government service, and two not so employed, who are qualified for the post." [30] When the President finally did fill the place, he appointed F. M. Thorn, his Buffalo crony who had personally led the investigation that got Hilgard's scalp. As sometimes happens with such an appointment, Thorn was "gradually compelled by the force of circumstances to conduct the office in accordance with long established custom, and to trust the men whom his predecessors have trusted." [31] But for the moment the friends of the Coast Survey were beaten, sore, leaderless, and afraid of dismemberment. Since Powell had at times given aid and comfort to the designs of the Hydrographic Office and had seemingly carried off the right to map the land, he continued to be the object of Coast Survey bitterness.

Congressman Herbert of Alabama, leading the attack on the Geological Survey in the Allison Commission, had the wit to see the interrelation of Agassiz's *laissez-faire* theories and the defense of the Coast Survey. The scientist was trying to rationalize the inclusion of his family's favorite agency within the narrow limits he set for government activity at the same time he seemed to exclude the Geological Survey. The congressman had no interest in such fine distinctions himself, but could he not use the renowned Agassiz against Powell? "I beg to inquire," he wrote, "whether in your opinion the work of the Geological Bureau could profitably be brought within proper bounds. To me, it is very clear that Major Powell is transcending the rule you lay down, that Government ought not to do scientific work which can properly be accomplished by individual effort." After attacking various studies on the Comstock Lode, he asked, "is it necessary that a geological survey of the United States should go at all into paleontology?" Should the government geologist do topography? "For myself, I am free to confess that . . . if the people must be taxed for the maintenance of all that machinery to complete the geological survey . . . the time has come when Congress ought to consider seriously whether it ought not to abolish the whole survey." [32]

Agassiz seized the opportunity. From the studies of the Comstock Lode "private individuals have learned nothing . . . and the scope of the investigation on mining industries . . . are all such as seem to me to fall within the limits of private investigation." [33] Ironically, Clarence King was at that time a partner of Agassiz's in mining ventures.[34] In general, the biologist did not see "why men of science should ask more than other branches of knowledge, literature, fine arts, &c. . . There is no end to that kind of interesting documents which the heads of bureaus could get printed at Government expense, and which few individuals or societies would print, even had they the means at their command." Publications were extravagant and editions too large. As to paleontology, that "is just one of the things which private individuals and learned societies can do just as well as the Government." They will do it cheaper. "There are always in the different universities plenty of people who will be too thankful to do such work for the sake of doing it." And they "will manage to get the gist of their results published before the scientific societies to which they belong." He himself declined to have the collections he made with the Coast Survey published by the government, putting up some $30,000 for private editions. As to topography, he agreed with Powell that a geologic map without it was impossible, but if "the States are not willing to go to that expense, it seems plain that they do not wish the Government to go to that expense for them." [35] The *laissez-faire* theories of an outstanding scientist, coinciding so closely with those of a congressional critic, were neveretheless pro-Coast Survey and anti-Geological Survey. The new-style bureau as administered by Powell, rather than all government science, was Agassiz's real target.

Powell's Rebuttal — Government Research Stimulates Private Research

John Wesley Powell met directly a challenge whose "power depends upon . . . the great name of Agassiz." Referring to the many

years his assailant had devoted "to the accumulation of a great fortune," Powell understood that Agassiz had a "plan to create by gigantic business enterprises a great fund . . . and that it has been his
ambition to make the museum over which he presides the American
center of scientific research, and the agency which should create,
control, and diffuse the increasing knowledge of the New World."
Agassiz perhaps would have fulfilled this dream if it had been a
possibility. "Due honor should be accorded him for the brilliancy of
his unrealized scheme . . . but a hundred millionaires could not do
the work in scientific research now done by the General Government." Shall "scientific research and the progress of American civilization wait until the contagion of his example shall inspire a hundred
millionaires to engage in like good works?" Powell's answer drew the
line between science for private groups, even learned societies, and
for the public. "Before that time comes scientific research will be well
endowed by the people of the United States in the exercise of their
wisdom and in the confident belief that knowledge is for the welfare
of all the people." Work paid for by the treasury "should be given
back to the people at large through the agencies of the public libraries
which they have established. To turn over all this material to private
societies and museums for publication would be to defraud the
people of that for which the money was expended." [36]

Of course the major was quite aware that Agassiz considered the
Coast Survey in a different light from the Geological Survey, pointing out that the biologist's studies "were based upon materials collected by the General Government, at great expense." Also "worthy
of remark" was the fact that "Mr. Agassiz's principal work in research has been published by the British Government." [37] But Powell
was not interested in fencing to take advantage of inconsistencies. He
wished to examine the question, "what scientific researches should
the government endow?"

His first principle in answering this question was that "the
government should not undertake to promote research in those fields
where private enterprise may be relied on for good and exhaustive
work, especially while vast fields where private enterprise cannot
work are still unoccupied by agents of the government." Powell thus
did not believe in indiscriminate government science any more than
Agassiz did, and he had used this rule against Hayden's botany and
zoology in 1878. But his standard of judgment was private enterprise's
actual present performance, not its theoretical potentiality. "Mr.

Agassiz seems to have adopted this principle, and then by some strange mischance used it to exclude the work of the Geological Survey . . . failing to understand that topographic and geologic surveying must be carried on by governments, or be so feebly prosecuted as to be of little value." [38]

Here Powell's appreciation of the fact that for "a long term of years the General Government has provided for general topographic and geologic surveys" provided him with effective evidence. Estimating the expenditures at $6,000,000 and those of the States at $4,000,000, he gave some slight credit to individuals, "notably some able college professors," but judged their contributions "very small when compared with the great works accomplished by the official surveys." Even in general geology and paleontology the "principal monographic works . . . are official." In surveying the literature of geology, he found that of 120,000 pages "not more than about 6,000 are the direct and independent results of private investigation." He also found "no instance . . . of the systematic investigation of any considerable area — such, for example, as the Comstock, Eureka, or Leadville district — by any private party, corporation, or nongovernmental organization of any kind." [39] Since he felt the government had historically an important share in geology, he proceeded to further principles which might more precisely define a dividing line from private science.

The government "should promote the welfare of the people" by providing for "investigations in those fields most vitally affecting the great industries in which the people engage." This was a statement of the problem approach. Not only mining profited from the Geological Survey, but also agriculture, as in the study of soils, forests, and irrigation.

"One national survey is more efficient than many state surveys," and "the plant for geological investigation is too expensive for private agencies." To engage in research in geology already cost a lot of money. Shall the "genius of only the wealthy be employed in advancing the boundaries of human knowledge, and does great wealth invariably inspire its possessors with love for research, insight into Nature's laws, and patience in long labor?" A reason for the increasing cost of geology is implicit in a further principle. "The results of local investigation are of general value to many districts, and a knowledge of the geology of one locality must be derived from an examination of many other localities." This statement of Mill's principle of the centralization of knowledge required that "a survey should be organized upon the broadest territorial base possible," because "one such integrated organization can accomplish far greater results than a score could accomplish with the same amount of money divided among them." [40]

However completely he had justified the Geological Survey and the principles on which a new scientific bureau must operate, Powell still had to contend with Agassiz's fear that private science in America would ultimately be destroyed by competition from the taxrich government. Under the motto "all governmental research stimulates, promotes, and guides private research" Powell in one reiterative paragraph sketched a theory of knowledge in which science, the power of government, and democracy all found full place.

Possession of property is exclusive; possession of knowledge is not exclusive; for the knowledge which one man has may also be the possession of another. The learning of one man does not subtract from the learning of another, as if there were a limited quantity of unknown truth. Intellectual activity does not compete with other intellectual activity for exclusive possession of truth; scholarship breeds scholarship, wisdom breeds wisdom, discovery breeds discovery. Property may be divided into exclusive ownership for utilization and preservation, but knowledge is utilized and preserved by multiple ownership. That which one man gains by discovery is the gain of other men. And these multiple gains become invested capital, the interest on which is all paid to every owner, and the revenue of new discovery is boundless. It may be wrong to take another man's purse, but it is always right to take another man's knowledge, and it is the highest virtue to promote another man's investigation. The laws of political economy do not belong to the economics of science and intellectual progress. While ownership of property precludes other ownership of the same, ownership of knowledge promotes other ownership of the same, and when research is properly organized every man's work is an aid to every other man's.[41]

The Investigation's Outcome

How might a congressional committee be expected to react to this fundamental debate in a decade when *laissez-faire* political theory and the gospel of wealth were dominant currents? Powell had distinguished between the economics of scientific knowledge and the laws of political economy. But Hilary Herbert explicitly tied his position to the formal philosophy of *laissez faire* then at its height, quoting the historian H. T. Buckle on the dire consequences to former civilizations of government interference. In his parting shot at Powell, on the last page of the last voluminous report, he reiterated, "Buckle is right. Government patronage shackles that spirit of independent thought which is the life of science." [42] Yet no doubt shadows the clear-cut answer. Powell won; government science won; the new-style bureau won; the Geological Survey won; even the Coast Survey won.

Herbert gained an important advantage by getting his bills ready first. His legislation for the Geological Survey was a masterpiece of consistency. It provided that "the Geological Survey shall not . . . except for the collection, classification, and proper care of fossils and other material, expend any money for paleontological work or publications, nor for the general discussion of geological theories, nor shall it compose, compile, or prepare for publication monographs, or bulletins or other books." The sole exception was an "annual report, which shall embrace only the transactions of the Bureau for the year, and the results thereof." Since current publications would be thus abolished, all works "which have not been ordered by Congress to be printed, may be published by the authors thereof at their own expense," an arrangement quite to the taste of Alexander Agassiz. With its empire destroyed and economy accomplished, the Survey would not need its physical plant. The secretary of the interior would "cause to be sold . . . all such laboratories and other property now in use . . . as shall no longer be needed." [43]

Herbert, who, unlike Agassiz, considered science of every sort tainted with extravagance, did not stop at the water's edge. His bill for the Coast Survey recommended its transfer to the Navy, because the "real scientists on this subject of nautical maps are educated sailor-men, naval officers," who in other countries control the coast surveys. To him, the transcontinental connection was a stretch of authority as evil as anything the Geological Survey had done. Indeed, the rivalry between "these creatures of Government" showed a tendency to make them "active legislative forces and to control Congress." This "deserves to be carefully scrutinized, and whenever discovered promptly repressed." [44] The traditions of fifty years in the Coast Survey meant no more to Herbert than Powell's bright dreams for the future. Agassiz can only have felt chagrined.

By reporting the bills first, Herbert managed to convey the idea that they came from the whole commission. Powell, facing disaster, marshaled his arguments and prepared another lengthy defense, in which many other scientists not hitherto connected with the controversy joined. *Science* claimed that "national aid in the publication of scientific works is absolutely necessary." [45] Even Herbert realized that his report "immediately became the subject of much criticism." Within a few days he had to back down, withdrawing the section on selling off the Geological Survey's laboratories. By this time his report clearly emerged as a minority one, speaking only for himself and John T. Morgan, a senator also from Alabama. The majority, consisting of a Democrat from Indiana and Republicans from Iowa, Connecticut, and Maine, finally asserted itself, issuing a report that disowned Herbert's and Morgan's whole approach. The two ex-Confederates had taken the position of John C. Calhoun without a trace of his constitutional dialectics, using arguments that reflected the alliance of many post-Reconstruction Southern politicians with business interests in the North.[46]

The majority recommended a bill that somewhat restricted Powell's use of lump-sum appropriations for publications without submitting detailed estimates beforehand. Otherwise, they left the Geological Survey intact.[47] They saw no reason to transfer the Coast Survey to the Navy Department. By considering themselves incompetent "to decide upon the methods to be adopted in a survey so highly scientific in its character and objects," they implicitly acknowledged the necessity of having scientists administer the new bureaus. They recognized on the side of the civilian Coast Survey "the sanction of the scientists of our country, and . . . of more than two generations of experience and criticism." [48] Thus the Allison Commission by its very lack of action affirmed the usefulness of science to the nation and recognized the growth of the new bureaus as accepted parts of the government.

So furious were the passions concerning the relation of science to government that the majority of the Allison Commission seemed to be taking up a long-forgotten issue when they pronounced upon the department of science. They felt that "no such duplication of work or necessary connection of these bureaus with each other . . . makes such an establishment essential to their efficiency." Where "one bureau finds it necessary to utilize the work of another a request for information and data is always complied with." [49]

In contrast to their glorious and successful defense of the new scientific bureaus, the experts had done a ragged job for a department of science. The National Academy had done nothing to push the brain child of its committee, which admitted political defeat in advance. Powell dealt the idea of a commission a heavy blow, only to try to make the Smithsonian bear a strain for which it was not designed. Newcomb hedged at the critical point of cabinet status. Indeed, their enthusiastic proof that each of the bureaus had a unique mission and its own traditions undermined the main argument for consolidation — that the present system duplicated wastefully. This performance before the commission raised a question whether a department of science was a good idea even if possible.

An unusually astute observer who was both within government science and outside the controversies of the surveys summed up the case against the department of science late in 1886. Dr. John Shaw Billings of the Army Medical Corps praised the majority of the Allison Commission for concluding that "the work is, on the whole, being well done, and that the people are getting their worth for their money." He had a realistic view of the problems of the scientific administration, describing them in much the same way as had Powell and Newcomb. But "as to the desirability of centralization and consolidation of scientific interests and scientific work into one department under a single head, I confess I have serious doubts." In essence, he claimed a department organized along the lines of the branches of science denies the problem approach. While taking advantage of every opportunity to increase knowledge, every scientific branch of the government should be tied to the "practical results" the legislators are trying to achieve.[50]

Billings did not place his restrictions on the government's science because he held knowledge in contempt. Rather, "we live in a fortunate time and place — in the early manhood of a mighty nation, and in its capital city, which every year makes more beautiful, and richer in the treasures of science, literature and art." But all government science must have a clear purpose; "we may not rest and eat lotus; we may not devote our lives to our own pleasures, even though it be pleasure derived from scientific investigation. No man lives for himself alone; the scientific man should do so least of all." Never had the world "more need of him, and there never was a time when more care was needful lest his torch should prove a firebrand and destroy more than it illuminates." [51] The concentration of all science in the government into one department, representing a special professional interest, might make control by Congress and the executive harder instead of easier. And the full potentiality of science in the government could be achieved only if it permeated the whole structure.

Merely by taking no action at all, the Allison Commission both affirmed the worth of government science and denied the validity of a separate department for it.

XII

CONSERVATION

1865–1916

IN the closing two decades of the nineteenth century Americans began to realize for the first time that geography and sky-rocketing consumption were threatening to place a limit on the natural resources available to the nation. Frederick Jackson Turner's emphasis on the disappearance of the frontier in 1893 symbolized a fundamental shift in attitudes toward the natural wealth of the nation, especially that in the public domain. Under the assumption that the resources of the public lands were inexhaustible, most people demanded and would tolerate government activity only in the routine measuring and disposing services of the General Land Office. However, with the increasing belief that resources were not only limited but fast being exhausted, the government as protector of the public's interest had a very different role to play in managing and regulating the preservation and use of the nation's wealth, especially that of federally owned public lands. In the first decade of the twentieth century the conservation movement, as it came to be called, urged national management of natural resources. The assumption that science would furnish the data on which classification could be made was basic to all major plans proposed. Scientists were characteristically early and ardent conservationists.[1]

Powell's Irrigation Survey

Long before conservation became a political plank and a part of the fighting faith of the progressives, the scientific establishment of the government had reacted to various aspects of the problem. The pre-Civil War explorations sometimes got close to the concept of land classification.[2] Powell, in his *Lands of the Arid Regions*, had laid down a fundamental program for land classification, irrigation, and water rights in the West which, although a part of the National Academy's report, had largely lost out in the organic act of the Geological Survey. In spite of his clear vision of the application of science to the problem, Powell at first had to content himself with making the Geological Survey into a purely informational scientific agency. He could complete its structure as a new scientific bureau only by getting some regulatory function, and he chose the public domain, with its pervading aridity, as the problem to attack. In 1888 he got his opportunity.

Western senators, realizing that most sites on small streams capable of supplying irrigation water had already been utilized by private enterprise, began to look to large dams and reservoirs on the principal rivers. W. M. Stewart of Nevada and Henry M. Teller of Colorado put through a joint resolution directing the secretary of the interior to make "an examination of that portion of the United States where agriculture is carried on by means of irrigation," to determine dam sites "together with the capacity of streams, and the cost of construction and capacity of reservoirs, and such other facts as bear on the question." [3] Naturally referred to Powell, the resolution became in his hands a wedge for his whole general plan for the arid regions. Interpreting the area covered by the resolution as every place beyond the twenty-inch rainfall line, he took in two-fifths of the United States, proposing a topographical survey, a hydrographic survey, and finally a preliminary engineering survey at a cost of five to seven million dollars.[4] To get money, the Western senators resorted to the usual trick of tacking a rider on the sundry civil appropriations bill. In the House, a Colorado representative, momentarily frightened by the prospect of speculators following the surveying crews, added an amendment reserving from sale, in addition to dam and canal sites, "all lands made susceptible of irrigation." In conference committee a further amendment giving the President the right to restore lands to entry supposedly smoothed over any damage the first provision might do. Powell got $100,000 and immediately had Dutton and Thompson of the old Powell survey in the field both for topographic surveys of likely reservoir sites and for training a corps of technicians in stream-flow measurement and hydrography.[5]

The effect of the casual, almost unconscious legislation which created the irrigation survey is a perfect example of the change that comes over a scientific enterprise when it gets the sanction of law to put the results of research into practice. After much delay, the General Land Office finally ordered the district offices to cancel all filings since the date of the legislation that were on the "sites for reservoirs, ditches, or canals." But until Powell completed his survey, who knew where these would be? The attorney general ruled that until the President had certified the land "entries should not be permitted . . . upon any part of the arid regions which might possibly come within the operation of this act." [6] The President would not certify until Powell had surveyed. Hence, in effect, all the land laws were suspended, and the General Land Office was out of business until the major gave the word.

In this unsought but conspicuously powerful position, Powell still hoped to change the pattern of Western settlement from haphazard private selection to an over-all plan soundly based on scientific information. But he had to hurry. Claimants of lands in the public domain were in no mood to wait several years, and the major's efforts to get enough sites opened up to relieve the pressure was complicated by the necessity of checking the titles of previous private claims at the General Land Office.

Meanwhile, the Westerners were appalled. They began to call Powell the "tycoon of many tails" and quote:

> Upon what meat doth this our Caesar feed
> That he is grown so great?

Senator Stewart now emerged as the major's leading antagonist, calling him "the King of the lobby." [7] The Western friends of irrigation felt that obvious sites should be opened immediately without waiting for a scientific survey or a plan for the use of the water of a whole river. Powell's science was simply an unnecessary delay which they would not tolerate. In April 1890, they showed their strength by setting up an artesian-well survey in the Department of Agriculture in the face of a large body of Geological Survey data that already showed this source of water to be of limited value. [8]

Organic acts tucked away in appropriation bills are as vulnerable as they are easily obtainable. In August 1890, Powell's enemies, having failed to move the whole irrigation survey to the Department of Agriculture, got to him by cutting the program from a proposed $700,000 to $162,500. All powers of reserving lands from entry disappeared and with them the possibility of shaping the pattern of settlement by scientific investigation. Reference to hydrography was omitted, cutting an essential step from Powell's program. Despite a record $719,000 for

the Geological Survey, the irrigation work was dead, and with it
Powell's great plan of orderly settlement based on the facts of environment as determined by science.[9]

The wreck did not stop with the irrigation survey. Powell, the
most brilliant scientific administrator since Bache and the personification of the new scientific bureau, had suffered defeat. The Geological
Survey, and beyond it the rest of government science, lay exposed.
The impending change of administrations, monetary difficulties, and a
gathering business depression gave hope to the trimmers of scientific
activity who had lost out in the Allison Commission. Stewart, joined
now by Hilary Herbert, moved against pure science. Because the old
rivalry between Cope and Marsh had by this time become a *cause
célèbre* in the pages of the New York *Herald*, Powell's enemies struck
hard at paleontology. Herbert claimed that "when the morning of
resurrection shall come, some paleontologist will be searching for some
previously undiscovered species of extinct beings, and some geologist
will be picking away at the rocks . . . There is no end to it." [10] In
1892, by a vote of 26 to 23, the Senate cut the Geological Survey's appropriation drastically and itemized sums for salaries. The final total
was only $430,000 and, with nothing specified for his salary, Marsh
had to be dropped. The Geological Survey was reduced to topography alone. The ax fell also on the Coast and Geodetic Survey, the
Lighthouse Commission, the Smithsonian Institution, the Naval Observatory, and the Bureau of American Ethnology.[11] Powell's precocious effort to harness the water resources of the public domain to
science had resulted in a sharp setback to government science.

The major himself retreated to the safety of the Bureau of American Ethnology as soon as he could arrange for the directorship of the
Geological Survey to go to C. D. Walcott, his assistant with the
hardiest exterior for political abuse. Yet the irrigation survey laid the
basis for the later conservation movement. It left a mountain of data
which would later prove useful; [12] it left the major's vision of aridity
as the key to Western environment. But mainly it left men. Powell
bequeathed to the conservation movement a group, trained and dedicated, that was ready to fill the posts and take the responsibility when
political fortune again made a program of reclamation possible. F. H.
Newell, A. P. Davis, Walcott, W J McGee, and many others got both
their training and their inspiration from John Wesley Powell.

The Department of Agriculture provided a caretaker outfit to keep the irrigation problem formally under government scrutiny. After the Powell survey collapse, an Office of Irrigation Investigations had appeared in the Division of Experiment Stations. Under Elwood Mead it worked for private irrigation through better water laws and gathered data on amounts of water needed for various crops. Although not closely related to other bureaus in the Department of Agriculture, the Office of Irrigation Investigations was also far removed in its approach from the Geological Survey.[13]

The Fish Commission

Besides the frontal assault on the water problem of the West, the government's scientific establishment responded to the looming exhaustibility of material resources in other ways. Characteristically quiet in his movements, Spencer F. Baird managed to bring research to the fisheries of the nation and a new bureau into the government establishment. While director of the National Museum and assistant secretary of the Smithsonian, Baird, beginning in the 1860's, often spent his summers along the New England coast, where he noticed the "great diminution in the numbers of the fish which furnish the Summer food supply to the Coast . . . & I found the same impression to be almost universal." Most people blamed the use of nets and "the capturing of the fish on or near their breeding ground before they have spawned," which led to a demand for "laws preventing or regulating the employment of nets or weirs." Baird believed that laws had to come in part, at least, from the federal government, and as a preliminary step he suggested a broad research program. "We must ascertain, among other facts, at what time the fish reach our coast, and during what period they remain, when they spawn and where, what is the nature of their food," and many other problems which added up not only to the biology of each species of fish, but to the ecology of life in the ocean.[14] While his personal reputation gave him the influence he needed in Congress, Baird was careful to keep his brain child away from the churning waters of politics. Envisaging a task of only a year or two, he expressly had included in the joint resolution of 1871 that set up the Fish Commission a clause providing that the President appoint "from among the civil officers or employees of the Government, one person of proved scientific and practical acquaintance with the fishes of the coast to be Commissioner of Fish and Fisheries, to serve without additional salary." [15] No money meant no patronage. After the first appropriation of $5000 came another of $15,000, and Baird was in the business for the rest of his life, even after he succeeded Henry as secretary of the Smithsonian.[16]

Baird, the true scientist, put his investigations on the highest plane. Using the same device on his assistants he used on himself, he got students from universities to serve without pay, developing many of the next generation of zoologists, among them C. Hart Merriam and George Brown Goode.[17] In 1881, he got $190,000 for the seagoing vessel *Albatross*, especially equipped for marine biology.[18] After using various places on the New England coast, Baird settled on Wood's Hole, Massachusetts, as the site for a permanent station. Characteristically, Baird arranged for purchase of the land by private subscribers, among whom were Johns Hopkins University, Princeton University, Williams College, and Alexander Agassiz. The institutions had the right to send a specialist to the station to carry on research. The government thus broke the ground for the famous research institution that Wood's Hole became after the establishment of the private Marine Biological Laboratory there in 1888.[19] Baird's broad and fundamental research program led an English scientist to say before the British Association that, while there expensive royal commissions visit the coast and question fishermen who have knowledge of only one small area, in America "the questions are put to nature and not to fishermen." [20]

However, almost from the beginning the fish commissioner had to concern himself with more than research. The advocates of artificial propagation became active through the American Fish Cultural Association in urging federal support. In 1873, Congress appropriated $15,000 which by 1887 had reached $161,000, as compared to a research budget of $20,000. Thus, hatcheries and fish culture became in quantity, at least, the main business of the new organization. By 1880, statistics also began to develop as a separate activity.[21]

With routine operations pressing him, Baird found himself spending six hours a day on Fish Commission business. He furnished office space in his own home, losing about $2000 a year besides drawing no salary. Yet when he asked for money to rent offices, the chairman of the appropriations committee refused because "he was opposed to anything that looked like fastening an additonal bureau upon the Government." [22]

Clearly, that was just what had happened, and the agency gradually conformed to the general administrative pattern. When Baird died, Congress voted his widow $25,000, which was hardly straight pay for the services rendered. The commissionership became a salaried post, providing both Cleveland and McKinley the opportunity to pay political debts. In 1903, the independent commission became the Bureau of Fisheries in the Department of Commerce and Labor.[23] By 1910, it had reached stability.

Wildlife Research — the Biological Survey

Less directly identified with a single commercial interest than fish, the land wildlife of the country began to get scientific attention at about the same period.[24] Although the concept of wildlife as an exhaustible natural resource had scarcely begun in the decades of the slaughter of the passenger pigeon and the buffalo, some dawning of the importance of the mutual relations of birds to insects appeared about 1850, when rude observations led to the disastrous introduction of the English sparrow, supposedly to eat caterpillars — an attempt to alter the balance of nature that itself created a problem. With the beginning of entomological work under C. V. Riley,[25] "economic ornithology" began to grow as a companion study. American scientists had, of course, been studying animals in nature since colonial times and, thanks in part to earlier government efforts, had a considerable start on the taxonomy of North American wildlife. With increasing specialization in scientific societies the American Ornithologists' Union became active in 1883, studying the sparrow problem and naming a committee to determine the distribution of each species of bird in North America. It set up a country-wide reporting system, including not only private collectors but the Lighthouse Board and the Department of Marine and Fisheries of Canada.[26]

Soon the volume of information, national and more in scope, swamped the committee, whose chairman, C. Hart Merriam, prepared a memorial to Congress and got a hearing with the help of Baird and C. V. Riley. The result was an appropriation of $5000, and a section of economic ornithology appeared in Riley's Division of Entomology in the Department of Agriculture. In 1886, the Division of Economic Ornithology and Mammalogy became separate, and, when Merriam was appointed its head, the government in effect took over the responsibility for the program originated by the Ornithologists' Union.[27]

Almost immediately Merriam began to play down the word "economic" in the division's title, by 1891 dropping it altogether. At the same time, he pushed the Union's interest in exact data of bird migration and distribution on a continental scale. In studying mammals and the smaller vertebrates, he introduced new techniques based on the use of the cyclone trap which quickly revealed many more species in North America than anyone had dreamed of before.[28] To know the wildlife of the continent with the accuracy now obtainable required a complete reëxploration. By 1894, Merriam was referring to his work as "biological investigations" and announcing comprehensive conclusions on the relation of the geographic distribution of plants and animals to temperature. The secretary of agriculture could now say that the name of the division "is unfortunate," for it "is in effect a biological survey, and should be so named, for its principal occupation is the preparation of large-scale maps of North America, showing the boundaries of the different faunas and floras, or life areas." [29] In 1896, the appropriations were made for the Division of the Biological Survey, and the scientists were called biologists. Out of the original problem approach had come a comprehensive research program in fundamental natural history only nominally held in the Department of Agriculture. Merriam's affinities were close to the Smithsonian and parallel to the Fish Commission, although unconnected with any routine program.

Scientists' Early Interest in Forest Protection

The exhaustibility of resources was perhaps most evident in the disappearance of the forests, the most striking change that man had made in American environment up to the end of the nineteenth century. Scattered scientists and nature lovers had noted some of the consequences before 1850. Henry had made a report on forest trees and their economic uses one of the first projects of the Smithsonian. At the AAAS meeting in 1873, Dr. Franklin B. Hough spoke on "The Duty of Governments in the Preservation of Forests," inspiring the appointment of a committee to memorialize Congress and state legislatures.[30]

When a House bill to set up a commissioner of forestry in the Department of the Interior had failed, friends of the measure used the familiar tactic of a rider to the free-seed clause of the Department of Agriculture appropriation. Two thousand dollars could be used to pay someone acquainted with statistics to "prosecute investigations and inquiries" into forest problems. No mention appeared of the public domain or the management of its forests. After a few years of irregular activity under Hough and others, the forestry work became a division in the Department of Agriculture in the mid-1880's under German-trained Bernhard E. Fernow.[31]

In his first annual report, Fernow eschewed all desire for regulatory powers, relying on "simply the example of systematic and successful management" and on "advice and guidance" to influence private owners. He undertook studies in forest botany, especially nomenclature, and some research in timber physics and the properties of wood that might aid in the utilization of forest products. Yet his program early showed some basic limitations. In the first place, he had no land on which to practice forestry and had bad luck in making deals with the Department of the Interior, the Army, or private owners that would give him access to forests even for research purposes. Although nominally in the Department of Agriculture, the division had little to do with farmers and drew no support from their organizations. The old custom of distributing free seeds accomplished nothing. In 1890, Congress pushed Fernow into some rain-making experiments which disturbed the neighborhood of Washington with explosions and brought ridicule on both the Division of Forestry and government science. By 1898, the agency consisted of eleven people and had an appropriation of $28,000.[32]

Meanwhile, the millions of acres of forest land in the public domain precipitated a crisis that would not wait for Fernow's cautious background research. As settlement of the West proceeded, private interests increasingly took title to tree-covered lands for grazing and lumbering, destroying the forest community in pursuit of immediate gains. Nature lovers such as John Muir, decrying spoliation of the wilderness, were able to get small tracts such as redwood groves set aside. Charles Sprague Sargent, representing the botanists' interest in the forests, made a comprehensive survey for the census of 1880, publishing a "Report on the Forests of North America." An American Forestry Congress gave organizational focus to these groups.[33] Much emphasis fell on tree planting, which the Timber Culture Act vainly attempted to extend to the West.[34]

Such activity, however, did not come to grips with the necessity of effective regulation if the forests of the public domain were to be saved.[35] The commission that grew out of the reorganization of the surveys in 1879 had pointed to a fundamental policy by suggesting withdrawal of all timber lands from sale and control of their use.[36] Although plenty of bills were introduced, nothing went through until 1891, when an insertion in an act concerning the public lands gave the President power to reserve forests from sale. In two years, over 17,500,000 acres had been withdrawn, checking the passage of title to private interests. But complete lack of any provision for managing or protecting the reserves left them to the hazards of fire and thievery, with unreserved forests still wide open.[37]

Although the most pressing problems were political, economic, or administrative, scientists had long taken an interest in the forests. They felt they had both something at stake in forest preservation and something to contribute. Botanists, led by Sargent, and naturalists such as Muir tried to arouse interest in preserving the woods unspoiled, but their appeal was not rooted in economic realities. Fernow was the leader of the handful of Europeans with forest training who had found their way here, but their attitude was largely passive.

The appearance on the scene of Gifford Pinchot, a young American who styled himself a "forester," added a new type of scientist-administrator interested in action. A graduate of Yale who went to Europe for forestry training, Pinchot decided after a year that the thorough and intensive methods used in the Old World forests had no relevance to American environment and that the opinion he heard everywhere of the impossibility of forest management in the United States was wrong. Without staying to become a technically qualified forester or, to his later regret, getting much education in the sciences that impinged on his problem, he came back to the United States determined to take management directly into the woods and to learn his methods empirically as he went.[38]

After getting some private land under management at Biltmore, North Carolina, Pinchot became a free-lance forest consultant. He at least sat in, while visiting Sargent in Brookline, Massachusetts, in 1895, on the hatching of a scheme to get some action on the problem of forests. Old Wolcott Gibbs, at last president of the National Academy he had helped to found, suggested that the quickest way to get a commission appointed was to arrange for someone to request it of the Academy. In February 1896, they chose the Department of the Interior, getting Cleveland's secretary Hoke Smith to ask the National Academy for a report on whether forestry on the public lands was desirable and on the effect of forests on climate, soil, and water. By asking for specific legislation the committee had the opportunity to set up a forestry program for the government.[39]

Besides Sargent as chairman, the committee was made up of Henry L. Abbott of the Army Engineers, Alexander Agassiz, Professor W. H. Brewer of Yale, Wolcott Gibbs, Arnold Hague of the Geological Survey, and Pinchot. Hague, who had worked on the King Survey and with Powell since the 1879 reorganization, brought with him the Geological Survey approach, fitting in the forests as one element in the control of aridity. Since Pinchot's ideas of practical forest management harmonized with this tradition — "Without the Survey, then and later," he said, "the Commission would have been up a very tall tree" [40] — the two worked closely together to develop a plan of action which they hoped to get into Cleveland's last annual message in December 1896. Sargent and H. L. Abbott, on the other hand, favored military control of the forests, placing most emphasis on policing and fire protection.

The commission toured the Western forests in the summer of 1896, concluding that large additional areas should be reserved. Although they missed the opportunity to get a statement in Cleveland's annual message, they did get him to set aside some 21,000,000 acres, more than doubling the reserves. The Western interests that had fought Powell's Irrigation Survey now took up the cry against the forest reform. The bitterness of the rift between the Eastern gold Democrats such as Cleveland and the free-silver enthusiasts of the West doubtless aggravated the issue.

Senator C. D. Clark of Wyoming immediately introduced an amendment to the sundry civil bill throwing the reserves open again. Although this did not pass, neither did the efforts of the friends of the forests to set up an administration for them. A compromise provision was finally included, which caused Cleveland in one of the last acts of his administration to veto the whole appropriation bill, giving the forestry problem a great deal of publicity, and insuring its immediate consideration in the first session of the new Congress under McKinley.[41]

Pinchot gave the credit for a way out of the forestry deadlock, so reminiscent of the fight that had ruined Powell, to C. D. Walcott of the Geological Survey, who lined up to support legislation for the reserves a South Dakota senator who had already announced against them. The result was a compromise allowing the Cleveland reserves, except those in California, to be open for entry for nine months, then to be closed permanently. Part of the law provided for a survey of forest reserves by the Geological Survey and gave the secretary of the interior power to "make such rules and regulations and establish such service as will insure the objects of such reservations, namely, to regulate their occupancy and use and to preserve the forests therein from destruction." The way was open both for full economic use of the reserved forests — the surest way to allaying Western opposition — and for the establishment of a professional forest corps within the government.[42]

The report of the National Academy committee, which came too late to influence the legislation, was a mixture of the ideas of the older forestry advocates and the newer principles of the professional forester–Geological Survey alliance. An elaborate account of European experience and a predilection for military control reflected the views of Sargent and Abbott, while Pinchot and Hague managed to insert a program of forestry management which in many ways foreshadowed later policy.[43]

Concerning the reserved lands, the report pointed out that in the "peculiar topographical and climatic conditions of Western North America" forests were "essential to the profitable and permanent occupation of the country." They "collect and in a measure regulate the flow of streams, the waters of which, carefully conserved and distributed artificially, would render possible the reclamation of vast areas of so-called desert lands." [44] Since private investment in the timber operations in the West could not make long-term plans, "silviculture . . . will only be really successful under Government control and administration; for, dealing with crops which often do not reach maturity until the end of one or two centuries, it can only be made profitable by carrying out, without interruption and under thoroughly trained officers, plans which must often be followed during the lives of several generations of men." [45] Ultimately the program could be self-supporting, especially since any investment in protection from fire and overgrazing would mean huge savings of timber resources. The outline for a system of forest administration, while influenced by Sargent's penchant for the Army and West Point graduates, recognized that "wise forest management calls for technical knowledge which must be based on a liberal scientific education." [46]

The legislation of 1897 and the National Academy report laid down a policy for the forest reserves and opened the way for their administration by the Department of the Interior. At best, the friends of the forests could hope for an agency modeled on the Geological Survey and working closely with it. At worst, they feared the reputation and practices of the General Land Office, which would put the power in the hands of political hacks instead of trained scientists.

Gifford Pinchot, full of misgivings, took a job as confidential forest agent for the secretary of the interior. Sargent, disgusted, wrote that while he had expected Pinchot eventually "to take a prominent place in National Forestry," the younger man had "gone over now to the politicians . . . without consulting his friends in the Academy, and his usefulness, I fear, is nearly at an end." [47] Yet Walcott and Hague of the Geological Survey had engineered the appointment, and Pinchot worked closely with their chief geographer, Henry Gannett, who undertook the accurate mapping of the reserves as provided in the 1897 legislation. [48] The Geological Survey, blocked from irrigation, nevertheless kept alive the vision of scientific planning in the West that it had inherited from Powell and here tried to extend it to a problem he had never taken up. With luck, the Interior Department could become the shelter for a whole group of scientific agencies concerned with natural resources.

Forestry in the Department of Agriculture

At this juncture, Fernow, whose Division of Forestry in the Department of Agriculture had had no part in the reserves, decided to resign and become head of a school of forestry at Cornell. Since his was the only permanent position in the government dealing specifically with forestry, Walcott suggested Pinchot, whose friends induced him to accept a job he feared had no future. [49] As chief of a division in the Department of Agriculture, Pinchot was placed under civil service without an examination and got a slight distinction from the title "Forester." Tama Jim Wilson gave him a free hand to develop his own program. Pinchot had huge contempt for Fernow's approach — that before the division "could manage a forest growth intelligently it must know first of all the biology, or life history, of all kinds of trees which composed it." While conceding this might be true in Germany, "forestry in the land of the ingenious Yankee could be built on a whole lot less information than that." [50] For Pinchot the object of research was what he called "Silvics" — rate of growth in height and diameter, amount of wood produced, coming of seed years, tolerance of young trees for shade and sunlight — the things a ranger needed to know to estimate what should be done to sustain yield in a given stand of trees. The data had to be translatable into practical instructions for selecting from the individual trees those eligible for lumbering.[51]

His first objective was to get support for the division by making it useful to private timber owners. *Circular 21*, issued only a few months after Pinchot took office, offered the division's assistance to anyone who wished to harvest his timber and still have a second crop. Working plans and full practical directions were to be given by agents on the ground. Owners of larger tracts paid the expenses but not the salaries of the division's parties. With all this emphasis on "getting forestry into the woods," research tended to be empirical and picked up by the way in actual practice. But the policy had important results. Requests for aid flooded the division. By 1904, 900,000 acres of private lands were under management, and applications for advice had come from the owners of 2,000,000 acres more.[52] Within a short time Pinchot forged an alliance between the lumbermen and the division. Because the larger operators had a greater margin with which to experiment on long-term methods, the Weyerhaeuser Lumber Company and other giants were foremost in adopting the recommendations and in praising Pinchot's work.[53]

Armed with more requests for aid than he could meet with the insignificant establishment he inherited, Pinchot steadily and successfully multiplied his requests for appropriations. The rate of increase between 1899 and 1905 is spectacular: [54] 1899, $28,520; 1900, $48,520; 1901, $88,520; 1902, $185,440; 1903, $291,860; 1904, $350,000; 1905, $439,873. Not since the early days of Powell had anyone built a scientific agency with such speed, and Pinchot looked well to the necessary exterior buttresses of his bureau. In 1900, he released his assistant chief, Henry S. Graves, to Yale to head a new forestry school whose initial endowment came from the Pinchot family. With quarters in the former home of O. C. Marsh in New Haven, the postgraduate program, especially designed to produce men for the forest service, set the standard for education in the specialty throughout the country.[55]

A purely advisory bureau was never Pinchot's ideal or intention. From the beginning he worked for a transfer of the reserves from the

Department of the Interior to the Department of Agriculture. Deploring the work of political appointees sent out by the General Land Office to administer the reserves,[56] he managed to get the secretary of the interior to call on the Division of Forestry for advice. Pinchot knew this was different from controlling the reserves, but "it was on the straight road to control." [57] In the first test, in the Black Hills, the foresters showed their ability to produce enough rough results in a hurry.[58]

One of the offshoots of the new forestry was research in grazing. All the natural-resource problems of the West were parts of a single pattern. Just as irrigation implied forests on the high watersheds, protection of timber required attention to grazing, next to fire the greatest danger to proper management. In 1897, the secretary of the interior closed the forest reserves to grazing entirely, thus precipitating a bitter controversy among different groups in the West. For instance, in the Salt River Valley in Arizona, the irrigation ranchers claimed that grazing in the mountains caused a water shortage, while woolgrowers protested the other way. Pinchot teamed up with F. V. Coville of the Division of Agrostology of the Department of Agriculture to study the effects of grazing on forests and water supply, their report suggesting regulation of grazing rather than flat prohibition.[59] Pinchot's whole approach envisaged wise use of natural resources on the reserved lands, not total exclusion from them.

The Progressive Era and Conservation

When William McKinley died in Buffalo on September 14, 1901, the government had already developed a considerable scientific establishment oriented to natural-resources problems, especially on the public domain. In the Department of the Interior, the Geological Survey was a major agency and in the Department of Agriculture, the Bureau of Forestry was well on the way. In lesser degree, the Biological Survey, the Fish Commission, the Division of Agrostology, and the Office of Irrigation Investigations were conservation agencies. The heirs of Powell had a reasonably coherent body of doctrine for dealing with the problems of the arid West; they had gathered many data and had the techniques for gathering more. Walcott, Hague, Gannett, and Newell of the Geological Survey and W J McGee of the Bureau of American Ethnology were only the most prominent of the long-time associates of Powell who stood ready in 1901. Pinchot, although completely unappreciative of Powell's personal influence, had long worked with the Geological Survey, his forest program fitting harmoniously into the general scheme. Agencies, experience, doctrines, and men were already a part of the government, but, as yet, there was no successful program.

The lack in 1901 was a fundamental policy on how to use these scientific agencies, giving them legal power to carry out their plans. Almost the whole body of law on which these agencies were based consisted of provisions written into appropriations bills. In the background was the smarting defeat of the Powell irrigation survey and the anomalous position of the forest reserves under a different department from the research agency. Regardless of their growth, the scientific bureaus had not won a single major test that might have given them legal sanctions to put a conservation program into active effect in the field.

Theodore Roosevelt's unexpected ascendancy brought into the White House for the first time since John Quincy Adams a man with a personal background for science. As an undergraduate at Harvard, Roosevelt had some ambitions to become a zoologist, and his sojourns in the West had given him direct experience with the essential factors of the conservation program. He was, for instance, able and willing to give his friend Hart Merriam copious advice about the direction that biological research should take.[60]

The beginning of Roosevelt's presidency is also the usual line of demarcation for the beginning of the Progressive Era, in which the prevailing mood favored an active policy toward the problems emerging from rapid industrialization. Big business combinations caused concern. The ethics and machinery of government on every level became material for reform. In their search for ways to relieve the weakness, inefficiency, and corruption which they conceived plagued democratic government in the face of mounting concentrations of economic power, the progressives had great hopes for science. Trained, impartial experts swayed only by scientific facts could make the decisions concerning allocation of natural resources now realized to be scarce. A decision based on science would of itself be in the public interest. This pattern had conspicuously appeared in the alliance for the regulation of foods and drugs between the muckrakers who investigated the packing industry and Harvey Wiley. Conservation was the broadest front on which the alliance of the progressives and science could operate, largely because here the government already had at hand the men, the institutions, and the clearly defined lines of action. The progressives had only to furnish the support where it had failed before — in Congress and on the high levels of the executive branch.

Pinchot and Frederick G. Newell of the Geological Survey had met Roosevelt while he was governor of New York. They lost no time getting to the new President to ask him to make a strong statement in his first message. When he gave them authority to prepare a draft, they huddled with others to produce a water- and forest-policy statement. The resulting message of December 2, 1901, proclaiming that "the fundamental idea of forestry is the perpetuation of forests by use," urged the consolidation of all activities, including the reserves, into Pinchot's bureau. In the West, where "water, not land" measured production, "irrigation works should be built by the National Government." [61] The doctrines developed by the bureaus now had a vigorous champion in the new President.

The immediate outgrowth of this executive interest was the Newlands Act of 1902, which provided for government construction of irrigation works, with repayment to come from the users of the water. The reclaimed lands were to be broken up into small holdings.[62] Roosevelt, regarding "the irrigation business as one of the great features of my administration," specified as the controlling agency the Geological Survey, "of which Mr. Walcott is the Director and Mr. Newell the Hydrographer. These men have been tested and tried and we know how well they will do their work." [63] The Office of Irrigation Investigations in the Department of Agriculture thus lost out, the new Reclamation Service growing up within the framework of the Geological Survey tradition.

The work from the beginning largely concerned the building of dams and administering of projects. The mapping and surveying that had so completely occupied Powell were now considered as only preliminary to the main task, which had little to do with the research of the parent agency. In 1907, the Reclamation Service became a separate bureau in the Interior Department, with Newell as director.[64] The plan of the Service more closely resembled a construction company than a scientific organization, research playing a minimum role. If the Geological Survey and its offspring were considered together, the spectrum from fundamental to applied research would be fairly continuous, but the Reclamation Service alone concentrated almost exclusively on engineering.[65]

Gifford Pinchot and Theodore Roosevelt

Meanwhile, Pinchot finally got control of the forest reserves in 1905, after a long and complicated campaign against the General Land Office. Working through outside groups, he engineered an American Forest Congress which crystallized the sentiment of lumber and business interests as well as professional and amateur foresters in favor of the transfer. Pinchot's objectives were in perfect accord with the concepts of the new scientific bureau. He lined up his outside help. He got regulatory functions with the power to make arrests for violations. He even arranged to retain income from the forests in a special fund that he could use for the Bureau of Forestry for five years without recourse to Congress for an appropriation. Although he lost this feature later, he was here trying to get the flexibility for shifting operations that Powell had had in his earlier years with the Geological Survey. He began calling the reserves National Forests and got the name of the bureau changed to the Forest Service, which assisted him in building up the *esprit de corps* of a professional career service.[66] Uniforms gave the rangers a status distinguished from ordinary civil servants. Pinchot instilled not only the determination to practice scientific forestry but also the doctrine that "all land is to be devoted to its most productive use for the permanent good of the whole people, and not for the temporary benefit of individuals or companies." [67] This ideal, which was social rather than scientific, was the guide for the new service.

With the Forest Service fairly established in the Department of Agriculture and reclamation under way in the Department of the Interior, conservation had become a big business in the government and an integral part of Roosevelt's program. The Forest Service appropriations jumped to $1,195,218 in 1906 and $3,572,922 in 1908.[68] Pinchot, with the responsibility of formulating policies, was closer to the seats of power in the executive department than any scientist before him, even Powell and Bache. Although technically only a bureau chief, Pinchot's personal relations with Roosevelt made him somewhat like a cabinet member for conservation.

One example of the extension of conservation policy was the reviving of the Geological Survey's function of "the classification of the public lands" which, although mentioned in the organic act of 1879, had not been practiced since the irrigation survey. In 1906, without any additional legislation, the Geological Survey began to segregate potential coal beds and later added water-power sites and lands bearing oil, phosphate, potash, and various other minerals.[69] On the basis of this research a reservation policy similar to that for the forests developed.

Power and success led to a further broadening of the program. Pinchot fancied he started the world conservation movement while riding in Rock Creek Park one afternoon in 1907 by conceiving all the problems of natural resources as one connected and interrelated whole.[70] Prominent among those who joined him in making the Conservation Movement the religion for a crusade was W J McGee, Powell's longtime associate in the Bureau of American Ethnology, who had left the government service in 1903 to work on the Saint Louis exposition and later at the Saint Louis Museum.[71] Pinchot called him the "scientific brains of the Conservation movement all through its early stages." [72]

These zealots, who generated the ideas to which Roosevelt lent his name and prestige, began to talk of treating the great river systems, especially the Mississippi, as single units for which plans would embrace purity of water, electric power, flood control, and reclamation of lands by irrigation and drainage.[73] Using a favorite device, Roosevelt appointed an Inland Waterways Commission made up partly of experts — Pinchot, Newell, and the chief of Army Engineers — and partly of members of Congress. Although it worked without appropriation and without a staff, McGee was given a minor job in the Department of Agriculture so that he could become a member of the commission and its secretary.[74] The report, strongly influenced by conservation principles, recommended a multipurpose approach to river development which strikingly anticipated the Tennessee Valley Authority of a quarter of a century later.[75] They even included a specimen plan for the Ohio River system, worked out by the chief hydrographer of the Geological Survey.[76]

By investigating the lower courses of rivers and taking up flood control and navigation, the conservationists were invading the traditional preserve to which the Army Engineers had retreated after the reorganization of 1879. With the planning dominated by congressmen who took care of their own localities, the Engineers confined themselves to improving navigation as their sole purpose. Floods were controlled by great levees such as those on the lower Mississippi. General Alexander Mackenzie, who served on the Inland Waterways Commission as chief of Engineers, sharply dissented from the conservationists, questioning whether many of their purposes "are as clearly and necessarily associated with the subject of channel improvement and interstate commerce as is assumed in the report." [77] In the Engineers, the General Land Office now had a rival as the chief enemy of the conservation ideas.

Out of the Inland Waterways Commission grew the famous conference of governors on conservation in 1908 and also the Newlands Bill, which provided for multipurpose river control. A National Conservation Commission began an ambitious inventory of resources, again with McGee as the leading spirit.[78] Thus, in 1909, the scientists concerned with conservation reached a peak of power to make and influence basic national policy. They were now the shapers as well as the agents of executive decisions, and the very machinery of the expert commission which Roosevelt so often appointed worked on the assumption that science could furnish answers superior to those arrived at by political negotiation. On March 2, 1909, two days before he left the White House, Roosevelt wrote to Pinchot: "I owe to you a peculiar debt of obligation for a very large part of the achievement of this Administration." [79]

The Legacy of the Conservationists of the Progressive Era

After Roosevelt went off to Africa to hunt large vertebrates for the Smithsonian, "probably the leading scientific institution in the world," [80] the conservationists never had quite such enthusiastic support from the chief executive. William Howard Taft as secretary of war had sympathized with the Army Engineers. The Newlands Bill never became law. The National Conservation Commission eventually expired. Pinchot's dismissal in 1910 after his controversy with Taft over Secretary of the Interior Richard Ballinger showed equally that conservation had become the very stuff of politics and that its former high position under Roosevelt had not lasted. Now sufficiently important as an issue to have a part in splitting the Republican party in 1912, conservation nevertheless ceased to endow the bureau chiefs concerned with it with extraordinary powers. Barred from direct access to the President, the conservationists turned to appeals to the public.

Implicit in the conservation movement even during Roosevelt's administration was a split between those who urged managed use of natural resources and the lovers of the outdoors who wished to preserve nature unspoiled. From time to time, ever since the Hayden survey had managed to get Yellowstone Park reserved, the aesthetic conservationists had induced Congress to set aside areas of the public domain of unusual interest. The proposal to build a dam in the Hetch Hetchy Valley, within Yosemite Park, to provide San Francisco with water, became the test that brought on a bitter struggle between the aesthetic conservationists and the practical conservationists led by Pinchot. Neither Sargent nor John Muir, the patron saint of Yosemite Park and the Sierra Club, received a bid to the governors' conference in 1908, a snub directly traceable to Pinchot.[81] The Hetch Hetchy dam was finally built, serving notice on a future generation that the National Parks were not above attack. Out of the activities of the aesthetic conservationists, however, gradually came an increased appreciation of parks and monuments.[82] Since no administrative system or central organization joined these various reservations, a movement gradually took shape demanding a national park service, comparable to the Forest Service but distinct both in principle and practice. Because of the dominance of the idea of conservation for use in the Forest Service and the Department of Agriculture, the nature lovers looked to the Department of the Interior. Working closely with the Geological Survey, Stephen T. Mather, the first director of the National Park Service, created an efficient central agency in 1916. By 1920, the new service began to consider science education as a part of its role, using the parks as outdoor living museums of nature.[83]

As conservation rose to the level of high policy, it gathered such momentous political consequences that the scientific bureaus on which it was based were forced ever more rigorously to relate their activities to immediate problems. It was well enough for Fernow, who had no power, to treat forestry as a complex science, but Pinchot had to have results quickly. Throughout the bureaus a shift of the emphasis of research from basic to applied accompanied the acquisition of power. The evolution of Pinchot himself from a forester to an administrator to a politician charts the course of this development.

Besides the Forest Service and the Bureau of Reclamation, which are the most obvious examples of the shift, the experience of the Biological Survey in the Roosevelt period forcefully illustrates the pressures in the direction of applied research. Theodore Roosevelt's friend Hart Merriam had exorcised almost completely the economic aspects of wildlife study and substituted the pursuit of basic research in North American wildlife. Yet the Lacey Act of 1900 to control the shipment of game birds in interstate commerce started the Survey on the road to becoming a regulatory agency. In addition, the first national bird reservation at Pelican Island, Florida, introduced responsibility for administering small but significant tracts of public land.

In 1906, the House committee on agriculture began to ask pointed questions about the practical value of Merriam's program. Why was most of the work being done on the Pacific Coast? Was a knowledge of the distribution of skunks in North America of any use to agriculture? [84] Although President Roosevelt sympathized with Merriam and his basic science program,[85] the appropriation bill of 1907 expressed misgivings by calling for a report on whether the Biological Survey duplicated the work of other agencies and on its "practical value to the agricultural interests of the country." [86] Because of the pressure the Survey itself changed its tune. Merriam justified his studies in geographic distribution as background for practice. He sent to Hawaii for Henry W. Henshaw to come back as his administrative assistant.[87] Another of the long-time associates of Powell who shaped so many bureaus in these years, Henshaw immediately proclaimed that the "pursuit of science solely for its own sake, however commendable it may be, is not the spirit that animates our government in its support of scientific research. In its aims and ambitions this is a practical age." [88]

From this time on the major emphasis of the Survey was on two lines of practical problems, the control of undesirable or harmful wildlife, and the protection of beneficial types. Links between the Survey and other conservation agencies increased. For instance, it coöperated with the Forest Service on investigations of wolves within the reserves.[89] After a decent interval Merriam, his personal research endowed by Mrs. E. H. Harriman, turned over the bureau to Henshaw.[90]

In 1913, after a long campaign, the Biological Survey gained an important bit of regulatory power when the appropriation act placed all migratory and insectivorous birds under national authority.[91] In

1916, this power got a more stable legal sanction in a treaty with Great Britain.[92] Such increased authority was among the forces that made the Biological Survey, while still maintaining a program of basic research, shift the bulk of its activities markedly in the direction of applications.

The pressure for shifting to practical studies that permeated the conservation agencies as they gained responsibility, had it run its full course, might have exterminated basic research and limited applied research to short-run projects looking toward immediate results. But the essential interrelation between basic knowledge and the desired practical gains made it inevitable that the shift would produce a countermovement. As the programs became more concerned with regulation and administration, the lack of a steady flow of fundamental information always began to attract attention.

Pinchot's scorn for Fernow drove government forestry out of the study and into the woods, only to find that effectiveness would not increase until a great deal more was known of several sciences and of the twilight zones between them.[93] Research was needed along two main lines, production and utilization. The latter, which dealt with new methods of handling and using forest products, was the easier to grasp and the first to revive. By 1907, sections devoted to timber tests and wood chemistry were on a permanent basis in the Service. In 1910, these activities became the Forest Products Laboratory in Madison, Wisconsin, which, in coöperation with the university there, grew into an outstanding research organization.[94]

Research in forest production came more slowly, partly because the time schedule for tree crops is so long. In 1908, an experiment station on Coconino plateau in Arizona began experiments that must run over periods of many years before results are possible. At this time, Pinchot decentralized the administration of the Service and research became the responsibility of each of the various regions.[95] After the Arizona station other regions followed suit, establishing a geographic pattern analogous to the agricultural experiment stations, but not financed by the states or rigidly confined by their boundaries.

The countermovement against the shift of forestry research toward practicality did not carry the full way to a reëmphasis on basic information only. It rather preserved and furthered a broad program of applied research. A forest is such a complex unit and the forester can apply controls to it so indirectly that much of his practice must continue to come from empirical data rather than from basic science. For instance, improvement of trees by the intensive use of the science of genetics as in agriculture is out of the question.[96] Pinchot considered all research conclusions as compromises between what was best for the forests and what was practical in logging.[97] Yet through its research the Forest Service became and remained an important scientific agency.

Conservation, which had its roots in the changed attitude toward resources once considered inexhaustible, added a whole new dimension to the government's scientific establishment. Only vaguely foreshadowed in 1865, the new bureaus concerned with conservation were both well developed and stable by 1916. Because of the tremendous responsibility of the public domain, their action had more profound and more obvious political, social, and economic repercussions than government science had had before. They had become more involved in the great issues of domestic politics. They had gathered more power for both regulation and administration. Their direct participation in the business of government meant that the ideals which animated and sometimes divided them often had political and economic as well as scientific roots. But they remained for the most part research institutions, representing the most ambitious attempt yet made to apply science on a general scale to a fundamental national problem.

XIII

MEDICINE AND PUBLIC HEALTH

1865–1916

ALTHOUGH the application of science to human beings made unparalleled strides in the period from 1865 to 1916, the government was more reluctant to impose the results of research on the citizens themselves than on their environment or their domestic plants and animals. Medicine was the preserve of a private profession, and public health depended on the police powers reserved to the states. Only the federal government's special wards led it to take an interest in health. The Army, the Navy, and merchant seamen were the major groups under federal medical care in 1865.

The Army Medical Library and Museum

On the surface, the Army Medical Corps suffered the same decline after the war as the rest of the service. Both appointments and promotions were frozen in 1869. Surgeon General J. K. Barnes stayed on until 1883, and his two successors had served in the Mexican War. Without even enough doctors to take care of routine, the corps used many contract surgeons.[1] Under such circumstances science could expect neither money nor equipment nor high-level policy support. Yet medical science in the Army did not share in the decline that beset general research in all other branches of the military. Although larger forces may have produced this paradox, the role of a very few outstanding individuals seems preponderant. Armed with two institutions which the chance of the Civil War happened to leave behind — the Army Medical Museum and the Surgeon General's Library [2] — John Shaw Billings and a few others not only kept alive the ideals of fundamental science but managed to make significant contributions. Billings was preëminent as organizer and administrator, but J. J. Woodward, George M. Sternberg, Walter Reed, and others added considerably to an atmosphere of broad learning and exciting discovery.

In 1868 the windfall of $80,000 left over from the war fell into Assistant Surgeon Billings's hands to use for the Surgeon General's Library, which to that point had been little different from dozens of other collections of books gathered by various bureaus in Washington for their own immediate use and convenience. Billings's plans included building a general medical library equal to the national collections in other countries and the preparation of a catalogue which would not merely identify his holdings but serve as a general bibliography. In 1880 he published at government expense the volume of the *Index Catalogue* for the letter "A." With the alphabet complete in 1895, a second series began immediately. To provide interim coverage, he established the periodical *Index Medicus* without government support. The library gradually became a national institution in the fullest sense and the peer of any in the world, and the *Index Catalogue* was unique in both scope and usefulness.[3]

Part of Billings's success with the library came from his standing in the medical profession as a whole. A leader in America, he also had a substantial European reputation. When the bequest of Johns Hopkins gave the chance for a new departure in medical education in the United States, Billings's plans became the basis for the design of the new hospital. His reports as adviser to the trustees on hospital construction and their relation to the training of doctors and nurses had an important effect on the institution that became the pattern for medical education in the twentieth century.[4]

The Army Medical Museum, even more than the library a child of the Civil War, had an equally illustrious career in the three decades following 1865. Dominated in the earlier years by the preparation of the *Medical and Surgical History of the War of the Rebellion*, the museum also became a center for experimentation in methods and equipment for photomicrography. J. J. Woodward began this work, with George M. Sternberg later applying the techniques to microorganisms. Billings added the museum to his duties in 1883, pushing his favorite policy of creating national medical institutions with close civilian ties and going beyond immediate military interests.[5]

The Marine Hospital Service

Meanwhile, the medical service for merchant seamen, theoretically paid for by their contributions from wages and administered by the Treasury Department, had reached such a low level of efficiency that some reform was imperative. With no central direction or uniform standards, the numerous hospitals did not even provide adequate care for their charges, much less advance knowledge in any way. In 1870, the secretary of the treasury called in Billings of the Army to aid Dr. W. D. Stewart, inspector of marine hospitals, in preparing a plan for reform.[6] Describing each port as "a law unto itself," they recommended a larger seaman's fee and an increase of expert supervision. Acting on this report the Congress passed a law which established the post of supervising surgeon. The first version of the bill had envisaged a military appointee such as Billings, but the Senate amended it to provide the appointment of a civilian.[7]

Since one of the great needs was competent personnel in a civil service that as yet had no merit system, Billings had recommended a commissioned corps of doctors along the lines of the Army. The first supervising surgeon of the newly-organized Marine Hospital Service, Dr. John M. Woodworth, who had served in the Civil War, built up a commissioned corps, recruiting by professional examination from recent medical graduates and promoting only through the ranks. Although a military system for hospital stewards gave way in the 1880's, the commissioned corps resisted all outside attacks and received specific legal recognition in 1889.[8]

The Marine Hospital Service, better able to meet its responsibilities to merchant seamen after the reorganization of 1870, still had no authority to undertake public health measures in the interest of the entire population. However, its personnel, stationed in the larger ports, found themselves strategically placed to observe yellow fever epidemics, and as early as 1874 they had orders to inspect local quarantine systems and to study local laws. Since Woodworth was an active advocate of centralized quarantine authority, the service gradually prepared itself to take on at least supplementary duties in times of crisis.[9]

The National Board of Health

The 1870's were years of ferment for public health movements. Massachusetts established the first modern state board of health in 1869. The American Public Health Association, organized in 1872, became a forum for opinion and the nucleus of a profession. A proposal at its first meeting for a national bureau of health led to a bill vainly introduced in the Senate. The American Medical Association also received a report on a national department of health in 1875 from a committee under Dr. Henry Ingersoll Bowditch of Massachusetts. While cautioning against premature action, the report outlined a plan for a central board of information and investigation modeled on the ideas of Jeremy Bentham. A secretary of health would have the surgeon generals of the Army, Navy, and Marine Hospital Service as undersecretaries. Representatives from each state would sit on a council. Although nothing came of its comprehensive scheme, the American Medical Association generally lent support to the American Public Health Association. Billings and J. J. Woodward of the Army tended to take the lead among these groups and to oppose both the simple reliance on quarantine and the expansion of the Marine Hospital Service into a general health agency.[10]

The yellow fever epidemic of 1878 gave both urgency and opportunity to the competing health groups. State and local officers, especially in the South, demanded drastic action. Although a quarantine act of that year gave some limited powers to the Marine Hospital Service alone, the swiftness of the crisis and the death of Surgeon General Woodworth prevented the full exploitation of this new authority. At the same time the meeting of the American Public Health Association in Richmond, at which Billings was influential, voted down a national quarantine and appointed a committee which proposed federal aid to state boards of health and a provisional national health commission appointed by the National Academy. Clearly these sanitarians felt that not enough was known about the diseases to be controlled to make quarantine effective and that the only hope was research. Their provisional commission would investigate the causes of yellow fever and also draw up a plan for a permanent national public health organization.[11] The reliance on the National Academy reflected its new role in the Geological Survey crisis which was unfolding at the same time.

Although the bills directly embodying the American Public Health Association's proposals lost out in Congress, the rough-and-tumble of the last day of the session gave their friends a chance to put through a hastily drawn measure. March 3, 1879, the same day Abram Hewitt was jamming through the consolidation of the surveys, saw the creation of a National Board of Health.[12] The National Academy idea having drawn too much opposition, the President, with the advice and consent of the Senate, was to appoint seven members, to be joined by officers detailed from the Army, Navy, the Marine Hospital Service, and the Department of Justice. A temporary group, it was supposedly to "obtain information on all matters of public health," and also, with the help of the National Academy, report "a plan for a national public health organization." An appropriation of $500,000 was to be used as grants-in-aid to state boards of health. The whole matter went through so haphazardly, however, that the last provision was omitted in the version sent from the House to the Senate, and as finally passed the only money involved was $50,000 for salaries and expenses.[13]

The board appointed by Rutherford B. Hayes was in effect chosen by the American Public Health Association, whose president, James L. Cabell, became chairman of the new body. The members were widely representative of the best public health authorities developing in the states, for instance Bowditch of Massachusetts. The Army designated Billings, who became vice-chairman and the leading member. As a group it was admirably suited to direct a research program, which was its genuine intention.[14] It immediately sent a yellow fever commission to Cuba, George M. Sternberg being one of the members. Billings took personal charge of a revision of standard nomenclature for disease, corresponding with the Royal College of Physicians on the subject. He also organized a vital-statistics program, in coöperation with the Tenth Census of 1880, by defining the federal registration area of mortality. Instead of building up a staff, the board gave grants to individual scientists, usually in the universities. Ira Remsen of Johns Hopkins investigated organic matter in the air. Some studied disinfectants. Others investigated sewers, soils, diseases of food-producing animals, adulteration of food and drugs, and the board directly carried on sanitary surveys around New York and Memphis, Tennessee.[15]

Of course the yellow fever commission sent to Cuba did not hit upon the factors that would lead to the control of the disease. (How much harder to find the keys in 1879, before Laveran, Theobald Smith, Sir Ronald Ross, and many who furnished negative data of importance.) Yet clearly ten years of such research would have markedly hastened the rate of advance in several promising lines and altered the timetable on which results were applied to benefit the health of the nation and the world. Although condemned in Congress as extravagant, the actual cost of research — $30,000 in all — was small compared to that of the quarantine, and much of its value came from the donated services of the investigators. Indeed, the board boasted that it "was able to accomplish so much at so trifling an expenditure of the public funds by reason of the fact that nearly all the eminent scientists . . . did not receive anything for their personal services, the only charge being for the pay of assistants and for other necessary working expenses." [16]

In the face of epidemic yellow fever, the Southern seacoast states, disappointed in a measure that held no hope for immediate results, demanded a quarantine act in the special session of the new Congress in the spring of 1879. Ironically, the National Board of Health consisted of the very men who had recommended against such action in the report of the committee of the American Public Health Association. They now again suggested federal grants-in-aid to state and local quarantine authorities. But eight years before the Hatch Act in agriculture, this principle seemed even stranger and more radical constitutionally than a national quarantine. The unwilling board thus either had to accept the power it had advised against or see some other agency overshadow it. In the act of June 2, 1879, Congress repealed the quarantine act of 1878 for a period of four years, gave the board authority to erect temporary stations along the coast, and appropriated $500,000 for the first year's expenses. The Marine Hospital Service was completely shut out from a function it coveted and toward which it had worked. Although the board put its faith in research, its destiny was to depend on its ability to administer a quarantine that could not be based on an adequate concept of the disease to be controlled. [17]

Once the responsibility was thrust upon it, the vigor with which the National Board attacked quarantine led it into a maze of difficulties with state and local authorities, whose powers were in no way diminished by the addition of federal inspection. Shipowners felt they had just one more barrier to clear. With the appearance of yellow fever in Memphis, the board found itself imposing interstate quarantine on the Mississippi, arousing the implacable opposition of the Louisiana authorities. By supplying funds only to actual yellow fever areas in the South and denying a quota to "every twopenny township," the board engendered further opposition. [18]

Against these forces a board of doctors attached to no department, with a term of four years on their organic act, had no effective defense. Their only recourse being the President, the board found itself completely blocked in 1882 when Chester A. Arthur transferred its appropriation to the Marine Hospital Service — cruel and contemptuous treatment, according to Dr. Bowditch.[19] Congress refused to renew the quarantine power in 1883, allowing it to revert to the Marine Hospital Service under the Act of 1878. When Billings resigned, the board was really dead, although it had a shadowy existence in law until 1893.[20]

The principals in this most egregious failure in the history of government science had their explanations for the wreck. Billings felt that the board's creation was premature. "Forced into existence in an emergency, it was only to be expected that as soon as this emergency had passed it would find itself without the support of an educated public opinion."[21] Dr. Cabell blamed four forces — the Louisiana State Board, the chief of the Marine Hospital Service, the great commercial communities whose business was delayed, and the politicians of both great parties who found no patronage in the appointment of scientists.[22]

As the exact contemporary of the Geological Survey, the National Board of Health may be fairly compared to the ideal new scientific bureau. It had certain of the characteristics that enabled Powell to succeed where it failed. An affirmative attitude toward research, a group outside the government solicitous of its welfare, imaginative leadership, and a thorough commitment to the problem approach all told in its favor. Yet, fatal flaws are also evident. Billings, the potential Powell of the outfit, was a military officer who never got rid of his usual duties, much less his obligations to the Army and the War Department. Neither he nor anyone else could work for the kind of political alignment that Powell built so carefully. Further, regulatory power attracted opposition before a research program could prove itself, as if Powell had had to fight his losing irrigation survey battle at the same time he built his research organization. A stable and dedicated personnel was lacking, up to and including a full-time permanent chief. The organic act was both clumsy and fickle in that it gave no assurance of permanence.

Even by pioneering the device of grants-in-aid to states and to individual investigators the board got into trouble. Both these techniques had great possibilities, but without a strong central organization to set standards the state grant ran the danger of logrolling while those to individuals were limited by the necessity of the principal investigator's donating his services. The National Academy had the same rule, which barred members from any but occasional advisory services. Baird's free services to the Fish Commission had in the end to be reimbursed, and the practice was abandoned. Powell, more realistic, had paid O. C. Marsh a salary. The National Board of Health was expecting a great deal in trying to found a sustained research program entirely on the gifts of time of the leading scientists.

Nevertheless, the impression persists that, with a little better luck, a little better administrative framework, and a little more time, the National Board of Health might have survived and, applying its wealth of talent to its own improvement, might have emerged as a permanent and comprehensive public health bureau for the government. As much was being accomplished in Germany at about the same time. The profound effects of the lack of such an emergence are patent in the latest attempt to revise and coördinate the public health establishment.[23] While efforts to revive a central body for public health within the government became endemic, the demise of the National Board ensured that programs impinging on health springing up in various agencies must develop in virtual isolation from one another.

The Army Medical Corps's Period of Great Accomplishment

Although quarantine went back to the Marine Hospital Service, any impulse to research that survived at all remained in the Army. The yellow fever problem, which had touched off the crisis of 1878 and 1879, fell to the small but brilliant group of officers in which Sternberg now became the leader. Billings's influence remained great, his library and museum providing shelter for those who felt the excitement of the tremendous new developments in bacteriology. But he was personally an old-school sanitarian, and most of the remaining chapters in his career, brilliant ones to be sure, lay outside the Army and the government. Sternberg's service on the National Board's yellow fever commission marked the end of more than a decade of frontier duty and the beginning of a period of research in bacteriology in which he matched paces with Koch and Pasteur. His work covered most of the disease bacteria then coming to light and a great many others to which overenthusiastic scientists attributed major maladies, especially yellow fever.[24]

Walter Reed was another young officer who surmounted eleven years of frontier isolation to enter the exciting new world of bacteriology. Assigned to recruiting duty in Baltimore in 1890, he studied at Johns Hopkins in the great early days under William H. Welch. In 1893, he went to Washington as curator of the Army Medical Museum, becoming a champion of diphtheria antitoxin and of government control of its preparation.[25]

A new era in Army medicine began in 1893 when Sternberg became surgeon general. By means of an administrative order from the secretary of war he set up the Army Medical School in Washington to give advanced and specialized training in military medicine and public health. Walter Reed became professor of bacteriology. The research tradition of the museum and library and the special emphasis on bacteriology now had a means of perpetuating itself.[26]

The Spanish-American War, although only a summer excursion as a military and naval campaign, thrust the United States, so long preoccupied with the affairs inside its own borders, into new regions and a position of prominence among the great powers. Of all branches of the government, none received a more severe jolt from this new venture than the Army Medical Corps. In the face of the fact that only 541 men died in battle while 3500 fell to disease, the protest was vain that the regular doctors, few in numbers to begin with, had no control over the health of the volunteers. Typhoid in the camps took most lives, while yellow fever and malaria forced a precipitous withdrawal of forces from Cuba. Even in glorious victory, the whole country realized that the Army had shown up poorly in the face of disease. The commission investigating the conduct of the war recommended sweeping reforms, nearly all of which received the approval of the surgeon general and were gradually introduced.[27]

The very calamities, however, that shamed the Army Medical Corps as an operating service in the field provided unparalleled opportunities for those few who had devoted themselves to research. Sternberg appointed a board on typhoid headed by Walter Reed. The data collected by these officers were the first of a series of steps which by 1917, combined with European research, had almost removed the disease as a military hazard.[28]

After this initial success Sternberg turned to the scourge that had so long held the attention of American investigators.[29] The results of the Yellow Fever Board in Cuba in 1900 are as famous as any episode in the annals of government science. The death of Dr. J. W. Lazear and the deliberate risks run by those who submitted to experimental mosquito bites made folk heroes. Reed's own untimely death shortly afterward gave him a legendary aura in the eyes of the American people. For a time these men appeared as the ultimate and eternal conquerors of yellow jack, which could within a generation be exterminated from the earth.

Their discovery was neither original nor final. Carlos Finlay had believed in the mosquito theory for a long time, and subsequent events have made the expectation of complete extinction seem hopelessly naïve. The spread of sylvan yellow fever from the forests of South America has emphasized that Reed and his brave colleagues scarcely began to understand the disease. They did not isolate the virus or develop a vaccine. Rather they gathered data on which they reconstructed the life cycle of the disease with just enough accuracy to allow them to interrupt it as the virus passed through the mosquito now called *Aëdes aegypti*. Their belief that this domestic species alone was the carrier, oversimple in view of recent discoveries, gave them a possible means of control by the administrative measures that Colonel W. C. Gorgas promptly adopted.

Considered from the point of view of developing government science, the yellow fever story comes into a new perspective. Sternberg had worked on yellow fever for years, clearing away a number of bacteria that had been supposed to cause it. Walter Reed had studied its pathology under Welch at Johns Hopkins, coalescing with the Army tradition the best of the new research spirit then infusing private medicine. The Army Medical School, Museum, and Library provided a headquarters. The overseas responsibilities that gave the surgeon general a chance to appoint a board provided an opportunity only for a thoroughly prepared team. After Reed reached Cuba he could and did call on other agencies for information. Henry R. Carter of the Marine Hospital Service contributed a discovery of great and timely importance. L. O. Howard of the Department of Agriculture supplied entomologic data otherwise beyond the reach of doctors of medicine.

Part of the spectacular result of the work of the Yellow Fever Commission stemmed from the military nature of the occupation of Cuba. Never before had a research unit of the federal government had the local police power to impose its findings immediately on a large population. Colonel W. C. Gorgas, chief sanitary officer in Havana, could go after mosquito breeding places by direct action. Leonard Wood, the military governor of the island, was a doctor of medicine who had shifted to the line only when he took command of the Rough Riders. He backed both the research and the control measures with a plenitude of power that would have been entirely unavailable to the President of the United States had the infected city been New Orleans instead of Havana.

The Panama sequel was almost entirely determined by the relation of scientific knowledge to authority. The method of control having been worked out in the comparatively simple military government of Cuba, Gorgas had to persuade a bewildering series of commissions to use them to save the Panama Canal from the disaster that disease brought on French enterprise there. Arriving on the Isthmus in April 1904, Gorgas brought as his quarantine officer Carter of the Marine Hospital Service. President Roosevelt refused the pleas of the American Medical Association to give Gorgas a seat on the commission that directed the American canal effort. Cut off from direct access to the chairman of the commission, Gorgas was for a time in 1905 in danger of losing out to doctors whom the second commission considered more practical. Only appeals to Roosevelt from medical authorities such as Welch saved him on the eve of his greatest accomplishments. As soon as his yellow fever program was underway, he attacked malaria. By 1907, when he was appointed a commissioner, his program was a generally acclaimed success.[30]

The Army's long-continued occupation of lands in tropical climates in the years after 1898 solidified the Medical Corps's interest and ability in public health and preventive medicine. By the outbreak of World War I the accomplishments were sufficient to overthrow disease as the primary cause of death among armies. The Army Medical Corps and companion work in the Navy made the record of 1917–1918, such a contrast with the Spanish-American War, possible.

An instance of the discoveries resulting from the Army Medical Corps's expanded horizon is Lieutenant Bailey K. Ashford's work in Puerto Rico.[31] A graduate of the Army Medical School who went in with the American troops in 1898, Ashford demonstrated the relation of widespread anemia to hookworm and over a period of several years developed a practical treatment. This important bit of applied research strongly illustrated the tendency of public health measures generated by Army research to escape into private hands as soon as their applicability to the general public had become apparent. Ashford's own Puerto Rican Anemia Commission gave way to his Institute of Tropical Medicine and Hygiene, which Columbia University took over in 1926, making Ashford a professor.[32]

On a larger scale, the control of hookworm, like that of yellow fever, had world-wide possibilities, and the attack on it, also like that on the more famous disease, soon gravitated to the new Rockefeller Foundation.[33] Charles W. Stiles was the major figure in this program. Not only was hookworm virtually eliminated as a threat in the Southern United States, but a wholesale attack on the great strongholds of the infestation in Asia went forward on a scale never approached by government action. The early twentieth century, alone of periods in the nation's history, found private agencies with the strength and will to embark on scientific programs involving whole segments of the population.

The great research and public health record of the Army Medical Corps cannot be explained in terms either of monetary support, which was small, or of the general medical service of the Corps, which was often inadequate and provenly not in control of its area of responsibility during the Spanish-American War. Although much credit is due the individual qualities of a small group — Billings, Sternberg, Reed, Ashford, and their less famous associates — their scientific brilliance alone does not account for everything. Rather, they managed with small means to develop research traditions embodied in institutions within their Corps. They were fortunate in orienting their work around problems the solutions of which were on the highroad that the science of bacteriology was then traveling. Their great accomplishment was in having prepared themselves before the opportunities of 1898.

The Evolution of the Public Health Service

Meanwhile, the Marine Hospital Service was slowly developing larger possibilities for research. As early as 1887 Dr. J. J. Kinyoun, fresh from Europe and the influence of the great bacteriologists, set up a laboratory at the Marine Hospital on Staten Island. After four years without legal sanction, Kinyoun's single-handed enterprise received mention in a congressional act and under the name of the Hygienic Laboratory moved to Washington. In the next few years

Kinyoun began programs of standardizing biologic preparations, examination of water supplies, modernization of quarantine equipment, and animal experimentation.[34]

Besides the Hygienic Laboratory, the Marine Hospital Service developed other new interests in the 1890's. Interstate control of diseases, medical inspection of immigrants, new quarantine powers, and new responsibilities brought on by the Spanish-American War, broadened the service's functions beyond the care of merchant seamen.[35] The result was mounting pressure to revise the service's scope and organization.[36] An act of 1901 showed a marked advance by appropriating $35,000 to erect buildings for "a laboratory for the investigation of infectious and contagious diseases, and matters pertaining to public health." [37] This wording gave the Hygienic Laboratory a specific research mission for communicable diseases and admitted a general health responsibility for the service. A Division of Scientific Research appeared in the organization in September 1901.

In the first year of Theodore Roosevelt, the Congress that passed the Newlands Act for the reclamation of arid lands took a favorable view of a broadened health service. Reflecting the changed attitude, the name became the Public Health and Marine Hospital Service. The Hygienic Laboratory developed regular divisions of chemistry, bacteriology, pathology, zoology, and pharmacology.[38] While depending largely on the commissioned service for research personnel, the surgeon general could if necessary appoint civilian scientists as heads of the divisions of chemistry, zoology, and pharmacology.[39]

As the Public Health and Marine Hospital Service moved toward a broader program, it had to reckon both with the lack of coördination among governmental agencies interested in health and with the increasing dominance of the foundations and universities. The 1902 law recognized these forces by providing for the Hygienic Laboratory an advisory board made up of the director, one member each from the Army, the Navy, and the Bureau of Animal Industry, and five others "skilled in laboratory work in its relation to public health, and not in the regular employment of the Government." [40] The board included such leaders of the profession as Welch, Simon Flexner, and W. T. Sedgwick, men who had much to do with the development of public health through private institutions.[41] A beginning of liaison with state authorities was also included in the form of an annual conference. A small but important regulatory function also came to the

Hygienic Laboratory in 1902 with the passage of the biologics-control act establishing tests and licensing for manufacturers of "viruses, serums, toxins, antitoxins" and the like.[42]

The progressives who came to see conservation as a unified scientific management of the national wealth also began to look upon health as a resource and a part of the general welfare in which scientific research could produce dramatic results. Some of the social legislation that produced the bitterest controversy and the most far-reaching court decisions was enacted in the interest of health. In 1906 Professor J. P. Norton of Yale read a paper to the AAAS suggesting a national department of health. Significantly, Norton was an economist rather than a medical man, his appeal stressing losses from preventable illness.[43] The resulting Committee of One Hundred on National Health of the AAAS began a crusade for the conservation of human resources which paralleled and coöperated with those whose main concern was environment. Irving Fisher, another Yale economist, became the chairman of the committee and the author of its manifesto, *A Report on National Vitality: Its Wastes and Conservation*.[44] Among other themes, he called on the federal government to "remove the reproach that more pains are now being taken to protect the health of farm animals than of human beings," and he urged the building of "more and greater laboratories for research in preventive medicine and public hygiene." [45]

The Committee of One Hundred had no trouble rounding up distinguished endorsements from both major parties, state governors, the Grange, and labor organizations, as well as boards of health.[46] Theodore Roosevelt favored the grouping of all health services into one bureau but did not wish to increase the size of the cabinet by adding a new department.[47] President Taft recommended an independent establishment.[48] In spite of these minor organizational differences, this peculiarly progressive scheme seemed to its advocates well on the way to triumphant accomplishment.

Throughout Taft's administration a general expectation of some health reorganization in the government persisted. But the task began to seem more and more difficult. In the first place, the minor differences over organization tended to grow, producing the same kind of split among friends that had so hampered the National Board of Health. The Committee of One Hundred continued to support a cabinet department, opposing all efforts to use the Public Health and

Marine Hospital Service as a vehicle for expansion.[49] In the second place, public health was now sufficiently important to draw the outright hostility of small but determined groups whose power of opposition medical men and health officers tended to underestimate. The National League for Medical Freedom, representing various sects, the Anti-Vivisection League, and the Christian Science Church appeared at hearings to protest all forms of federal health reorganization.[50]

By 1912 the progressives had clearly failed to get their plan through Congress. Their long campaign, however, resulted in a law which, shortening the name of the Public Health and Marine Hospital Service to the Public Health Service, empowered the organization to "study and investigate the diseases of men and conditions influencing the propagation and spread thereof, including sanitation and sewage and the pollution . . . of the navigable streams and lakes." [51] While implicitly emphasizing communicable diseases and evading entirely the issue of reorganization, this law had the great importance of opening the whole field of public health to research by the government, thus recognizing it as a legitimate sphere of federal activity.

The Public Health Service began to get large dividends on its research program when Dr. Joseph Goldberger went to work on pellagra, which had been first recognized in the United States in 1906. So great was the reputation of bacteriology that the early efforts of the Hygienic Laboratory involved the assumption that this disease was infectious. In 1914 Goldberger became head of a team of forty-one investigators who attacked every aspect of pellagra and soon concluded that the cause was dietary deficiency. The way was open not only to controlling this one disease but to attacking a wide range of problems connected with diet rather than bacteria.[52]

Compared with other scientific agencies, the research money available to the Public Health Service was small. While total expenditures had long been sizable, ranging from $414,000 in 1875 to $2,785,000 in 1915, only a small fraction was available for research.[53] Both agriculture and the great conservation agencies completely overshadowed it. Yet the legal and structural accomplishments of the progressive era laid the foundation for the dominance of government research in public health in the middle of the twentieth century.

XIV

THE COMPLETION OF THE
FEDERAL SCIENTIFIC ESTABLISHMENT

1900–1916

BY the early twentieth century large-scale industry, the dominant force in the United States, was moving toward a closer relation
with science. The first industrial research laboratories had just begun
to make headway, and the great captains of industry had begun
to favor science in their philanthropies. Unlike agriculture, whose
friends had already directly called on the government for a research
establishment, large-scale industry felt no such need. Agitation for a
Department of Industry to match the Department of Agriculture
met with no enthusiasm. However, the swelling scientific needs of
industry did not entirely ignore the government, their pressure
pumping new life into ancient functions.

The four agencies most dramatically aroused by the stimulus of
industry were the National Bureau of Standards, the Bureau of the
Census, the Bureau of Mines, and the National Advisory Committee
for Aeronautics. The first two had a history as old as the republic.
The government's concern with minerals went back at least to the
middle of the nineteenth century. The godmother of research in
aeronautics was the Smithsonian Institution. In spite of their ancient
roots, these agencies all began new careers close to industrial technology soon after 1900.

The National Bureau of Standards

The government's responsibilities for maintaining standards of
weights and measures, enshrined in the Constitution itself, had rested
in the Coast Survey since the time of Hassler. Charles Sanders Peirce,
head of the office of weights and measures, though better known to

posterity as a philosopher than as a civil servant, testified before the Allison Commission that the "office of weights and measures at present is a very slight affair, I am sorry to say." [1] He emphasized that he had authority only to make standards for the states and for agricultural schools, lacking power to issue certificates of verification. Often his office had no means to verify instruments referred to them, an inadequacy that led to a demand for government action. Instruments needing careful calibration had to be sent abroad, usually to Germany.

Meanwhile, outside the Coast Survey, various agencies had made piecemeal responses to demands for standards. The Treasury Department, sometimes with the aid of the National Academy, had long investigated meters for measuring alcoholic liquors. The Navy had adopted a standard gauge for bolts, nuts, and screw threads. The Army had developed a machine for testing iron and steel which, set up at the Watertown Arsenal, was available "for all persons who may desire to use it, upon the payment of a suitable fee for each test." [2] The Department of Agriculture was setting standards of purity for foods and drugs. Although no coördination existed between these activities, they early showed that, as technology became more complex, other standards than simple weight and measure were called for both inside and outside the government.

Perhaps the most dramatic development in the demand for standards accompanied the rise of the electrical industry. Congress appropriated $7500 for a national conference of electricians in Philadelphia in 1884. Among the recommendations framed there was a request for "a bureau charged with the duty of examining and verifying instruments for electrical and other physical measurements." Although nothing came of the 1884 effort, an international conference in Chicago at the time of the Columbian Exposition in 1893 adopted standard units of electrical measure, among them the henry. Since no bureau existed to work out the details, the act defining these units for the United States called upon the National Academy. This was not, however, a workable permanent solution, and by 1895 preliminary steps were under way to set up electrical standards in the Office of Weights and Measures of the Coast and Geodetic Survey. [3] With the addition of electricity, the rationale of combining surveying with standard measures — a union that went all the way back to Hassler — broke down completely. The Coast and Geodetic Survey had the resources in neither money nor trained personnel to go off on this new line.

Henry S. Pritchett, when he became superintendent of the Coast and Geodetic Survey in 1897, found weights and measures under the charge of a field officer of the survey, who had two scientific assistants, an instrument maker, and a messenger. Believing that a physicist was needed in the job, Pritchett got $3000 from Congress, with which he hired assistant professor S. W. Stratton away from the University of Chicago.[4] The first duty of the new director of the Office of Weights and Measures was to recommend a plan for enlargement.

Some of the impetus toward reorganization came from the sting of foreign superiority. Germany, so much admired in this period for scientific research, led in the field with the Normal-Aitchungs-Commission in 1868, and the later Physikalische-Technische Reichsanstalt provided the equivalent of a national physical laboratory, to which many American industries had to apply for service. The Kew Observatory in Great Britain began a parallel evolution in 1871 which produced the National Physical Laboratory in 1899.[5] Pritchett had the Reichsanstalt in mind when he asked Stratton to draw up a plan.[6]

The campaign for a new bureau showed careful planning. Stratton consulted first others in the Coast and Geodetic Survey, and then a number of private physicists, chemists, and manufacturers. The result was not "a copy of the Reichsanstalt, but a standardizing bureau adapted to American science and to American manufacture." [7] Secretary of the Treasury Lyman J. Gage gave his approval before the draft of the bill went to Congress. By clearing with established bureaus Stratton insured himself from opposition within the executive. These consultations, adjusting the bill in advance to the pressures from interested parties, rendered appropriation-rider stealth unnecessary. The Congress wrote an organic act out in the open. After presentation, the bill was endorsed by the National Academy, the AAAS, the American Physical Society, the American Chemical Society, the American Institute of Electrical Engineers, and other organizations, making it a real scientists' proposal. When passed on March 3, 1901, the act created a National Bureau of Standards which besides full powers over custody, preparation, and testing of standards had as its responsibilities "the solution of problems which arise in connection with standards; the determination of physical constants and the properties of materials, when such data are of great importance to scientific or manufacturing interests and are not to be obtained of sufficient accuracy elsewhere." [8] The problem approach — the limitation of the new bureau to questions relating to standards — was here preserved, but even in the original wording the possibility of a coverage as broad as the physical sciences themselves is inherent.

Besides service to the federal, state, and municipal governments, the new bureau was to provide, for a fee, standards "for any scientific society, educational institution, firm, corporation, or individual pursuits" requiring their use. The act also created a visiting committee similar to that later provided for the Hygienic Laboratory of the Marine Hospital and Public Health Service. Five members, "men prominent in the various interests involved, and not in the employ of the Government," were to report on the efficiency of the scientific work.[9] This device, plus the smooth and straight-forward course of the legislation through Congress, was a sign of a reviving ability of Congress as a whole to grapple adequately with the problems of government science and to provide administrative forms adapted to its needs.

Since Stratton was appointed by McKinley as the first director, preparations for the shift began before the law went into effect. The new bureau remained under the Treasury Department until 1903, when it became a part of the new Department of Commerce and Labor. Besides money for buildings, appropriations rose by 1905 to a level slightly under $200,000.[10] Several divisions appeared almost immediately: electricity; weights and measures; thermometry, pyrometry, and heat measurement; optics; chemistry; and engineering, instruments, and materials.[11] Ten years later the Division of Metallurgy and the Division of Structural, Engineering, and Miscellaneous Materials had been added, while appropriations advanced to $696,000.[12] The problem approach was maintained since the broad aim of determining standards prevailed in all the sciences represented.

In this apparently smooth evolution two countertendencies appeared. On one hand, the work required to develop testing instruments, to set standards accurately, and to determine physical constants led directly toward basic physical and chemical research. On the other hand, the demands of other government agencies for acceptance standards of the supplies they bought and the requests of industry for processes of all sorts led directly to applied research of very practical and sometimes limited scope. For instance, to standardize government purchases of electric lights, the bureau investigated the relative properties of plain and frosted bulbs, and, after conferences with manufacturers, adopted specifications for purchase. In 1905 the chief of the electrical division described the bureau as "the American National Physical Laboratory, using the word physical in a liberal sense, as its work includes both chemistry and engineering." Its function was "to contribute something to the advancement of human knowledge and to serve the public." [13] In general, the urge to undertake basic research — to become a national physical laboratory — more than held its own, even though the secretary of commerce and labor deleted the word "National" from the title of the bureau in 1903. The original name was restored in 1934.

Since the need for standards research had been felt in many parts of the government for a long time, the bureau naturally rubbed shoulders with others when it broadened out beyond the traditional concern with weights and measures. The Bureau of Chemistry in the Department of Agriculture especially reacted, both because Wiley was trying to do chemistry for other agencies through his contracts laboratory and because pure-food-and-drug problems usually involved standards of purity and the like. Besides, the Bureau of Standards began to experiment with the properties of sugar. The father of the food-and-drugs act and of the beet-sugar industry fulminated that the Bureau of Standards "has broken deeply into the activities already started by the Bureau of Chemistry and some of the other Bureaus of the Department of Agriculture, violating the fundamental principle of ethical standards." [14] However, Wiley failed to show that the Bureau of Standards had any serious intention of taking over enforcement of the pure-food-and-drug laws or any extensive program in agricultural chemistry.

A somewhat parallel friction arose with the Geological Survey, which had developed several separate programs of investigating structural materials for the use of the government. Cement, lime, steel, and ceramics had been especially stressed. In 1910, when the Bureau of Mines was created, these activities were at first transferred to it, but almost immediately a repeal went through which gave the Bureau of Standards a $50,000 appropriation and both the equipment and personnel that had been under the Geological Survey. [15]

In contrast to the electrical industry, for instance, the building trades were the most completely decentralized and atomized business group with which the bureau had to deal. Private contractors were seldom in a position to conduct research on either materials or methods.[16] Hence the bureau found itself doing applied research for the building business in the same way the Department of Agriculture helped the farmer. Requests from industry and the responsibility for saving the government money in its own building operations forced the expansion of applied research on a very practical level, in contrast to the basic problems under attack elsewhere in the Bureau.

The Bureau did not always successfully invade standards activities in other departments. For instance, it did not try to get the determination of time away from the Naval Observatory.[17] Congress continued to place under the secretary of agriculture many standards of particular interest to the farmer, for instance, the grading of cotton.[18] In 1908 an effort was made to transfer the iron-and-steel-testing machine of the Army's Watertown Arsenal to the Department of Commerce and Labor, but Congress almost immediately changed its mind.[19]

Compared with the old Office of Weights and Measures, the new Bureau of Standards was an aggressive and expansionist outfit. In terms of its ideal of a national physical laboratory, its growth was no more than proportional to the rapid penetration of science into technology in the United States in this period. While insistent demands always ran ahead of the bureau's programs, it was often alert to new opportunities. For instance, experimentation in radio communications was attracting ever-increasing attention in widely separated agencies of the government. In 1915 Stratton pointed out in his *Annual Report* that it would "not only be more economical, but productive of more efficient work to concentrate the laboratory work of the Government at one place in a small laboratory especially designed for it." He had secured agreements from the War, Navy, Treasury, Post Office, Agriculture, and Commerce Departments that "the location of the laboratory at the Bureau of Standards would prove of great benefit both as to economical performance of the work and by its close proximity to the scientific work of the Bureau, especially that of the electrical division." [20] With a generosity characteristic of its treatment of the bureau, Congress voted $50,000 for a radio laboratory the next year.[21]

The Bureau of Standards had established itself and its basic policies firmly by 1916. It had expanded the meaning of weights and measures to research in physical standards of many sorts. It became a direct link between the government and industry, usually staying in the background but occasionally, as in building materials, ranging into applied research. Although apparently uninterested in regulation, many of its activities were in effect research programs for the commissions then proliferating in both federal and state governments. For instance, research in gas and electrical measurement was of direct aid in the regulation of public utilities.[22] It was a new scientific bureau, but with a difference. Its expansive tendencies grew from the pervasiveness of the problems of physical standards themselves. Even though its divisions tended to approximate the lines of disciplines in science, it never went all the way to indiscriminate physics and chemistry for their own sakes. After 1916, the bureau began to give away research programs it had fostered at as great a rate as it took them in.

The Census

Like weights and measures, the census had a formal place in the Constitution. Political in its main objective, its first six decennial enumerations markedly expanded the variety of the information it gathered without improving its crude administration. The actual count was in the hands of federal marshals, and the organization died completely after a spasmodic effort each decade. The census of 1850 attained much higher standards, and the returns received more sophisticated analysis, but the basic difficulties of the marshals and the ephemeral organizations still dogged the census of 1870.[23]

Francis A. Walker, superintendent of the ninth census in 1870, learned first-hand the limitations of the old arrangements. Furthermore, even at this time it was clear that an increasingly urban and industrial United States had to have more extensive statistical information. Walker took the lead in securing new legislation which took the tenth census away from the marshals and enabled the superintendent to organize his own staff. During the interim years of the 1870's, Walker exercised only a general and gratuitous supervision over the publications of the ninth census while teaching political economy and history at Sheffield Scientific School.[24]

In 1879 Walker again took active command, not only organizing the enumerations but assembling an impressive array of scientific talent. This was the pregnant period of the reorganization of the surveys, the National Board of Health, and the beginning of a new approach in agriculture. Most of the key men of these new developments participated in the preparation of the voluminous reports. John Shaw Billings in the *Report on the Mortality and Vital Statistics of the United States* [25] virtually established vital statistics on a national scale, an achievement that far outlasted the National Board of Health. Powell at the Bureau of Ethnology launched an ambitious classification of Indian languages under the cloak of an appointment from Walker.[26] C. S. Sargent's *Report on the Forests of North America* [27] was an epochal study in plant taxonomy. Clarence King directed the work that led to *Statistics and Technology of the Precious Metals*,[28] by S. F. Emmons and G. F. Becker. Raphael Pumpelly prepared a *Report on the Mining Industries of the United States* [29] which ran off into special investigations of iron resources and of the coals of the Northwest. Other reports included ambitious studies of the technology of petroleum, coke, and building stones.[30] The agriculture volume contained a number of monographs, such as W. H. Brewer's on cereal production.[31]

Walker aimed at more than counting heads, demography, or even the nascent social sciences which were his main concern. The tenth census was a general-purpose scientific organization which placed special emphasis on natural resources. Since statistical problems pervaded all the sciences and since the staff assembled by Walker was so many-sided and illustrious, the census had an open path to many areas later occupied by other agencies. But the policy, which appeared stupid even at the time,[32] of setting up the organization every ten years and then allowing it to fall into ruin without even providing a nucleus, doomed the census to losing its opportunity. Organic legislation for a permanent bureau and the removal of the census from party politics was beyond the Congresses of these years. Walker, a potentially great scientific administrator, left in 1881 to become president of Massachusetts Institute of Technology. The census of 1890, following the pattern Walker laid down, covered more subjects than any before or several thereafter.[33]

By 1900 the growth of the Department of Agriculture and conservation agencies limited the possibility of doing big things in those directions. The future of the census lay in statistics. Also, an industrial and urban society demanded more than an orgy of figures once every ten years. Schedules once a part of the main census only were increasingly turned over to experts for continuous investigation. This tendency only added cogency to the arguments for a permanent organization.

Important as was permanency of organization, the friends of the census also attempted to create a central statistical agency for the whole government.[34] While not as inclusive as Walker's wide-ranging investigations, the plan for such a centralized agency was suggested by the same considerations as produced the drive for a department of health — the desire to collect organizations doing the same work into one administrative structure.

The aim of permanency was accomplished in the first year of Theodore Roosevelt, when Congress passed an act establishing the Bureau of the Census in the Department of the Interior.[35] The second aim seemed on the way to fulfilment when in 1903 the Bureau of the Census was transferred to the newly created Department of Commerce and Labor, whose secretary was given power to consolidate all statistical work. The same law gave the President power to transfer to the Department of Commerce and Labor any office engaged in statistics except those in the Department of Agriculture.[36] Thus the legal basis was complete for the coördination of most government statistical research into one bureau. Since the need for statistics was multiplying unprecedentedly in these years, the new bureau had a high growth potential.

However, the Bureau of the Census failed to become a central statistical organization serving the whole government.[37] Vital statistics did remain under the Bureau of the Census until 1946, when they were transferred to the Public Health Service. But in other great areas, such as agriculture, coöperation became the rule,[38] and most of the greatly expanded statistical establishment of the government grew up in individual bureaus close to operating and regulatory functions. Perhaps, like a general chemistry laboratory, a central statistical organization was as quixotic as a bureau of typewriting.

Nevertheless, the establishment of the permanent Census Bureau gave demography a stable place in the government, by implication establishing the social sciences as well. The year 1902 thus marks the change of an ancient activity, hallowed by the Constitution, into an agency capable of providing the continuous flow of information necessary in the Progressive Era.

The Bureau of Mines

The mining industry increasingly felt the impulse of three forces in the early years of the twentieth century.[39] First, the products of mines were the basis of the new industrial dominance. In the second place, minerals doubly attracted the attention of the conservationists because they were nonrenewable natural resources which were being dissipated at an appalling rate. Third, the hazards of working in the mines made them among the most spectacular targets of that strong urge among the progressives to improve the industrial environment of workingmen.

Since the conservationists had one of the government's best-organized research outfits in the Geological Survey, they were the first to tackle the problems of mining. Although Clarence King had begun studies back in 1880, Powell had largely concentrated on other things. The major's successor, Walcott, after 1894 began again to publish studies of technological processes, and in 1899 he proposed a division of mines and mining to make a systematic inquiry into the value of economic minerals. In 1904 Congress finally appropriated $60,000 for testing the coals of the United States at the Louisiana Purchase Exposition in Saint Louis.[40]

The leading member of the committee appointed to set up the coal-testing plant was Joseph A. Holmes, conservation-minded state geologist of North Carolina who many years before had urged to Gifford Pinchot the need for federal forestry in the Southeast.[41] N. W. Lord of Ohio State University had charge of the chemical laboratory, bringing with him a number of graduate students.[42] As the law provided, all machinery and all coal tested came from various private companies without cost to the government. After the close of the exposition the plant continued to operate, forming the nucleus for an organization, which, when moved to Pittsburgh in 1907, became the Technological Branch of the Geological Survey.[43]

The year 1907 also saw an unusual number of mining disasters, the eight hundred miners who died in December alone dramatizing a steadily worsening trend. The "people spoke through the medium of the newspapers and demanded that the Government stop this slaughter." [44] Almost immediately an appropriation followed of $150,000 to the Technological Branch of the Geological Survey "for the protection of lives of miners in the territories and in the District of Columbia, and for conducting investigations as to the cause of mine explosions with a view to increasing safety in mining." [45] Although actual regulation within the states was thus carefully excluded, the federal government had here turned to scientific research as a means of meeting an urgent problem.[46]

An important branch of the Geological Survey had shifted its program in the direction of applied research, and in the field of mine safety had wandered far from the original interests of the parent organization. In a similar and parallel movement the Bureau of Reclamation had separated from the Geological Survey completely, remaining in the Department of the Interior. A Bureau of Mines now seemed the natural and inevitable evolution. In 1908–1909 three programs — testing of fuels, testing of structural materials, and safety investigations — had a total appropriation of over $500,000, all under Joseph A. Holmes. In 1910 Congress took the final step to create the new bureau in the Department of the Interior.[47]

The organic legislation was less than ideal from the point of view of Holmes, who became the first director. The Bureau of Standards made away with the testing of structural materials.[48] More significant, the provisions of the act, coupled with the size of the early appropriations, actually curtailed research in emphasizing mine safety.[49] Hopes for expanding beyond the coal industry into the metal mining of the West and for energetic studies on the conservation of mineral resources seemed dim.

A new organic act in 1913 described the field of the Bureau of Mines as "mining, metallurgy, and mineral technology." Holmes could now describe as his purpose the conduct, "in behalf of the public welfare, of such fundamental inquiries and investigations as will lead to increasing safety, efficiency, and economy in the mining industry." He now put prevention of waste on an equal plane with mine safety.[50] Following closely this clarification of its position, the bureau developed research programs for new fields.

The burgeoning petroleum industry had attracted some attention even in the Geological Survey period, and in 1914 a Petroleum Division began with limited funds to provide production research for an industry that had in its rapid growth provided little research of its own. From the beginning the division stood for conservation and elimination of waste, even though it did not pretend to any direct regulatory power.[51]

In 1915 the bureau began to reach out toward research in non-ferrous metals, which had been the subject of some of the Geological Survey's early studies and which carried the work into the West. While conceivably one large Western establishment would have sufficed, Congress provided for ten regional experiment stations, scattering both the bureau and the local benefits over the face of the West.[52]

As the experiment-station system emerged over the next few years, it took on some of the characteristics of the agricultural establishment. The laboratories were normally located on the campuses of universities, in close coöperation with their engineering departments. In serving regions other than states, however, they more closely resembled the Forest Service research stations, drawing on federal rather than state funds. Instead of aiming at a general program for each geographic area, the bureau concentrated research in appropriate problems at each of the regional stations.

The comparison between the Bureau of Mines and the Department of Agriculture is instructive both for the similarities and the differences that it reveals. Joseph Holmes was aware of the essential similarity between the mining industry and agriculture. He saw that mining and farming were the foundation industries of the economy, and that the producing units of both were small and local. He attributed the large growth of agricultural productivity in the early years of the twentieth century in part to the expenditure of federal money and suggested that the limited amount of research under the Geological Survey and early bureau furnished "specific evidence of the larger benefits that may be expected to result from larger expenditures for mining investigations." [53]

In spite of the parallel, the federal government had relatively neglected mining, which was the "more difficult to understand in view of the hazards . . . that should appeal to the humanitarian as well as the commercial instincts of the American people." Holmes attributed this neglect to misapprehensions concerning mining. The usual assumption was that the industry was controlled by a few large corporations who demanded federal support. Actually the large companies were few, and "at the request of the Bureau of Mines, a number of them have expended considerable sums from their own funds for investigations that promise to be useful not only to them but to other less important mining developments." Most people did not appreciate that the large number of small enterprises depended on government research to provide methods profitable enough to keep them from "being helplessly transferred to a few large corporations which alone may have the funds for developing the processes." The humanitarian appeal was at a disadvantage because it "comes from employees working under hazardous conditions, a majority of whom are unfamiliar with our language, our laws, or our institutions." Finally, the benefits of conservation in mining actually went not to producers but to "the consumers or users of mineral products who are distributed throughout every part of the country." [54]

Holmes's analysis shows that the Bureau of Mines, a generation younger than the Department of Agriculture, was a child of the twentieth century in its relation to the great corporations, to small business, to labor, and to the consumer. It was not able to establish direct regulation; the explosives it approved for mine use were marked "permissible." Yet it established a true mission for itself in an age when both the universities and industrial laboratories were already on the scene. The career of Frederick G. Cottrell, a University of California chemistry professor who simultaneously worked for Holmes and the Anaconda Smelter Smoke Commission in the years between 1911 and 1914, symbolizes the bureau's many-sided relation with other research enterprises.[55] The Bureau of Mines was a midget among giants, but the impartial ideal of science gave it added stature.

The Smithsonian and the National Advisory Committee for Aeronautics

In the midst of the bureau-building and the unprecedented application of science that occupied the government as it entered the twentieth century, the crowning administrative invention of pre-Civil War American science — the Smithsonian Institution — not only survived but entered dynamic new fields. The National Museum had gained major momentum with the Centennial Exposition in Philadelphia in 1876, inheriting many of the collections, for which Congress provided a new building.[56] Under the perceptive leadership of George Brown Goode, the museum extended its range and functions. More significant for basic research, the Smithsonian's abiding concern was the flood of material coming into the National Museum from the Geological Survey and the Department of Agriculture. Along with the Bureau of American Ethnology, the museum became a great research center in anthropology, zoology, and botany. The large new building unit that rose on the Mall in the early 1900's conspicuously indicated the continuing importance of the museum in the age of the new scientific bureau.[57]

Meanwhile the Smithsonian Institution proper continued to seek ways to serve science on an endowment that was relatively dwindling in importance. The original endowment had been twice as large as Yale's, larger than those of Princeton, Columbia, and the University of Pennsylvania, and equal to Harvard's. In 1901 the secretary stated that the "Institution's endowment has in . . . fifty years increased but from $600,000 to somewhat less than $1,000,000, but the *average* endowment of the five universities named is now about $8,000,000, indicating that in this regard the Institution's fund for scientific purposes is relatively unimportant compared with what it was fifty years ago." [58] For such small means to make a measurable impression on the course of scientific research required ever more difficult selectivity in the choice of objectives.

With much depending on the personal interests of the secretary, the regents fell into the rhythm of alternating between physical and biological scientists. Henry had been a physicist and Baird a zoologist. Hence the secretaryship went to Samuel Pierpont Langley, an astronomer and specialist in the analysis of the spectrum of the sun. He soon organized the Astrophysical Laboratory, which became a permanent part of the Smithsonian. Even before Langley left the Allegheny Observatory in Pittsburgh for the Smithsonian, however, he had developed a side interest in the theory and practice of heavier-than-air flight which opened a significant chapter in Smithsonian history.[59]

After publishing his *Experiments in Aerodynamics* in 1891, Langley began to construct model flying machines to test his data. These were years of heightening interest in the possibility of flight, as Senator Henry Cabot Lodge attested when he unsuccessfully introduced a bill in the Senate in 1894 to offer a $100,000 prize for an aerial machine.[60] Langley lent prestige to research in an oft-ridiculed field both by his own reputation and by that of the Institution.

During the Spanish-American War an Army-Navy board investigated his experiments to determine the possibilities "of developing a large-size man-carrying machine for war purposes." As a result the War Department allotted Langley $50,000, and the Smithsonian $20,000 more, giving aeronautical research a promising start, with funds, a strong military motivation, and a renowned scientist. All this talent was concentrated immediately on the finished product — a large, self-propelling "aerodrome" that would carry a man.[61]

Langley was ready by 1903 to stake all on the flight of his machine. An attempt in October failed with only minor damage to the equipment, but in December the plane plunged into the Potomac River, putting an end to Langley's career in aeronautics. The newspaper ridicule closed his laboratory at the Smithsonian and drove the War Department out of the field in the critical years immediately following. The Wright brothers' successful flight had occurred just nine days after Langley's Folly, but the government and the Smithsonian, after a bold start, now shied off from aeronautical proposals.[62]

Langley's ghost continued to hover in the Smithsonian after his death in 1906. The new secretary was C. D. Walcott, who had guided the Geological Survey as Powell's chosen successor. By fostering the programs that became the Bureau of Reclamation and the Bureau of Mines, he made a place for himself as one of the leading architects of government science in the era of Theodore Roosevelt. The secretary and Alexander Graham Bell, a member of the board of regents, kept up the Smithsonian's interest in aeronautics largely out of respect for Langley.[63]

Because of the proposals of individuals in the government who had become interested in aviation and who realized that the lack of fundamental information was hampering development, President Taft appointed in 1912 a commission to study the need for a research laboratory.[64] Walcott was a member, and the report recommended a laboratory within the Smithsonian supported by appropriations. When Congress did not act, the Smithsonian on its own initiative resolved to revive Langley's laboratory to study "the problems of aerodromics, particularly those of aero-dynamics, with such research and experimentation as may be necessary to increase the safety and effectiveness of aerial locomotion for the purposes of commerce, national defense, and the welfare of man." The secretary was to secure the "coöperation of governmental and other agencies in the development of aerodromical research under the direction of the Smithsonian Institution."[65] As "a private organization having governmental functions and prerogatives," the Smithsonian hoped to coördinate the research of the Bureau of Standards, the Weather Bureau, and the War and Navy Departments as well as conduct the Langley Aerodynamical

Laboratory, which it envisaged as a building on the Institution's grounds with a flying field in nearby Potomac Park. An advisory committee of fourteen members was organized, consisting of the director of the laboratory, members designated by the secretaries of the Navy, war, agriculture, and commerce, and others appointed by the secretary of the Smithsonian who "may be acquainted with the needs of aeronautics." [66]

This experiment in coördination almost immediately ran afoul of a ruling by the comptroller of the treasury that it was unlawful for any government employee to serve on such an advisory committee without authority from Congress.[67] Regardless of the legal technicalities, the Smithsonian was pushing into a field requiring more aggressiveness than the Institution had usually shown. To provide both basic and applied research for a whole industry, even a new and amorphous one, was more in the tradition of the Department of Agriculture. To add the task of coördinating several agencies with one another and the industry with the government immensely complicated the problem for an institution whose strength normally came from standing slightly aloof from the government.

Walcott and Bell accordingly memorialized Congress to take over their advisory committee. By this time the war in Europe had added urgency to the pleas of the friends of aviation. "The United States is the only first-class nation that does not have an advisory committee for aeronautics and suitable research laboratories placed under its direction." Perhaps partly a result of this lack, the outbreak of war found the United States with 23 airplanes, while France had 1400, Germany 1000, Russia 800, and Great Britain 400.[68] Nevertheless, the legislative outlook was unfavorable, and President Woodrow Wilson reputedly feared the effect on American neutrality.[69] To Walcott, the heir of John Wesley Powell, the proper tactics in a tight place involved a rider to an appropriation bill. Since the House Committee on Naval Affairs proved sympathetic, the Smithsonian's advisory committee was attached to the naval appropriation bill. Franklin Delano Roosevelt, acting secretary of the Navy, approved the idea but felt that fourteen was too large a number and that the public membership should be cut from seven to three so that "the Government should always have a controlling interest." He feared that the "interests of private parties must be more or less commercial and influenced by such considerations." [70]

As finally passed by Congress in 1915, the act provided for a committee of twelve, all appointed by the President of the United States. Two members from the Army, two from the Navy, one each from the Smithsonian, the Weather Bureau, and the Bureau of Standards, gave the government a majority. The five public members were to be men "acquainted with the needs of aeronautical science, either civil or military or skilled in aeronautical engineering or its allied sciences." The duty of the committee, which served without pay, was to "supervise and direct the scientific study of the problems of flight, with a view to their practical solution, and to determine the problems which should be experimentally attacked, and to discuss their solution and their application to practical questions." If laboratories were placed under the direction of the advisory committee, it might "direct and conduct research and experiment in aeronautics." [71] Thus the National Advisory Committee for Aeronautics, later regularly referred to as NACA, was in itself an independent executive body with more than advisory power.

After a few early changes, Walcott was chairman until his death in 1927. His invidious part in the unseemly squabble between the Smithsonian and Orville Wright over the Langley aerodrome has tended to overshadow his and the Institution's role in the organization of the committee and the impress they gave to its structure. The government as a whole seemed unable to respond to the demands of a new industry which had unusually strong links to national security and a desperate need for research. The ancient and impoverished Smithsonian provided the shelter for early experiments and the form of the committee. Several of the first appointees had worked under the Smithsonian, and the history of the Langley Aerodynamical Laboratory merged into the activities that eventually concentrated under NACA at Langley Field, Virginia.

Government Science and Large-Scale Industry in the Progressive Era

The Progressive Era, with a wide-ranging concern for the problems emerging from a complex society, had seen the creation of scientific bureaus shaped by the relations of government and large-scale industry. The National Bureau of Standards, the Bureau of the Census, the Bureau of Mines, and the NACA, were all the servants of industry as well as of the government. Although less famous and less tied to politics than the conservation agencies which dealt with the public domain, these new organizations were a part of the revival of the vigor and efficiency of the federal government, which in the late nineteenth century had often been dwarfed by the great corporate domains of the captains of industry. The bureaus renewed the ability of the government to conduct its own business in a society dominated by a complex technology that increasingly depended on research for guidance.

The industry-oriented bureaus also did more. In the name of the general welfare, they sought answers to those problems that industry needed to have solved but was unable or unwilling to answer for itself. With industrial research still in its infancy, the new bureaus performed its functions for the building trades and mineral and petroleum production in the same way the Department of Agriculture served the farmer. After the sudden birth of the aviation industry, the government had to provide most of the research. Even in industries that early set up their own laboratories, such as electrical manufacturing, the abstruse research involved in defining and maintaining standards was left to the government. Without the efforts of these bureaus the use of science would have penetrated more slowly into technology. The presence of the large number of trained scientists in the civil service, subject to a constant turnover, enriched private industrial research indirectly by giving experts highly specialized experience which they carried with them when they left the government.

In the first years of the twentieth century a government without science was already unthinkable. Excepting military applications, the government's scientific establishment was virtually complete in 1916. The NACA set up a laboratory for the study of a subject that to most people seemed like a problem for the future. The Bureau of Standards was already working with cathode-ray tubes and radioactive minerals. From the point of view of 1940, the foundations of every important scientific institution within the government were already in place in the last bright days of peace that the United States enjoyed before entering the first World War.

XV

PATTERNS OF GOVERNMENT
RESEARCH IN MODERN AMERICA
1865–1916

THE bewildering multiplicity of the government's research establishment in 1916 was the result of its history. An agency, once established, usually stayed in existence although its original purpose might change many times. Indeed, the failures usually reappeared in another form after a seemly retirement. Creations of the pre-Civil War era did not vanish even in the twentieth century. The Smithsonian Institution, the National Museum, the Coast Survey, the Naval Observatory, the *Nautical Almanac*, the Corps of Engineers, the service academies, and the Army Medical Department were humbled only by the size of later establishments. Many of them still performed important research and embodied respected traditions after the geographic explorations that gave them their initial impetus had ceased to be a major concern.

The Civil War had seen a start toward research technology in military production, but this largely disappeared after Appomattox. Only the National Academy and a few accidents such as the Army Medical Museum and Library commemorated the single great war effort in a century of peace.

A Survey of Fifty Years of Bureau Development

The creation of the Department of Agriculture and the land grants for colleges, a result of secession but not of the war, was the opening of the great period of bureau-building essentially unhampered by the constitutional questionings that had shaped earlier development. In a sense, the Department of Agriculture became a laboratory experimenting on the nature of a scientific bureau, gradually perfecting a standard form adapted to its particular problem, its position within the government framework, and its need for communication with the world of science outside. As the new scientific bureau grew, it became both a tool for research and, through regulation and persuasion, an instrument of power. By 1890 the department showed its strength by taking over the Weather Bureau from the Army.

Meanwhile the wilderness beyond the frontier, which had provided the main motive for government science during the first seventy-five years of the republic, was disappearing. As geographic problems receded in importance because of the very success of exploration, the *ad hoc* machinery that had grown up within the Army and Navy gradually became an anachronism and was replaced by a new, vigorous, and permanent organization, the Geological Survey. Although beset by enemies and doubters, Powell's bureau weathered the Allison Commission hearings and the defeat of the irrigation survey to become a nucleus of research in the government second only to the Department of Agriculture.

The emergence of the United States as a world power opened new horizons for the Army Medical Corps, which in defiance of all the rules had been readying its research tools for just such an opportunity. Sternberg, Walter Reed, and Leonard Wood had served for years at frontier posts in the West. When the focus of the nation shifted overseas, they and their colleagues went to Havana, Puerto Rico, and Panama. Their success reflected the long preparation sculptured by the Army Medical School, the Museum, and the Library.

With the settlement of the West and the disappearance of a frontier line, the assumption that natural resources were inexhaustible gave way to concern for their conservation. The Interior Department, building on the Geological Survey, became the shelter for the Reclamation Service and the Bureau of Mines. Meanwhile in the Department of Agriculture Gifford Pinchot built the Forest Service into a powerful administrative agency which increasingly applied research to conservation problems.

By 1900 America had become both urban and industrial, generating forces that put new life into ancient scientific functions of the government — the census and standards of weights and measures. A permanent Bureau of the Census conducting continuous studies mitigated the decennial orgy-and-famine cycle of the nineteenth century, providing demographic and statistical information of increasing subtlety and reliability. The Bureau of Standards, reacting to the demands of industrial technology, especially in new fields such as electricity, became at least partially a national physical laboratory. The Bureau of Mines by 1916 had begun to furnish safety and technologic research to the mining and petroleum industries.

Although the National Board of Health proved a false dawn of the federal government's support of research on infectious diseases, sanitation, and hygiene, the Public Health Service eventually filled the gap. It evolved from the Marine Hospitals, limited to simple medical service for a particular group, into a research organization which besides routine duties had some responsibility for the general welfare by improving the health of the whole people. The progressives who assisted this shift saw public health as the human side of the conservation movement.

The National Advisory Committee for Aeronautics was the capstone of the federal establishment. Even it was a product of earlier trends. The ancient Smithsonian sheltered its embryonic stages, and the long experience of the Geological Survey provided the trick that accomplished its creation. Yet the NACA was a new kind of organization for a new problem. It was the last product of a profoundly peaceful and fertile period of bureau-building and also the first war-research agency of World War I.

The galaxy of bureaus existed in law through organic acts. In them Congress had spoken, occasionally straightforwardly, usually by the devious method of appropriation-bill riders. This stratagem, however, was not wanton political immorality. Legislating scientific bureaus into existence was a technically difficult problem with which the machinery of Congress could not cope directly. Most of the people's representatives in the late nineteenth century had little background for science or appreciation of its results. Indeed, 1865 to 1900 was not a great period for legislation on any subject. Hence the appropriation-bill rider allowed a small number of congressmen, adequately coached by experts, to legislate in the interest of science. Theirs was a constructive achievement, their circumventions detouring ignorance and lethargy, not the rights of a vigilant people. One of the great changes of the Progressive Era was a wider appreciation of the use of science in the public interest. The organic act for the Bureau of Standards and the Newlands Act showed a new ability of Congress to deal with science directly. The difficulties encountered by the few unfortunates, such as the Army Medical Library, that did not have organic acts only underline their importance. These agencies were condemned to explaining their reason for existence over and over again without gaining a secure position.

The handling of scientific personnel in the government underwent great changes. The rise of the merit system in the civil service gradually took in most scientists in the government, working upward from the bottom. As political patronage in the scientific bureaus declined accordingly, the ideal of using trained people regardless of party affiliation came closer to realization. The great *esprit de corps* of such new bureaus as the Forest Service, comparing so favorably with the old General Land Office, for instance, is impressive evidence that the new way brought to the government incomparably more scientific competence than it could possibly have achieved otherwise. Yet the abiding and fundamental gain of the merit system was offset both by the difficulty of applying it to higher scientific positions and by the fact that it put a premium on seniority within the service rather than on scientific brilliance. In 1916 the ranks of government scientists contained a small number of congressmen's nephews and also no Simon Newcombs.

A special personnel problem was the bureau chief, who most reflective people realized must ideally be both scientist and administrator. Was he to be an expert, chosen solely for competence and on the recommendation of scientific bodies such as the National Academy? Or was he a policy-making administrator who should be chosen by the President, his political advisers, and the senators of the applicant's state? A few, such as Pinchot, were politician and scientist at the same time. Most Presidents, especially those whose predecessor belonged to the opposing party, yielded at least some of the time to political needs, appointing unqualified party men as chiefs of scientific bureaus. Cleveland's appointee, Thorn of the Coast and Geodetic Survey, turned into an acceptable administrator, as did G. M. Bowers, to whom McKinley gave the Bureau of Fisheries in return for political support in West Virginia. Others did not learn so easily.

By 1916 the principle of professional scientists as bureau chiefs had gained sufficient ground for Republican nominee Charles Evans Hughes to make an issue in his campaign of President Wilson's appointments. He attacked two specifically, pointing out that in the Coast and Geodetic Survey an eminent scientist "was displaced to make room for an excellent stock breeder and veterinary surgeon."

Science heralded the raising of the issue as a recognition of the importance of science but observed on Wilson's behalf that the two appointments "are the only ones in which the President is open to criticism." [1]

Central Scientific Organizations

As a part of her emergence as a world power, the United States had an opportunity to create a government scientific establishment *de novo*. The Bureau of Science in the Philippines, springing up promptly after American occupation, demonstrated that research was as much a part of a government in 1901 as a post office or a revenue service. With its beautifully centralized organization it dramatized by contrast that the home government's scientific establishment was the product not of logic but of history, the interplay of institutions and their environment through many changes.

Originally called the Bureau of Government Laboratories, the Philippine Bureau of Science was a microcosm of the home establishment with most of the historical irregularities absent.[2] By 1903 research already had started in public health, chemistry, biology, and weights and measures. In the next few years the Philippine Bureau of Mines was absorbed whole, and much research was transferred from the Bureaus of Agriculture and Forestry. Thus when the United States government had a *tabula rasa* and a simple situation, it created the equivalent of a department of science. This is not, of course, proof that such an organization was superior, especially since the Philippine Bureau of Science began to decline after a few years.

At home, however, the impulse toward a centralized, rational organization sank to a low ebb. The bureaus were too strong and too engrossed in their own development. Spectacular progress quieted apprehensions about coördinating the directions that science might take. After the high point of the Allison Commission in the 1880's, the discussion of a cabinet department of science dwindled almost to nothing.

The idea of a national university enjoyed only a pale revival. After a congressional resolution placing the collections and resources of the government in Washington at the disposal of qualified students, several attempts were made to create a university which would utilize the libraries, laboratories, and museums of the capital for advanced training. Great educational leaders of the day — Daniel Coit

Gilman, David Starr Jordan, William Rainey Harper, and Nicholas Murray Butler — lent at least lip service to the idea. Government administrators — notably C. W. Dabney and C. D. Walcott — also worked for it.[3] But the several schemes proposed never came close to obtaining financial support, government or private. Even if a university had emerged, the day was long past when it would have coordinated government science as Joel Barlow had envisioned back in 1807. Some federal institutions did arise in the District of Columbia, notably Howard University and St. Elizabeth's Hospital, but their purpose was to benefit groups with special problems rather than to become national organizations.

The National Academy of Sciences had the dignity of an honorary society and a legal position as adviser to the government, but little else. After the forest commission in 1896 it had a negligible influence on policy. Its attempts to control the appointments of bureau chiefs accomplished only occasional successes. All the Presidents of the period listened to the Academy's recommendations only when they chose.[4] Simon Newcomb commented that one "hardly knows where to look for a spectacle less befitting our civilization than that of such a body of men searching through Washington to find a suitable place for their meeting . . . grateful to one of their officers when he has a spare corner in which to keep their records; wondering what shall be done with an invitation from a foreign organization."[5]

Theodore Roosevelt, in the early days of his administration, when scientific bureaus were coming to life all around, made a gesture toward the problem of central organization. In 1903 he appointed to a Committee of Organization of Government Scientific Work a general, an admiral, and three conservationists — Pinchot, Walcott, and James R. Garfield.[6] For four months this group met, working up a series of reports on government scientific bureaus which if it had been published would have given a cross section of the establishment such as the science committee of the National Resources Committee did in 1938 and the Steelman Report in 1947. Unfortunately, the reports were never published.

The aim of the committee was the avoidance of duplication. They looked both to consolidation of bureaus doing similar work and to the coördination of those with mutual interests even across departmental barriers. But research agencies to them were so many indestructible atoms to be moved around in different combinations without essentially changing their nature. As in the movement for a department of health and a central statistical agency, coördination and consolidation were popular slogans during the Progressive Era which could always attract adherents but seldom got results. The Department of Commerce and Labor, formed while the committee met, gave a new administrative shelter for the Coast and Geodetic Survey, the Bureau of Standards, the Bureau of the Census, and the Bureau of Fisheries.[7] But these transfers simply added another cluster of atoms to the executive.

In their conclusions the committee recognized that the development of scientific work had been enormous since 1890, that this had largely come through the development of special bureaus to meet special needs, and that it often broke over departmental barriers. They estimated a total expenditure for science of over $11,000,000 in fiscal 1902. Although looking for duplication, they admitted they found little. Rather they saw lack of efficiency and coördination, and their remedy throws much light on their real concerns. Most of the transfers aimed to put all conservation agencies under the Department of Agriculture and to consolidate statistics in the Bureau of the Census. Pinchot treated the committee largely as a move in the fight for the forest reserves.[8] Clearly these men, of whom Walcott, Pinchot, and Garfield were bureau-builders without peer, saw even a committee to organize government scientific work as an opportunity for furthering construction jobs already under way. In their exuberance for science it never occurred to them to question the direction in which government research was trending. The committee's work and its unpublished reports soon sank into oblivion.[9]

In 1908 Congress gave the National Academy a similar chance by requesting it in an appropriation bill "to consider certain questions relating to the scientific work under the United States Government." The committee set up by the Academy was distinguished, including the heads of Cornell University, the University of Wisconsin, Massachusetts Institute of Technology, the Carnegie Institution, and the Lick Observatory. This group, leaning heavily on the unpublished reports of the 1903 committee, came to parallel conclusions.[10] Actual duplication was "relatively unimportant," but organizations and plants overlapped making "consolidation of some branches of work now carried on in several organizations . . . probably advisable." Since "anything like a rational correlation of allied branches of scien-

tific work" did not exist, the committee recommended a permanent
board to consider "the inauguration, the continuance, and the interrelations of the various branches of the scientific work." [11] Nothing
ever came of this suggestion, which had no more to offer than the
thoroughly aired and rejected proposals of the Allison Commission
era. The National Academy committee, which managed only to get
a five-page report printed, confessed its own helplessness when it
recognized that many of the bureaus "have been so long established
as to become integral parts of the departments to which they are assigned." Any consolidation or redistribution "should take into account
their origin and historical development as well as their present
status." [12]

The remarks of the 1903 committee as they pondered the organization of science in the government incidentally revealed prevalent
attitudes about the relation of basic to applied science. Pinchot, Walcott, and their colleagues emphasized that research should be organized around a problem, and that "the individual sciences and arts
should not be segregated in the separate bureaus and offices." [13] The
problem approach had won so completely and the shift toward
applied research had proceeded so far that the committee now left
research on broad and general grounds to private institutions. Science
"on the part of the Government should be limited nearly to utilitarian
purposes evidently for the general welfare." [14] Henry, Powell, and
Alexander Agassiz had each had versions of this precept, but none
of them had dared reduce government science to such unmitigated
practicality. The sublime faith of John Quincy Adams that the government owed humanity the support of science as a necessary element in
civilization had almost completely disappeared. To seek a reason is
to realize that private scientific institutions outside the government
were profoundly changed by the early years of the twentieth century.

The Estates of Science

In 1900 the universities, grown in one generation from colleges
with narrow courses of studies, seemed to have become the natural
homes of disinterested, pure science. The broadening of the curriculum, the introduction of the German seminar and its ideal of research, the creation of graduate schools, and the rapid accumulation
of endowment either created new centers of learning or remade old
ones.[15] With Johns Hopkins setting the pace, such universities as

Harvard, Cornell, Chicago, Columbia, and Michigan became the headquarters of fundamental research in the country.

The result was a division of labor which gave rise to the assumption that basic research belonged to the universities, leaving only applied research to the government. The difference heightened between the disinterested, cloistered seeker for pure knowledge and the grubby civil servant chained to mundane, grinding routine investigation. Although the split between basic research and the common concerns of society was noticeable fairly early in the nineteenth century,[16] after 1900 it became institutionalized in the division of functions between government and the universities.

Yet the universities and the government did not lose all touch with one another. The bureaus remained a major employer of men from the scientific departments of the graduate schools [17] and wanted more than they could get.[18] Those institutions that took some interest in applied science — Massachusetts Institute of Technology, Yale School of Forestry, the land-grant colleges — remained closely linked to the government service throughout this period. The Smithsonian and especially the National Museum kept the ideal of basic research alive in Washington. The visiting committees of the Bureau of Standards and the Hygienic Laboratory and later the NACA brought eminent university professors into touch with government research. But no Bache, Henry, or Newcomb represented the government in the ruling circles of science. Billings, Theobald Smith, and Francis A. Walker were there, but in private institutions.

Deepening the shadow cast over government by the universities was the dramatic rise of the foundations of men wealthy enough to rival the government itself. Andrew Carnegie, the most reflective of the great captains of industry, felt that "it might be reserved for me to fulfill one of Washington's dearest wishes — to establish a university in Washington." [19] Other counsels, however, prevailed, pointing to competition with existing universities and the enormous sum required. The Royal Institution of London became the model instead. The resulting Carnegie Institution of Washington "shall in the broadest and most liberal manner encourage investigation, research, and discovery." [20] To the $10,000,000 of 1902 Carnegie added $2,000,000 more in 1907 and yet another $10,000,000 in 1911. A giant compared to the Smithsonian, the Institution by 1915 developed three main lines of activity. It established departments for collaborative research, made grants to individual investigators, and provided support for publications.[21] The Department of Experimental Evolution, for example, was doing significant work in the new science of genetics, while the Department of Terrestrial Magnetism supported a program that had been a particular specialty of the Coast Survey in the time of Bache.

Many of the trustees of the Carnegie Institution were prominent figures in the development of government science. John Shaw Billings, Abram Hewitt, Samuel P. Langley, Henry S. Pritchett, Theobald Smith, and C. D. Walcott helped make the policy of the purely private organization.[22] At its outset some anxiety existed that it might compete with official agencies. George M. Sternberg feared "there would be a strong disposition to refuse or cut down appropriations for scientific work in various departments if it was believed that the funds of the Carnegie Institution could be made available for such work." But, fighter that he was, Harvey Wiley pointed out that government appropriations for science were already ten times the income from Carnegie's original gift and that the Institution would find "keen competition if they undertake any line of investigation already carried on under the auspices of the Government." [23]

Wiley proved correct that government science was already too big to be challenged by an institution even of the opulence of the Carnegie Institution. But the keen competition he envisaged did not materialize. Because of the shift toward applied research within the government and the care with which the Institution concentrated on fundamental problems, no serious conflict arose. The existence of the Carnegie Institution in the capital city did, however, emphasize the division of labor between basic and applied research.

The great Rockefeller foundations also cast a shadow by working in fields which, although unoccupied by the official agencies, were directed to aiding the public welfare on a scale usually possible only to governments. The General Education Board's early support of Seaman A. Knapp's agricultural extension program actually led to a new function for the Department of Agriculture. The Rockefeller Foundation took the discoveries concerning hookworm and yellow fever, both developed within the government, not only into the Southern United States but to other parts of the world. In addition the Foundation put large sums into basic research in many fields.[24]

The great foundations were a monument not only to the peculiar distribution of wealth that produced enormous fortunes but also to the effectiveness of science, which the donors now saw as a leading contributor to the well-being of mankind. With the government, the universities, and the foundations all in the field, room for fruitful investigation still remained in so many directions that they all expanded greatly in the same years. Since communication is a component part of science, the increased activity of public and private agencies, instead of producing unhealthy competition, made a denser matrix of new lines of discovery which aided all investigators. The growth of both universities and the foundations with their emphasis on basic research relieved the government of many responsibilities. This new partnership made government science seem to lose out in what was really a period of outstanding accomplishment.

Comparison with the Rest of the World

Science's lines of communication could not stop at national boundaries for government bureaus any more than for any other kind of researchers. The presence of Europe, fount of basic discoveries and models for scientific institutions, loomed large even as late as 1900. Throughout the first century of the republic, colonialism in basic discoveries had persisted, although in the forms of scientific institutions, including government agencies, direct borrowing had long since diminished markedly. In some fields of applied science America even began to export men and methods. One measure of the accomplishments of the government in science is the extent to which its techniques became exportable. Soon after the Civil War a trickle of agricultural missions began — to China, Japan, and Latin America. Men trained in the Geological Survey took up mining for the Chinese government, and Bailey Willis tried to apply its concepts concerning arid regions to Argentina.[25] After 1900 such missions went out in a steady stream.

Another sign that America's scientific dependence on Europe was diminishing showed itself in the large number of laments about the New World's inferiority. The beating of breasts over a condition long taken for granted implied a change. Carl Snyder, writing for a general audience in the North American Review at the turn of the century, added up the great discoveries of the nineteenth century, stressing the preponderant role of European investigators. Admitting the occasional appearance of a Franklin or a Henry and the large sums of money spent on science, he indicted the absence of both fundamental research and any institutions to compare with the Collège de France, the Royal Institution of London, the Pasteur Institute, or the German university system.[26]

Simon Newcomb, while generally corroborating Snyder on the paucity of creativeness in American science, felt that this unflattering view was nearer the truth in 1880 or even 1890 than in 1900. In balancing the various factors he pointed out "that no government is more liberal than our own in enterprises for the promotion of science." [27] Nevertheless, the government had a share in the responsibility for the lack of esteem in which real research was held. It had done nothing for the National Academy and never honored outstanding individual scientists. This was the "natural outcome of that gap between the world of politics and the world of learning which is so marked a feature of society as it exists today, both in the country at large and at the National Capital." [28]

Even though the accomplishments of the Progressive Era were yet to come when he wrote, Newcomb was unwilling to blame democracy itself for the shortcomings he found. Another commentator, who ranked the United States fourth in scientific attainment after Germany, Great Britain, and France, also refused to blame democracy.[29] But the question of the influence of democracy looms above bootless national comparisons as a central and fundamental dilemma.

Joseph Priestley's ardent belief in the congeniality of free institutions to science had not had an easy victory. On the other hand, the gloomy prediction that only monarchy could patronize science adequately had remained unfulfilled. Every generation of Americans had trouble adapting scientific institutions to their form of government. The masses and their representatives often lacked both interest and understanding for a task always delicate and complex. The absence of explicit rules in law and custom provoked endless controversy. Yet every generation did make the attempt, leaving its mark both on science and on the political framework. Heroes emerged — Hassler, Bache, Powell, Billings, Pinchot — to fight the battle for their time. Whether they were radicals or conservatives politically is irrelevant to their common belief that democracy would serve science, which in its turn would serve the people.

In the Progressive Era the trend culminated in the belief that science fostered democracy, that its freedom, devotion to truth, and objectivity were necessary ingredients of economic and political justice. W J McGee concluded, "America has become a nation of science. There is no industry, from agriculture to architecture, that is not shaped by research and its results; there is not one of our fifteen millions of families that does not enjoy the benefits of scientific advancement; there is no law on our statutes, no motive in our conduct, that has not been made juster by the straightforward and unselfish habit of thought fostered by scientific methods." [30]

The optimism, the sense of achievement, that reigned in both political and scientific circles before 1914, benefited more than most people then realized from the ninety-nine years of peace between Waterloo and Sarajevo. The Civil War had been a peculiarly American interlude. Only in peace could nationalism and universal science grow up together. After the outbreak of the war in Europe the United States had two more years to complete her scientific establishment designed for a peaceful world, and two years to realize that the horror overseas concerned her deeply. Beginning in 1916, the most pressing problems of government and science in the United States stemmed from wars of unimagined complexity and from fearful threats both to democracy and to the freedom of science.

XVI

THE IMPACT OF WORLD WAR I

WHEN the Allies and the Central Powers, in late 1914, settled into a stalemate of trench warfare, a new age had arrived. It was a total war of production and the attrition of the fiber of nations, a war of seemingly endless indecisiveness. The main factors were massed manpower and materiel. Only dimly a few men began to see that the way around the trenches lay in technological unconventionality — in using weapons that, although known before, had seemed fantastic. Although no nation had accurately foreseen the nature of the conflict, the Germans with their background of technological efficiency promptly introduced poison gas and the submarine. The Allies gradually groped toward an adequate use of the tank and the machine gun. But the United States, which in the early days of the war could imagine the new nightmare only vicariously, reacted very slowly. Not until the eve of our entry into hostilities did the federal government, in a tremendous flurry, attempt to reorder the whole relation of research to both the military and mass industry. In less than two years the Armistice as abruptly terminated the experiment.

Two related but distinct trends reached their culmination with America's entry into the war. The first was the tendency to large-scale mechanized industry which geared the whole economic and social life into a common effort of total war. The second was the application of scientific knowledge and methods to the technology both of weapons and of industry. Both these trends had been present in relatively undeveloped form during the Civil War. The intervening years of peace had brought them to completion even though the military had not participated very largely either in assisting or in understanding them.

The Military Research Establishment in 1914

Except for some enclaves such as the Naval Observatory, the old scientific spirit of the early nineteenth century had disappeared from the armed services. This loss was somewhat less serious because science was now called on for a different job. Exploration with its requirements of general information had given way to the application of science to the weapons of war themselves. The Corps of Engineers, for instance, had little left but civil works from its earlier activities. Hence the survivals of the old interests were of less importance than either the first rustling of weapons research in some of the services or the beginnings of a mechanism of evaluation by which new weapons could be developed, selected, and adapted to actual field use.

The Army had made a start at setting up a general staff and breaking the autonomy of the old service bureaus, thus clearing a possible channel for the introduction of new weapons and the revision of tactics and strategy to exploit the changes. In actual research, however, only the Signal Corps showed much energy and imagination. After losing the Weather Bureau to the Department of Agriculture, this branch had established itself in the field of military communications, and, beginning about 1907, became the repository of Army interest in radio and aviation. Although unable to get appropriations from Congress, the small Aviation Section had done some testing of planes, machine guns, and bombsights when in 1914 the need for pilots required the shift of all their planes to training.[1] The concept of aviation as a part of signaling equipment was to have important repercussions on its use and design during the war. The Signal Corps's work on radio was also tied closely to its earlier essential reliance on telegraphy.

In general, however, the Army's attitude toward both science and technology was far from dynamic. Even though most of the nearly two million automobiles in the world in 1913 were in the United States, the Quartermaster Corps was still testing mule wagons as well as trucks.[2] A Board of Ordnance and Fortifications had been more or less active since the 1880's, spending some money for research, but it had lost most of its control over the work of the technical services.[3] The Springfield rifle in use by the Army had been adopted in 1903, and no light artillery piece measured up to the French 75-millimeter gun of 1897.[4]

The Navy was far behind the Army in its command structure, the Office of Chief of Naval Operations being created only in 1915 with very modest powers. Yet by 1914 the naval revolution which had been in its early stages during the Civil War had culminated in the battleship of the *Dreadnaught* type, and the United States had reflected the changes. The *Nevada* and the *Oklahoma*, commissioned in 1916, compared well with any battleships in the world and completely overshadowed the ships that had fought the Spanish-American War only eighteen years before. David W. Taylor had made a brilliant name in ship design, and reforming officers such as William S. Sims and Bradley Fiske had introduced changes in gunnery procedure that had increased both accuracy and rate of fire. The use of directors and mechanical computers was beginning by 1914. The shift from coal to oil as fuel brought important advantages.[5] The Navy had shown an early interest in radio, and its contracts had been an encouragement to some of the pioneer inventors such as Lee De Forest.[6] Profound as these changes were for naval warfare, they seldom proceeded directly from scientific research on the part of the Navy Department, but rather were wholesale borrowings from abroad, especially from the British. Sims was emphatic that his new system of gunnery was "taken bodily from Percy Scott," the reforming British admiral.[7] In the second place, although some of the early development of the submarine had taken place in the United States, the Navy had made no adequate assessment of the role of this new craft nor had it foreseen the need of developing new means of combating it. The main asset of the Navy was a familiarity with technological change and, in at least some quarters, an open attitude toward it.

Although spawned in a period of peace and directed toward nonmilitary problems, the civilian scientific bureaus of the government had as much to offer in research resources for a total war as had the military itself. Especially the new bureaus oriented to industrial problems — Standards and Mines — possessed skill of high importance which was much more completely developed than in the military. The Department of Agriculture by minor conversion had research resources for such important military supply problems as nitrates. The National Advisory Committee for Aeronautics, although still in embryo, had obvious relevance to the military. Since total war was to extend back to the sources of basic raw materials, the conservation agencies such as the Geological Survey and the Forest Service could bring their facilities to bear. The great drawback to the civilian establishment was its orientation around peacetime problems, the difficulty of dropping its usual work, and above all the lack of leadership from the military in the selection and priority of problems. No adequate administrative mechanism existed to mobilize the government's scientific establishment for war.

The Evolution of a Central Research Organization

Although the European battlefields began to teach clear lessons in total war almost immediately, the diplomatic position of the United States militated against quick response by increasing research. All moves to step up military activity seemed to threaten the policy of neutrality that Woodrow Wilson pursued. In December 1914, the President made no recommendation for an enlarged budget for the Army and Navy. Sentiment for "preparedness" began to appear and when espoused by Republican leaders such as Theodore Roosevelt and Elihu Root became a live political issue.[8] Only in 1915, after the unrestricted submarine warfare had definitely moved the administration toward a position of enforcing strict neutrality with regard to Germany, did Wilson become a convert to the policy of preparedness.[9] As the attitude of the President and his advisers changed, the way opened for the development of a central scientific organization geared to the war. Because the existing bureaus were entirely inadequate to a task of such magnitude, a mobilization of the nation's total scientific resources outside as well as within the government was essential.

The government as a whole responded to the emergency by creating a great apparatus entirely separate from existing machinery. These temporary organizations, ill-defined in both powers and functions, often went through many changes before attaining proper balance.[10] An act of August 1916 set up the Council of National Defense, made up of the secretaries of war, Navy, interior, agriculture, commerce, and labor, to "coördinate industries and resources for the national security and welfare." [11] Looked upon at first as an investigating and research body preparing for a "future war of defense inferentially far distant," the Council of National Defense soon found itself plunged into an executive position for which it was not well fitted.[12] After many mutations the War Industries Board and the Food Administration emerged under the strong leadership of Bernard Baruch and Herbert Hoover to control large segments of the economy.

Of course, the War Industries Board often found itself pushed directly into research activities by its own momentum. A technical and consulting staff gradually developed which had the full use of the Mellon Institute in Pittsburgh.[13] Especially active in the study of nitrates and strategic minerals, the War Industries Board sometimes clashed with the Geological Survey and the Bureau of Mines, the points at issue revolving around the stimulation of industries for the domestic production of metals such as chrome and manganese.[14] But in general the emergency organization was more interested in immediate large-scale results. Research was only a last resort, and its organization was left to those with a direct interest in it.

The Navy was first in reacting to the need for a central research organization. Secretary Josephus Daniels, whether or not he had any conscious knowledge of it, picked up precisely where Gideon Welles had left off in 1865. Then the Permanent Commission had done the most effective work accomplished during the Civil War by screening hundreds of unsolicited inventions. Now Daniels appointed a Naval Consulting Board whose chairman, Thomas A. Edison, was the very embodiment of the spirit of the inventor. On July 7, 1915, Daniels wrote to Edison, "one of the imperative needs of the Navy . . . is machinery and facilities for utilizing the natural inventive genius of Americans to meet the new conditions of warfare as shown abroad." He wished to set up "a department of invention and development, to which all ideas and suggestions, either from the service or from civilian inventors, can be referred for determination as to whether they contain practical suggestions for us to take up and perfect." [15] After Edison had sent an emissary to Washington indicating his willingness to serve, Daniels went to East Orange, New Jersey, and worked out the details of the board. Except for the chairman and one other, the membership was chosen by the eleven largest engineering societies of the United States. The resulting list was widely representative of engineers and inventors. Several of the pioneers of industrial research served — Willis R. Whitney, Frank J. Sprague, L. H. Baekeland, and Elmer A. Sperry, among others.[16] Significantly, no representation was given to the National Academy of Sciences, and only a few of the men chosen had any close connection with either university or government science.

Starting off without congressional authority or appropriation, the Naval Consulting Board divided itself into committees on almost the whole range of scientific and technological problems of interest to the military: chemistry, physics, aeronautics, internal-combustion engines, electricity, mines and torpedoes, ordnance and explosives, wireless and communications, transportation, production, ship construction, steam engineering, life-saving appliances, aids to navigation, food and sanitation, and public works.[17] Vigorous research programs along all these lines would have blanketed not only the Navy's but the whole nation's research effort.

Edison's position was as crucial to the whole enterprise as his name was to its renown. He made it clear to Daniels that he took up the work as an inventor and not as an administrator.[18] Because of his sixty-eight years and the impairment of his hearing, his decision was natural, but it left the board without vigorous executive leadership, a lack that strongly influenced its subsequent career.

The members of the Naval Consulting Board early realized that the inventors scattered through the country could not be expected to answer the problems of a kind of warfare that even the professionals hardly understood. They were also aware that the Navy had almost no facilities that could be used for research. They, therefore, proposed a naval laboratory under the command of an officer with a staff of "civilian experimenters, chemists, physicists, etc." They believed that "secrecy should be a governing factor." Besides the staff, an inventor should find there facilities for developing "the idea he has presented, provided he is a practical man." They envisaged an investment of $5,000,000 and an annual appropriation of $3,000,000. In March 1916, Daniels, Edison, Baekeland, and other members of the committee testified before the House Naval Affairs Committee, securing an authorization for the laboratory by way of the time-tested route of an appropriation bill. The initial sum was $1,000,000.[19]

This step could have placed Navy research on a new plane. Indeed, it provided the legal basis for the establishment of the Naval Research Laboratory in 1923. But in 1916 the members of the board were unable to capitalize their splendid opportunity. They split on whether to locate the laboratory at Annapolis, Sandy Hook, or on the Potomac near Washington. When war broke out in April 1917, the whole project dropped and the money remained idle.[20]

The major activity of the Naval Consulting Board became then the screening of the inventions with which the public in this as in other wars deluged the government. This work, the province of the Permanent Commission during the Civil War, was handled during

World War II by the National Inventors Council of the Department of Commerce.[21] Culling 110,000 suggestions was an onerous task for which the members of the Naval Consulting Board, busy executives of industrial concerns, were not entirely suited. Only 110 inventions had enough merit for detailed examination by the subcommittees and only one went into actual production. No clearer proof is needed that in time of total war random ingenuity is no alternative to the problem approach by teams of highly trained men thoroughly aware of both scientific theory and the needs of the services.[22]

The National Academy of Sciences provided in theory a legal vehicle for a broadly based organization of the nation's scientists. Yet it was so unaccustomed to action of any kind that it did not enter the field until several months after the Naval Consulting Board. Most people had forgotten the tribulations of the Civil War period and believed that the Academy had once been an effective war organization for science. Everyone equally recognized that in the moribund state which had overcome it in the early twentieth century the National Academy could not automatically take up the burdens of a central scientific organization. That the Academy finally was able to do anything was largely the result of the efforts of a group of reformers within its own ranks.

George Ellery Hale, as a friend pointed out, was an astronomer who regarded "himself as doing his best work as an initiator and promoter of scientific enterprises." [23] Director of the Mount Wilson Observatory, he was a thoroughly competent and respected scientist. He had shown great enterprise in raising money for expensive telescopes, establishing close relations with the Carnegie Corporation and the president of its board, Elihu Root. As early as 1910 Hale had become the active leader of a movement to rehabilitate the National Academy when he had become its foreign secretary. Looking across the Atlantic, he wished to make it a true counterpart of the Royal Society and the French Academy.[24]

With the outbreak of war in 1914 Hale's ideas of reform merged with an intense desire for preparedness, an enthusiasm for the Allies, and a critical attitude toward strict neutrality. His friend Elihu Root was, of course, one of the eminent Republicans who followed the lead of Theodore Roosevelt in attacking Wilson on these grounds. On July 13, 1915, Hale said in a telegram to Dr. William H. Welch, the president of the Academy, that it "is under strong obligations to offer services to President [of the United States] in event of war with Mexico or Germany." This stand was doubly bold because the Academy usually waited for its invitations and because it implied that war was imminent. Welch, while he could "imagine no objection to the Academy offering its services to the President in the event of war," considered such a thing improbable "while Wilson is president." [25]

Nothing further happened until April 1916, when Hale moved a resolution to offer the resources of the Academy to the President of the United States "in the event of a break in diplomatic relations with any other country." [26] The vote was unanimous in favor, and Welch blandly fell into line, taking a committee which included Hale and C. D. Walcott to see Wilson. After the President accepted the offer, the Academy began the slow process of evolving a new mechanism for dealing with the emergency. An organizing committee under Hale as chairman was made up entirely of relatively young men, including Robert A. Millikan, a physics professor at the University of Chicago who had a burning belief in the cause of the Allies. [27]

By common consent, membership in the National Academy was abandoned as the basis for participation, the organizing committee reaching out in all directions for investigators who could actually do the work. In contrast to the Naval Consulting Board, which represented only inventors and the engineering societies, Hale tried to include all the great estates of science in the country. The committee recommended that "there be formed a National Research Council, whose purpose shall be to bring into coöperation existing governmental, educational, industrial, and other research organizations" in strengthening the national defense. [28] Government, the universities, the foundations, and industry all had a genuine place. On July 25, 1916, President Wilson gave his blessing in a letter to Welch, promising coöperation from the departments and agreeing to appoint representatives of government bureaus as members of the Council. [29] The fact that this letter was published, in effect announcing the project, eased Hale's way considerably. [30] Both Massachusetts Institute of Technology and the Throop College of Technology (later California Institute of Technology) began expanding research facilities in anticipation of coöperation with the Council. [31] This broad base of coöperation was the great strength and one of the great accomplishments of the new organization.

Although the desire for preparedness and strong pro-Allies emotions were the immediate spur to Hale and his group, they did not abandon the long-range aim of reforming the Academy and through it the research institutions of the country. "It was recognized from the outset," wrote Hale, "that the activities of the committee should not be confined to the promotion of researches bearing directly upon military problems, but that true preparedness would best result from the encouragement of every form of investigation, whether for military and industrial application, or for the advancement of knowledge without regard to its immediate practical bearing." [32] Accordingly, the NRC reflected some of Hale's characteristics. It relied initially on the support of the Carnegie and Rockefeller Foundations for its money. It tried to foster permanent research institutions. And it emphasized the international aspects of science.

Hale went to Europe in August of 1916 to establish liaison with scientists in the Allied governments and laid the basis for that flow of information from the war itself which was necessary if American scientists were to comprehend either the magnitude or the nature of the problems. Hale's enthusiasm for the Allies and his interest in international exchange of scientific information made this duty especially congenial. [33]

In September 1916, the National Research Council held an organization meeting in New York. Hale became permanent chairman. Gano Dunn of J. D. White Engineering Corporation and John J. Carty of American Telephone and Telegraph Company played a major role, bringing an important segment of the research engineers into line. Carty buried the hatchet between "pure and applied research, pointing out that they do not differ in kind but merely in the objects to be accomplished." [34] Columbia's Michael Pupin also stressed the alliance between the scientists and engineers, speaking "at length on the value of coöperation in industrial research, as evidenced by the work of the Research Laboratory of the General Electric Company." The Engineering Foundation, a new private organization, placed funds at the disposal of the NRC and also the services of its secretary. [35] Government scientists such as C. D. Walcott and S. W. Stratton received prominent posts. The first list of members showed both a breadth and a practicality that the Academy itself could never have achieved.

The executive committee met several times in the fall of 1916, setting up a military committee of high-ranking army and navy officers. A beginning was made on committees representing scientific disciplines. Chemistry led off, followed by mathematics, astronomy, physics, geology and paleontology, geography, botany, zoology and animal morphology, physiology, medicine, hygiene, agriculture, psychology, and anthropology.[36]

However, a full-time working organization did not immediately follow. Hale went back to California without providing a permanent director for the NRC. Only in February 1917, with the severance of diplomatic relations with Germany, did Hale and R. A. Millikan throw up their private jobs to go to Washington. Millikan, as chairman of the physics committee, was already being inundated with requests from the military committee to do something about the submarine. In addition, Hale soon created for Millikan the post of "Third Vice-Chairman, Director of Research, and Executive Officer of the National Research Council." [37] Thus, the Chicago physicist, not yet fifty years old, became the key man administratively while still retaining his earlier post, which entailed active investigation in physics. Hale had some hopes of making Millikan the permanent counterpart of the secretary of the Royal Institution — an investigator making a career for himself in the capital and controlling American science largely by his influence and example. Millikan himself, however, subordinated everything to "helping in the war." [38]

With the outbreak of hostilities in April 1917, the Millikan point of view gained quick ascendancy, and Hale, whose particular talent was getting an organization started, soon withdrew into the background. In August 1917, he left for Pasadena for several months to see to the mounting of the telescope at Mount Wilson Observatory, and soon his restless energy was redirected toward forming an International Research Council.[39]

Meanwhile a small full-time staff gathered in Washington, and the NRC began reaching out in many directions for the power and support it needed to become an effective war-research agency. The main early personnel, besides Hale and Millikan, were a public health expert, a chemist, a second physicist, and an aeronautical engineer.[40] During several months, late in 1917, Millikan held down the office virtually alone.[41] Only in 1918 did the NRC get an active central office.

An early headache was money. Although the establishment of the NRC got around the statutory prohibition against members of the

National Academy receiving reimbursement for their services, no government appropriation was immediately in sight. Accordingly, the private sources most concerned with the NRC's activities came to its rescue. The Engineering Foundation had given small but significant support in the fall of 1916. However, the first substantial sum came when the Carnegie Corporation, of which Hale's friend Elihu Root was chairman of the board, made $50,000 available. An additional $100,000 followed in May 1918. The Rockefeller Foundation gave $50,000 in February 1918 for the Division of Medicine and Related Sciences.[42] A number of smaller gifts added to the total of private support. While it is impossible to estimate the total amount of government expenditures through the NRC during the war, the private funds remained a considerable percentage of the total financial support, and without them the operations of the central office in particular would have been drastically curtailed.[43]

In February 1917, the Council of National Defense requested the NRC to act as its department of research, responsible for "the organization of scientific investigation bearing on the national defense and on industries affected by the war." [44] At almost the same time the Council of National Defense appointed the Naval Consulting Board as its board of inventions.[45] Such arrangements did not fall into place automatically. For a time the NRC feared that the Council of National Defense would set up a scientific committee under a man generally considered incompetent. Welch, as president of the National Academy, exerting his influence on the Council of National Defense, secured the desired orders while resisting pressure from his own people to appeal directly to Woodrow Wilson.[46]

Basically, the NRC was fortunate not only to get the blessing of the over-all emergency body at the outset but also to have its sphere of authority clearly marked off from that of its Navy rival. Nevertheless, money from the Council of National Defense did not appear on the NRC's books until January 1918, when President Wilson authorized a grant of $29,250. The total sum received during the war through this channel was $128,650, somewhat less than the combined private support.[47]

Part of the funds from the Council of National Defense made possible an NRC activity important out of all proportion to its cost. The Research Information Service was a direct outgrowth of the chilling realization that American science was profoundly dependent on European science in general and on Allied military research in particular. This was apparent as early as Hale's visit to Europe in 1916. A mission to the Allies under Joseph S. Ames went over in the opening days of the war, and return missions of scientists from the major Allies were accredited to the NRC.[48] By early 1918, the funds received from the President through the Council of National Defense made possible a regular branch of the NRC known as the Research Information Service. With a central committee made up of the chief of military intelligence, the director of naval intelligence, and S. W. Stratton for the NRC, the new organization opened offices in London and Paris. Two leading scientists sailed in February 1918 to direct these offices from the advantageous posts of scientific attachés to the British and French embassies. A similar arrangement was later made in Rome.[49]

This type of activity had an especial appeal to those such as Hale who had the permanent ideals of the NRC at heart. "Properly regarded, this Information Service may be considered as the pioneer corps of the Council, surveying the progress of research in various parts of the world, selecting and reporting on the many activities of interest and importance . . . and disseminating it to scientific and technical men and to institutions which can use it to advantage." [50] The results of the Information Service, although by nature indeterminate, form a bright chapter in the annals of the NRC.

The general usefulness of the NRC had to be measured in the final analysis by its relations with the military. If science was to take a new place in the conduct of war, it had to do it in the field of weapons research, and the armed services were jealous guardians of their own preserves. NRC scientists such as Millikan were able to establish cordial but informal relations with the Navy through the Military Committee, while the existence of the Naval Consulting Board discouraged too close a legal tie. But the chief signal officer took a more aggressive attitude. General G. O. Squier was a man described by Millikan as "a strange character who . . . was in no sense an organizer nor a man of balanced judgment, but he had one great quality much needed at that time, namely, a willingness to assume responsibility and go ahead." [51] A graduate of West Point but also a Johns Hopkins Ph. D. and an electrical engineer,[52] Squier had been so impressed with the value of research for war that as early as 1916 he "predicted that the United States will probably find it desirable to appoint as a Cabinet officer a Secretary of Sciences in the not distant future." [53]

At the outbreak of war, Squier began to envisage the large-scale use of research. He wrote to Hale in July 1917 that in "the Signal Corps questions involving the selection and organization of large numbers of scientific men and the solution of research problems are constantly arising." Considering the NRC "the one agency in a position to meet the present needs," he requested "the research council to act as the advisory agent of the Signal Corps." To accomplish this he resorted to a very direct and personal scheme. "I would suggest that Dr. Robert A. Millikan, vice chairman and executive officer . . . apply for a major's commission in the Officers Reserve Corps, for detail in charge of this work." [54] With this beginning, Squier put virtually the whole physics committee into uniform and hence under orders. By capturing the executive officer himself, he acquired a certain military control over the whole NRC. Rather ironically, one of the most pressing needs of aviation in the Signal Corps was a meteorological service, a development entirely unforeseen in 1890 when the Weather Bureau was transferred to the Department of Agriculture.

Millikan was "not keen" about going into uniform, "for I was quite as active in connection with the Navy as with the Army, and in addition had had thus far free access to the offices even of the Secretaries of both War and Navy, which men in service in general did not have." [55] While thus recognizing at least partially the consequences of this step, Millikan admitted that throughout the government civilians chosen for large responsibilities were being taken into the armed service. Even Dr. William H. Welch, the portly president of the National Academy,[56] appeared on the streets of Washington in uniform. This direct action was the World War I way of bringing civilian resources into the service of the military. No mechanism existed for directing large funds for military purposes into civilian hands. The only alternatives were either to work out some new procedure or to take the civilians into uniform. In the confusion of the summer of 1917, the latter seemed to most of those involved the only course. Soon the practice spread to other activities of the NRC. Sound ranging for artillery began under the Signal Corps when the NRC recommended Augustus Trowbridge and Theodore Lyman as scientists to work out the procedure. The chief of ordnance then got cognizance and issued commissions to the same investigators. When they went overseas they were transferred to the Engineers, with whom they served at the front.[57] A. A. Michelson went to work on range finders at the behest of the NRC, but eventually the Navy's Bureau of Ordnance took control, Michelson receiving a commission.[58] Yet, this trend made it hard for the NRC to develop a balanced program for all the branches of the services and also rendered almost impossible the role of impartial critic and initiator of ideas that the Office of Scientific Research and Development during World War II so cherished.

Enough has been said of the activities of the wartime NRC to indicate that it was far from a tightly organized bureau, omnipotent within its sphere, actually directing all phases of the research effort. Millikan, commenting on the vagueness of early NRC records, emphasized that the "confusion was in the *situation*, which never got sharply into the records, rather than the way such records" appear.[59] His list of organizations with which the NRC had trouble is perhaps instructive in hinting at the actual boundaries of authority. He claimed it took longest to "work the right relation to . . . the Naval Consulting Board, the National Advisory Committee for Aeronautics, the Bureau of Standards, and the engineering societies." [60] Of these the NACA and the Bureau of Standards were especially significant because they represented new and vigorous research enterprises which tended to develop their own programs even though all their leaders had places in the NRC.

The Wartime Research Effort

Any given research program was likely to be the shared responsibility of several agencies with varying objectives, resources, and levels of activity. To trace the ramifications of the major research efforts would be impossible in a small compass even if much more were known about them than appears in available literature.[61] All that is possible is to indicate by example some of the major programs with a few aspects of their scientific and administrative setting.

The position of medical research clearly reflected the stresses of mobilization and war. A whole generation of medical accomplishment made it possible to hope for unprecedentedly good results in the care of battle wounded, in the protection of troops from disease, and in checking epidemics through the population as a whole. Virtually all the wartime agencies had some relation to medicine. The Council of National Defense had its General Medical Board which had a committee on research.[62] The NRC set up a Division of Medicine and Related Sciences under Victor C. Vaughan, an authority on public health, with a number of committees gathering information on medical activities.[63]

The Army, however, quickly became and remained the dominant force in medical research. W. C. Gorgas as surgeon general gave commissions both to Vaughan and to Welch, who became the "liaison man between America's medical laboratory men and the army."[64] In February 1918, the NRC reorganized its division under the chairmanship of commissioned medical officers. "The general plan has been to follow the advice of representatives of the War, Navy, and Labor Departments in determining urgent problems, and then to find the proper workers to investigate them."[65]

With the exception of its futile efforts to check the influenza epidemic, the record of medicine in the war was so outstanding that it introduced a new era of warfare in which the diseases that had once ravaged armies and civilians alike were kept under control. Yet how much of this gain was the result of research conducted during the war is hard to determine. The accumulated discoveries of the years since the last previous war had much to offer, and improved administration and hospital facilities doubtless also contributed. Most authorities point out, in addition, that much basic research was interrupted. Immediate remedies rather than knowledge of disease tended to take precedence. The stimulating effects of the war are to be found in the mass clinical opportunities and the general shaking up of the whole profession rather than in the research program and resultant discoveries from it.[66]

An activity originally organized under medicine in the NRC unexpectedly grew into an independent science during the war. Psychology was still struggling to find its own first principles when the great masses of draftees requiring classification gave it an unprecedented opportunity. As one of the scientists most responsible for the program said, "Fortunately alike for the science of psychology and the army, the practical work overrode the disadvantages of its name and ultimately converted psychology into a word to conjure with in the United States Army."[67] Both the Alpha and Beta tests, one of which was given to every soldier, and procedures for selecting men for specialized tasks convinced the most skeptical officer of the practical value of the new science.[68]

Weapons research for the armed services had to contend with two deterring factors which appeared in every program. The first was the overwhelming necessity of producing large quantities of materiel and the consequent desire to standardize early, usually by borrowing battle-tested designs from the Allies. The second factor was the relation of time lag in research to the probable end of the war. Since weapons research did not really begin on a large scale until 1917, results that could be developed, produced, and shipped across the Atlantic for use at the front in 1918 were not to be expected. Only the assumption that the war, which already seemed an eternity to the European participants, might last to 1919 or 1920 gave research any priority at all.

Aviation was so new as a weapon that the need for more knowledge assailed the government from all sides, quickly forcing those responsible to resort to wholesale borrowing in an effort to get into large-scale production even before the possible uses of aircraft in war were clearly understood. Hence much effort went into groping attempts to evolve an air doctrine on which requisitions could be based. A series of organizational shifts involving the American Expeditionary Force, the chief of staff, occasional technical missions, and many of the emergency agencies eventually set in motion a massive program of airplane building on British and French designs which was just beginning to roll in the fall of 1918.[69]

As aviation mushroomed, the chief signal officer gave official encouragement to an increased initiative on the part of the scientists he had taken into his outfit from the NRC. But early in 1918 the Signal Corps lost all control of aviation, throwing its Science and Research Division into a kind of no man's land between its former superior and a production-minded organization called the Board of Aircraft Production.[70] Near the end of the war, with the creation of a separate Air Corps, Millikan "was instructed that my whole physical science group . . . was to be transferred." The only difference it made was that "all of us . . . took off Signal Corps insignia and put on the 'wings.' "[71] Nevertheless, these shifts of authority meant that the NRC, in addition to falling under military control, had its position within the army structure confused. Even though the research group tagged after aviation through all these wanderings, it failed to become a sufficiently implanted part of the military to achieve permanence.

Within the military and production framework, the NACA and the NRC attempted to carry on a research program. The NACA started building its facilities at Langley Field, Virginia, but at the end of the war its first wind tunnel was still unfinished.[72] Army research centered at Dayton, Ohio, where Millikan's uniformed NRC personnel did most of their work. At the time of the Armistice the NRC people concerned with aviation were 22 officers, 121 enlisted men, and 16 civilian scientists.[73] The closest student of the problem concludes that the "achievements of the United States in creative design and experimental engineering, as contrasted to the result of production engineering, were important only insofar as they marked the growth of a new industry and developed a body of experience to guide the War Department in the postwar era." [74]

The need for a means to combat the submarine was the most pressing one facing the Allies in 1917 and also the one most insistently calling for scientific research. Since the British and French were as baffled as anyone, borrowing designs for production was impossible. Sir Ernest Rutherford, who came over with one of the early British missions, described it as "a problem of physics pure and simple." [75] Of several possible approaches, detection of the submarine's position by listening devices seemed to offer the most immediate hope of success.

The Naval Consulting Board entered the submarine-detection business in February 1917, when it set up a Special Problems Committee under which Willis R. Whitney of the General Electric Company had responsibility for detection by sound. He soon secured the coöperation of the Submarine Signal Company of Boston and later of the Western Electric Company in setting up a station for experiments at Nahant, Massachusetts.[76] At this point the NRC physics committee under Millikan was also becoming so active that the secretary of the Navy created a special committee with an admiral as senior member and with civilian advisory members from the Naval Consulting Board (Whitney), the NRC (Millikan), the Submarine Signal Company, and the Western Electric Company. Of the group of scientists drawn from industrial research working at Nahant, Irving Langmuir was the most notable.

Millikan soon moved, however, to get university physicists to work on sound detection in an independent project. He managed to get facilities at New London, Connecticut, for a group of men drawn from Yale, Chicago, Rice, Cornell, Wisconsin, and later Harvard. With the advantages of the naval base there, New London soon became the main center for research, merging the university and industrial teams. From these groups emerged a listening device for determining the bearing of a submarine by picking up its noise.[77]

In this field as in others the first glimmerings of a workable device led immediately to problems of development for service use. Although Max Mason, who produced one of the more promising sound-detection devices, was part of the university group, he turned to General Electric for development and production. By 1918, the Navy took over the financial responsibility for the New London station which the NRC had assumed in the early days.[78]

The physicists soon found themselves enmeshed not only in the development and production of listening devices but in the design and production of escort vessels. By Christmas 1917, Millikan was in touch with Henry Ford, who eventually began the mass production of eagle boats, designed to carry, among other things, listening gear.[79] Here again the factors of time and mass production invaded research programs and tended to dominate them.

In retrospect, the World War I research on submarine detection seems scarcely to have dented one of the most intractable problems of twentieth-century warfare. The depth charge, the convoy system, the mine, and old-fashioned seamanship counted for much more during the crucial days of 1917, while both submarines and detection gear moved into new ranges of performance before the outbreak of World War II. Yet the very approach to the problem as one that could be solved only by massed and coördinated scientific resources demonstrated clearly that a new era of warfare had arrived and that science had an essential place in it.

No weapon of World War I shocked the world public as did poison gas, which seemed by the very process of its generation somehow the machination of scientists. Although gas had been in use on the Western front for two years before 1917 and although the substances used had been well known in the laboratory, the United States' chemical establishment, extensive as it was, had almost no experience with the particular problem. The NRC had a committee on the uses of gas in warfare which appears not to have been very active as such.[80] The Army had almost no experience at all. The real center of knowledge and the possible vehicle for swift action was the Bureau of Mines, which had been working on gas as the cause of mine disasters for several years. In February 1917, the Bureau of Mines, calling attention to its abilities through the NRC, embarked on a program of adapting oxygen-breathing apparatus used in mining for use as a gas mask.[81] The problems of offense and defense being closely intertwined, research under the Bureau of Mines quickly burgeoned on gases as well as masks, leading to a marked expansion of personnel and, in the fall of 1917, the creation of a central laboratory at the American University in Washington. Out of this effort came a new gas, lewisite.

By this time groups interested in separate phases of gas warfare had cropped up in the Medical Corps, the Ordnance Department, the Signal Corps, and the Corps of Engineers. The AEF, as usual somewhat ahead in its organizational adjustments to the realities of war, set up a separate Gas Service, and General John J. Pershing cabled in September 1917, "Send at once chemical laboratory, complete equipment and personnel, including physiological and pathological sections, for extensive investigation of gases and powders." [82]

Two trends had now become quite clear, one in the direction of a major gas-warfare command within the Army, and the other in the direction of embracing all activities, including research and production as well as tactical use, within the military framework. Few chemical companies had any real interest in such toxic materials, and the Bureau of Mines with its other responsibilities could hardly expect to carry on the whole research program indefinitely. Branch laboratories were springing up in various universities and the problems ramified far beyond those that the Bureau of Mines had formerly cultivated. Van H. Manning, chief of the bureau, became apprehensive that his "gas work" would go to the Army even though he "had planned the work and gathered the force." [83]

The inevitable shift came in July 1918, with the creation of the Chemical Warfare Service. The American University Experiment Station went over intact to the Research Division of the new service. Not only were the chemists hired from the universities given commissions in accordance with the usual World War I practice, but also the key Bureau of Mines personnel. G. A. Burrell, who had headed the work for Mines, became a colonel and chief of the Research Division.[84] Production centered in Army hands at Edgewood Arsenal.[85]

Great as was the impact of science on the Army in creating a whole new service for gas warfare, chemistry played on a much larger stage than weapons research alone. Indeed, chemical production was one of the avenues by which military considerations merged into economic problems and eventually penetrated deeply into the fabric of American society. Science, caught in these ever-widening influences, was carried into all parts of the economy. Since chemistry was at just the proper stage to feel these impulses and to contribute most, this was in a real sense a "chemist's war." [86]

The need for helium to inflate airships and balloons touched off a major production program of an element that had before been a rare laboratory curiosity. As in the case of poison gas, the Bureau of Mines possessed some background information which made Burrell's group there the first center of activity. F. G. Cottrell of the bureau was already interested in processes that might make possible the separation of helium from the natural gas of certain Oklahoma and Texas fields.[87] Meanwhile, as a result of a British mission to Washington, the Army, Navy, and NRC became very enthusiastic about helium, and out of a bewildering series of conferences [88] emerged two plants at Fort Worth, Texas, and one near the Petrolia field in Texas, source of the helium-bearing gas. All three used different processes and got into production only in the last stages of the war, with the Navy and the Bureau of Mines backing rival methods. When the Armistice was signed, the first 147,000 cubic feet of the gas was awaiting shipment in New Orleans. The failure of aviation to develop far on the lighter-than-air principle has rendered the helium story less fascinating than it was to a generation that saw in it great potentialities. It remains, however, an excellent example of an industrial process in government hands rushed in a few months from laboratory to production under the stress of war.

The chemical problem that unfolded on the most gigantic scale was the necessity of securing a supply of nitrates, basic to the manufacture of explosives. This had been a cause for concern in the Civil War. Accustomed to depend on nitrates shipped from the natural deposits in Chile, the United States had little experience with the synthetic processes that had already been worked out in Europe, especially in Germany. The task, then, that faced the government and the American chemical industry was the evaluation of various processes and their adaptation to unprecedentedly large-scale production.[89] In spite of great accomplishment, including the building of the plant and power installations at Muscle Shoals, Alabama, the American supply during the war continued to depend on Chile. The greatest effect of the program, then, was on the potential production capacity of the chemical industry, plus the political problem of whether the government should remain in the nitrate business at Muscle Shoals.

The unaccustomed demands on science by the war were in many cases compounded by the abrupt breaking of lines of communication with Germany. Throughout the nineteenth century the easy flow of ideas and instruments in the international world of science had made it easy for Americans to rely on Europe. And the much-heralded borrowing of basic ideas was only a part of the debt. Whole technologies which supplied science with necessary services had grown up almost without American representatives. The most obvious of these was the manufacture of high-quality optical glass, which had before come from Germany to the exclusion of an American industry. Even the basic formulas on which the Germans worked were tightly held secrets. With wartime military demands for more and better optical instruments of all kinds, American scientists had to create an industry while replacing German data by research.[90] The two organizations with the background and facilities to attack the problem were the Bureau of Standards and the Geophysical Laboratory of the Carnegie Institution.

The Bureau of Standards set up an experimental furnace in their Pittsburgh station in the winter of 1914–15 to work out a system of production from the beginning.[91] Working with the various optical companies who undertook production, the chemists of the Geophysical Laboratory did much to determine the composition of various kinds of optical glass and to supervise the delicate controls necessary for its successful production. By early 1918, the American companies were making large quantities of glass which filled military needs as satisfactorily as the pre-war German product.[92]

The impact of total war wrought a subtle change in the concept of conservation. With great new demands on the supply of both minerals and food, the term came to mean maximum efficiency in production and minimum waste at the consumer's level. Conserving material resources for future generations rapidly dropped out of sight as a goal. The pressure for production greatly stimulated research in metallurgy,[93] and the need for food for the Allies brought the war home even to those biological sciences not closely related to medicine.

The Department of Agriculture, although it had already been doing for years the kind of research on production now called for, entered into the food-conservation programs of the emergency agencies.[94] The NRC's Division of Biology established itself as a part of the war effort in spite of some early scoffing. One observer doubted "if any other nation ever responded to the prospect of war with a scheme of national defense which included a Committee on Zoology and Animal Morphology." [95] Nevertheless, the biologists proved their relevance to many aspects of stepped-up food production.[96]

Effects on American Science

Although the entire period of the upheaval lasted no more than three years and hostilities less than twenty months, World War I had profound effects on every part of American science, whether supported by the government, by the universities, or by the foundations. The first major result was the infusion of research into the economy, especially into production, so thoroughly that industrial research as a branch of the country's scientific establishment dates its rise to eminence almost entirely from the war period.

The second major result was the use of coöperative research on a large scale. American scientists became accustomed to working together for the quick solution of an immediate problem. Not only did specialists learn to work with others like themselves, but they rubbed shoulders across the lines of the accustomed disciplines, often much to their own enlightenment. The government had pioneered in the problem approach from 1880 onward. Now it became the common experience of a whole generation of scientists — the ones who shaped institutions not only in the 1920's and 1930's but during the second World War as well.

The wartime NRC became a central scientific agency to an extent never dreamed of by the National Academy. It performed a real function as a clearinghouse of information and a focus of scientific personnel. Most of the great research efforts of the war fell at least nominally within its sphere. Yet it showed equally definite limitations. It never developed an adequate full-time administration to direct all phases of its program as a unit. Millikan, the executive officer, was deeply immersed in various special projects.

It never became the dispenser of large funds, and much of what money it had was from the private foundations. The only effective way it had to get military research funds was to have its scientists commissioned in some particular branch of the Army. As the war went on, more and more of the NRC's program went over to military control. It was a spawning ground of much-needed military scientific laboratories more than an independent agency supplementing the military programs. It became also less capable of initiating projects, depending increasingly on the assumption that the military knew what to ask for. In this respect, the Office of Scientific Research and Development of World War II started from a position immensely stronger than that held by the NRC in 1917–1918.

Basic science, according to most observers, did not fare well during the war years. Long-range programs suffered not only in government bureaus but also by the absorption of investigators from the universities. As in medicine, the need for immediate remedies took precedence over the quest for knowledge. Frank B. Jewett, himself prominent in war research, claimed that "in setting up the machinery to accomplish these [recent scientific wartime] achievements we at the same time set up the machinery for the destruction of advances beyond a certain point." By robbing the colleges, universities, and industries of trained scientists for "war's sweat-shop, it was inevitable that stupendous results should be obtained," but at the expense both of basic research and of training new men. "While I am not in a position to know the exact situation elsewhere in the world, I do know that we in the United States had early in the summer of 1918 arrived at the state where scientific man-producing machinery no longer existed." [97] This trend did not become more obvious because the sudden arrival of the Armistice did not give it a chance to run its course.

Indeed, World War I was a fragmentary experience for the American people. The Armistice caught the war effort just as it was gaining momentum. The first helium on the pier in New Orleans was a symptom of potential results that were on the point of fruition. Research and production programs alike had just begun to shake free from their organizational difficulties when the Armistice stopped them, and scientists as well as others showed great dispatch in dropping their wartime tasks and getting home again. Millikan was discharged from the Army on December 31, 1918. [98]

The autumn of 1918 marked more than just the cessation of hostilities. It also signaled a revulsion, a positive rejection of the war and all its works by the American people. Wilson lost the Congress and then the League of Nations. The structure of interallied coöperation fell to pieces. Appropriations for the Army and Navy plummeted. The wartime research structure, a target of emotional rejection, could not in any event have continued long into peacetime.

XVII

TRANSITION TO A BUSINESS ERA

1919—1929

LONG before the Armistice, the necessity of transition to a peacetime research pattern began to reassert itself. The decade 1919–1929 was to have a peculiar flavor of its own, compounded of disruptions left over from the war and new forces just beginning to become powerful. It was not a great period for heroic action on the part of the government. At the same time, science became a more conspicuous force in American life than it had ever been before. The government's research policy was a combination of these two basic conditions.

The Peacetime National Research Council

The National Research Council's main hopes for a peacetime role were the ideals with which George Ellery Hale had begun it. Even in the middle of the war these did not quite completely disappear, and as peace approached the NRC leaders began to talk in terms of stimulating basic research, coördinating the nation's scientific societies, and becoming the agency for representing the United States in international scientific affairs.

In February 1918, Robert Millikan received a letter from George Vincent, president of the Rockefeller Foundation, asking if there were some "device by which the scientific personnel and resources of the country could be better organized" for national service. "Is the National Research Council, which has been created out of the war emergency, likely to take on permanent form?" Is the federal government in a position to create a separate institution "on the analogy of certain research units in the Department of Agriculture and in the Geological Survey? Is the Bureau of Standards capable of extension into a national research institution?" [1] He then suggested the idea of a research institution to deal with physics comparable to the Rockefeller Institute for Medical Research.

Millikan responded by gathering sixteen physicists and chemists who, after discussing the proposal, voted against it by nine to seven. Millikan himself voiced the prevailing view that too much "centralization, even in the pursuit of science in this country, is a dangerous tendency." He would have preferred "not a central Institute of Physics and Chemistry, but the stimulation of at least a dozen such creative centers scattered all over the country . . . to be associated . . . with effective educational centers." [2] Since a private research institute appeared too centralized, Vincent's alternative of a government agency evidently received no serious consideration. What did emerge from these deliberations was a group of postdoctoral fellowships in physics and chemistry, administered by the NRC with funds supplied by the Rockefeller Foundations.[3] Handling fellowships was for the NRC far from as imposing as its wartime activity, but with the inception of this program the NRC had a solid peacetime function.

These conferences about Rockefeller Foundation plans led Hale, Millikan, and others to bestir themselves about the permanent legal status of the National Research Council. Hale, back in Washington in the spring of 1918, took the lead in preparing a draft which, on the advice of Elihu Root, was submitted to President Wilson as the basis of an executive order. Millikan lists the participants in the preparation of the draft besides Hale and himself as C. D. Walcott, J. C. Merriam, A. A. Noyes, J. J. Carty, and Gano Dunn. This group, the same that dominated the wartime NRC, was in effect an alliance of influential scientists from government, universities, foundations, and industry. Wilson passed the draft on to the Council of National Defense, from which the rumor soon emerged that the plan was disapproved. John J. Carty then went to New York to put it up to Colonel E. M. House, Wilson's confidential adviser, who agreed to present it to the President. Wilson fell in with the spirit of the plan, making a few verbal changes, which, according to Millikan, "improved upon our formulation." [4]

The resulting Executive Order of May 11, 1918 created a permanent National Research Council, deriving congressional sanction from the National Academy's Act of 1863. The new organization was thus a part of the National Academy and a possessor of its powers to advise the government. Yet its membership was not limited to the select few in the Academy, and in practice most of the scientists on it were representatives of the various learned societies. The duties listed reflect both the long-range ideas of the Hale group and the wartime atmosphere of the spring of 1918. They are:

1. In general, to stimulate research in the mathematical, physical, and biological sciences, and in the application of these sciences to engineering, agriculture, medicine, and the other useful arts, with the object of increasing knowledge, of strengthening the national defense, and contributing in other ways to the public welfare.

2. To survey the larger possibilities of science, to formulate comprehensive projects of research, and to develop effective means of utilizing the scientific and technical resources of the country for dealing with these projects.

3. To promote coöperation in research, at home and abroad, in order to secure concentration of effort, minimize duplication, and stimulate progress; but in all coöperative undertakings to give encouragement to individual initiative, as fundamentally important to the advancement of science.

4. To serve as a means of bringing American and foreign investigation into active coöperation with the scientific and technical services of the War and Navy Departments and with those of the civil branches of the Government.

5. To direct the attention of scientific and technical investigators to the present importance of military and industrial problems in connection with the War, and to aid in the solution of these problems by organizing specific researches.

6. To gather and collate scientific and technical information at home and abroad, in coöperation with governmental and other agencies, and to render such information available to duly accredited persons.[5]

In addition to the many private members, the President of the United States was to designate government representatives to the NRC on the nomination of the president of the National Academy. The wartime functions mentioned in points 4 and 5 thus gained firmer legal basis along with the broader activities of points 1, 2, 3, and 6. In this order, the scientists accomplished on a small scale what the diplomats failed to do. They secured their charter while the war was still on and the climate of opinion favorable. Had they waited, the NRC might well have gone the way of the League of Nations.

As soon as the Armistice was signed, the wartime NRC began to liquidate itself as rapidly as possible while a new peacetime NRC struggled to get itself in motion. By June 30, 1919, the NRC "passed out from under its more direct relations to the National Government through the Council of National Defense . . . and we may look forward to an early conclusion of all our more direct responsibilities to the Government." A special division was set up in the NRC to "continue our contacts with the Government, in the hope that they will lose nothing of their practical value because of the change in their external forms." [6] Despite this optimistic sentiment, nothing was done immediately to organize the Division of Government Relations.[7]

Finally, in December 1919, the personnel was appointed by the President of the United States, one representative from each bureau of the government interested in science. The list was extensive and often distinguished. Walcott was chairman, and among the representatives were Major General Squier for the Signal Corps, Rear Admiral Taylor for the Bureau of Construction and Repair, most of the bureau chiefs in the Department of Agriculture, and Stratton for the Bureau of Standards. However, the list was evidently too extensive and the names too distinguished, for reports of the NRC in the 1920's show that meetings were few, plans were nebulous, and action was not forthcoming.[8] This grandiose house of delegates of government science never really convened. Instead, the effectual separation of the NRC from the government that took place in 1919 became a settled policy.

The active part of the new peacetime NRC came to center around the representatives of the various scientific and engineering societies. Members were organized into divisions roughly paralleling the lines between scientific disciplines. By 1933 some 79 societies had named representatives, the memberships of all divisions and the executive board totaling 285.[9] Salaried chairmen provided continuity in the Washington staff. Thus, the NRC became a focus of scientific activity which had no precise counterpart before 1916.

For money the peacetime NRC relied exclusively on private sources, largely the great foundations. Rejecting the policy of building up an endowment, the council viewed itself as "an agency for the exercise of the maximum stimulation of research men and research agencies." [10] Thus, the NRC, with its wide connections in the American scientific public, became the dispenser of Rockefeller and Carnegie money for a variety of projects. The fellowships and grants were the heart of the peacetime program.

The Carnegie Corporation agreed to put up $5,000,000, of which one-third was used for a building to house both the Academy and the NRC. Architectural planning and raising the additional money to buy a site in Washington became the major preoccupation of the officers during the early 1920's. In 1924, President Coolidge made the dedication speech and George Ellery Hale received a special ovation. Some speakers considered it the opening of a temple to science — "pilot of industry, conqueror of disease, multiplier of the harvest, explorer of the universe, revealer of nature's laws, eternal guide to truth." [11]

The main carry-over from the war period was the peacetime NRC's position in international science. The Smithsonian had always been a universal institution, and the National Academy had had formal relations with its counterparts in other countries, but the wartime alliance and the United States' new position as a world power demanded something more. Hale was, of course, a strong advocate of international scientific organization and set in motion a series of conferences among the Allied nations which began before the Armistice. A plan prepared by the National Academy of Sciences became the basis for a new organization called the International Research Council. The very name marks the extent of American influence. The NRC became the agency designated for the nation as a member of the International Council, and the United States led in the formation of many of the constituent unions.[12] Again the scientists took a course different from the prevailing temper of the times, because these memberships persisted unbroken through the period of the rejection of the League of Nations. Indeed, Congress paid the small dues throughout the 1920's by a provision in the State Department appropriation bill.[13] With the growth of the power of the constituent unions, the name was changed in 1931 to International Council of Scientific Unions.

The NRC weathered the postwar transition by changing its nature and by limiting its operations to a scale set by its private sources of money. Within its reduced sphere it did effective work and began to gather a backlog of data on the administration of science, especially of grants and fellowships, which would ultimately be of more general use to public and private agencies alike. It kept alive a spark of formal American participation in international scientific organization. But it was not able to become a really forceful central organization for American science. It so nearly lost touch with the government that it was neither a coördinating center for science in the bureaus nor an active adviser.

The Federal Scientific Establishment

The military research agencies suffered an even more drastic shock at the end of the war than the NRC. While service research did not go all the way back to prewar innocence, the combination of whole-sale cuts in appropriations and rapid demobilization dealt heavy blows. The civilian scientists with commissions fled back to the campuses, leaving few regular officers with any experience in direct-ing research.

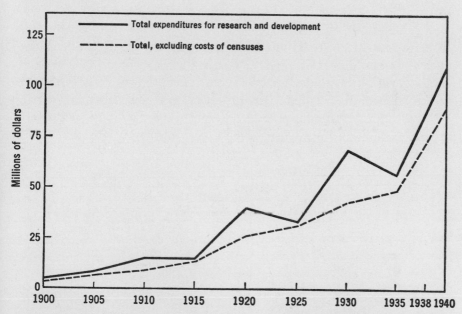

Fig. 2. Total government expenditures for research and development, 1900–1940. Estimates for the censuses have been separately identified to provide more meaningful comparisons for the selected five-year intervals. [Source: National Science Foundation, courtesy of Mrs. Mildred C. Allen.]

Besides lack of money, the services had to battle other discourag-ing trends. In the War Department reorganization of 1920 research was largely geared to procurement, under the direction of an assistant secretary of war. Procurement presupposed standardization, making the object of research a single item of equipment capable of mass production. All specifications tended to emanate from the General Staff, which set tactical and strategic requirements largely without the advice of scientists in scientific agencies. A War Department

Technical Committee made of representatives from the General Staff and the technical services had a shadowy existence until 1931, when it was abolished.[14] A further deterrent was the huge pile of World War I equipment, enough to supply the needs of the peacetime army for many years. Not only did procurement itself tend to suffer, but change in design became a luxury too expensive to be encouraged.[15]

The Chemical Warfare Service offers a good example of the vicissitudes of the postwar period. One critic described its demobilization as "one of the quickest operations of the war." [16] Personnel was turned out wholesale and Edgewood Arsenal "came near going to wrack and ruin." [17] Within the Army the service had to fight off schemes to reduce it to a branch of engineers or ordnance, and in the eyes of public opinion it had to combat the argument that even if another war occurred gas warfare would be outlawed.[18] By 1921, the worst crises were past, the service consisting essentially of a small cadre of officers and enlisted men, the plant at Edgewood, and a few civilian chemists engaged in research. Appropriations for the whole service plummeted below $1,500,000.[19]

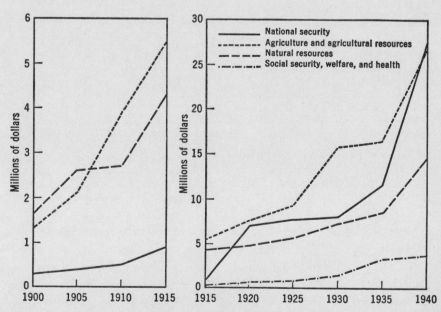

FIG. 3. Government expenditures for research and development distributed by budget function, 1900–1940. For definition of functional category, see *The Budget of the United States, 1955*, 1100. (Some functions are omitted here.) [Source: National Science Foundation, courtesy of Mrs. Mildred C. Allen.]

In the Navy, the trend might have been much the same except for the seed of a laboratory planted in 1916 by the Naval Consulting Board. By 1923, the dust had settled sufficiently for a Naval Research Laboratory to come into existence on the basis of the former congressional authorization. Here A. Hoyt Taylor and other civilians in navy employ began a series of experiments which eventually led to the devices known in World War II as radar.[20] Although under the command of a regular officer, the Naval Research Laboratory developed a certain *esprit* among its civilian scientists and a taste for fundamental work. This was, however, a very small exception to the general lack of research in both Army and Navy.

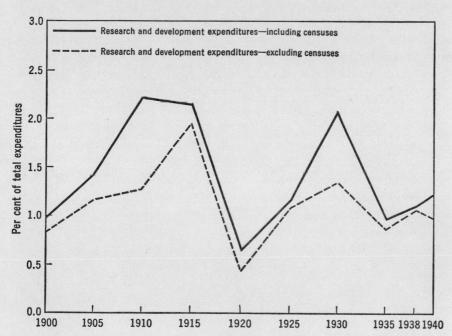

FIG. 4. Research and development expenditures as percentages of total government expenditures for all purposes, 1900–1940. [Source: National Science Foundation, courtesy of Mrs. Mildred C. Allen.]

By the early 1920's scientists had, in general, turned away from warfare as a field for application of their knowledge.[21] Indeed, with the American people deprecating military activities, the scientists became rather defensive about admitting that they had been "an agent of evil and a promoter of human capacity to do grievous things."

They denied responsibility for the war itself and said that they "hope from their hearts that they will never again" have to convert to a war footing.[22] Significantly, the nitrate program survived in the transfer of the Fixed Nitrogen Research Laboratory from the War Department to the Department of Agriculture in 1921.[23] Swords into plowshares — in this case explosives into fertilizer — was the trend of the day.

One agency that had been a part of the wartime research establishment had an atypical development in the postwar period. Thanks to its birth at the Smithsonian before the war began, the National Advisory Committee for Aeronautics had an organizational resilience not available to straight military outfits, and aviation had peacetime as well as wartime applications. Although military men represented the services on the committee, the civilian scientists were in practice predominant. Serving without pay, the members managed with the help of a permanent office staff to set basic policy and actually to administer the ever-growing research establishment at Langley Field, Virginia. The committee served as a focal point for aeronautical research not only of government agencies but of the private centers, especially at Massachusetts Institute of Technology.[24] The budget was not large — approximately $200,000 in 1923 to $500,000 in 1927.[25] But by clinging more and more closely to its fundamental mission of the scientific study of the problems of flight, the NACA made a distinguished record of achievements during the 1920's.[26]

The civilian scientific agencies of the government lived out the decade of normalcy without major incident. A slow general rise in funds available came largely from the fruition of programs already started. The systems of regional laboratories of the Bureau of Mines and the Forest Service filled out. A new act in 1928 established a generally broader policy of forest research. The National Park Service began to consider itself a purveyor of scientific information to the public. The demands of the automobile led to research in public road construction. In general, activity and expenditures increased during the decade without essentially changing the government's research establishments.[27]

The Public Health Service offers a good example of the doldrums of the 1920's. Stimulated to enter research in venereal diseases during the war by the Chamberlain-Kahn Act, the service salvaged a Division of Venereal Diseases when Congress cut off funds.[28] Numerous plans to reorganize and coördinate health services agitated the whole decade without marked result.[29] In practice, the commissioned corps had great difficulty covering its many duties. Of the 180 officers, many were examining immigrants in the ports of Europe in 1927. Twenty had to provide disaster relief in flood areas. Twelve had to go to Los Angeles to deal with an outbreak of bubonic plague. The hospitals were always in straits. Naturally, research moved slowly.[30] Nevertheless, accomplishments continued, such as the vaccine for Rocky Mountain spotted fever.[31]

Some of the grand veterans of government science had trouble keeping their places. The Smithsonian, dwarfed now both by bureaus and by the great foundations, tried to lift itself by the only way available in a period of tight federal budgets — a drive for increased endowments. The general prosperity of the country should spare at least a little for an old and honored servant. "Thou shalt not muzzle the ox when he treadeth out the corn," quoted Dwight L. Morrow at the climax of a meeting of notables at the Institution in 1927. The consensus of the group seemed to be that the Smithsonian had a place "as the inspirer and coördinator of research in pure science as it had been in the past, and that both governmental and private support should unite in making available more adequate means." [32] The meeting failed, however, to move those who had the power to help, leaving the Smithsonian impoverished while its museum costs continued to rise.

One change during the 1920's affected the government's scientific establishment out of all proportion to the money spent. The social sciences for the first time appeared as a distinct entity animating whole research activities. The Bureau of the Census had led the way, to be followed by such statistical agencies as the Bureau of Labor Statistics. More analytical bureaus also now became sufficiently effective to claim a place in the federal hierarchy. From this time forward the natural sciences had to reckon with a younger brother who clamored, usually feebly and ineffectively, for attention and funds.

The clearest example of a new social-science agency was the Bureau of Agricultural Economics, created in 1922 out of a miscellany of earlier statistical and analytical activities, some of them as old as the Department of Agriculture.[33] The new aspect after this time was a greater interest in general economic conditions outside the fences of the individual farm. In contrast to the prewar period, the 1920's found the farmers in a deep depression with an ever-lessening share of the nation's wealth. Increased production was less urgent than seeking a cure for a chronic economic malaise. Great organizations of farmers sprang up shouting of parity and wielding formidable political influence.[34] In this atmosphere the Bureau of Agricultural Economics became a key agency in the department, and economic problems began to become a factor in other types of research as well. The influx of men trained in university departments of economics and sociology also tended to weaken the homogeneity of the personnel of land-grant college origin within the department, paving the way for markedly changed attitudes in the 1930's.

A second effect of the rise of the social sciences on the federal establishment was the beginning of the systematic study of that structure as an object in itself. In the past, natural scientists closely associated with the government had speculated in a random way — sometimes with great acuteness — on research as a function of the state and on the forms in which it might be organized. The social scientists had begun to make their appearance among those interested in parts of the federal research structure by the time Irving Fisher issued his report on *National Vitality* in 1909. But only after World War I did the social science of the country muster enough personnel, technical ability, and interest to tackle the federal government as an institution worthy of systematic study. One of the most notable attempts during this decade was the series of *Service Monographs* of the Institute for Government Research, later the Brookings Institution. Running to more than sixty volumes, these studies analyzed the federal government's executive branch, bureau by bureau. Although soon out of date as handbooks, they provided a comprehensive cross section which displayed the research activities as well as other operations. Late in the period, President Herbert Hoover appointed a research committee on social trends, under the chairmanship of Wesley C. Mitchell, whose final report included an analysis of the growth of governmental functions.[35] Behind the whole enterprise was a new belief in the ability of the social sciences, which "might supply a basis for the formulation of large national policies looking to the next phase in the nation's development." [36]

Industrial Research and the Department of Commerce

Outside the federal establishment itself, a force that affected every phase of scientific activity in the United States came into its own —

industrial research. Appropriate to an era in which business was the dominant and most active sector of the nation, the laboratory of the great corporation finally completed the fusion of research and technology. In 1789, these two had been far apart, embodying separate traditions. Beginning with the Civil War, they moved toward one another hesitantly and intermittently. The great upheaval in science during World War I thoroughly mixed the two. In the 1920's the corporation that adopted research as an integral part of its business operations became normal where earlier it had been the rare exception. The 300 laboratories in 1920 had become 1625 by 1930, employing a total research personnel of over 34,000.[37] As the number and size of industrial research laboratories grew, it became apparent that a new estate of science in America had arisen, a companion to the government, the universities, and the private foundations.

The rise of the new estate had repercussions throughout the existing fabric of science. The universities had an important outlet for their products and a demand for the results of their research. The Division of Engineering and Industrial Research was one of the most active parts of the NRC.[38]

The government itself might have taken the lead in organizing the movement, as did that of Great Britain. One of the pioneers of industrial research in America specifically recognized the precedent on which it might be done, pointing out in 1913 that "through the combined efforts of the Department of Agriculture, the Experiment Stations, the Agricultural Colleges, and our manufacturers of agricultural machinery, there is devoted to American agriculture a far greater amount of scientific research and effort than is at the service of any other business in the world." [39] Besides agriculture, the Bureau of Standards, Bureau of Mines, NACA, and others had already entered the field and pioneered the techniques of coöperative research that were to prove so useful in private industrial laboratories. But the tide was set the other way. The railroads were returned to private owners in 1920, and economy and tax reductions were the desire both of business men and of the directors of policy. Frank B. Jewett, speaking of the need of industrial research for small business units, admitted that it "is, of course, conceivable that this service might be rendered by a government supported research organization with one or many laboratories." He saw in this course, however, only hazards, "the principal among which would be the difficulty of maintaining as high a standard of scientific and technical ability" as could be maintained in a nongovernmental organization.[40] In such a climate of opinion the government left the bulk of industrial research to the great corporations, who generally limited themselves to the application of science that might possibly yield some profit.

With business setting the tone in the country and with industrial research the wonder of science, the Department of Commerce was a natural center of new activity inside the government. Also, the secretary of commerce who served both Warren Harding and Calvin Coolidge was the one major political figure of the decade with an active appreciation of science. Herbert Hoover had begun his worldwide career as a mining engineer whose interests ranged beyond his immediate profession to such unusual activities as translating a sixteenth-century treatise on metallurgy. He had emerged from the war with a tremendous reputation as a humanitarian and an efficient administrator.[41]

An indication of the vigor that Hoover brought to his new job was his effort to confine the Department of Agriculture to production on the farm, allowing the Department of Commerce to "take up its activities where manufacturing, transportation, and distribution begins." [42] Although not successful in that particular interpretation, he quickly showed himself the most active of the cabinet members. The Bureau of Standards, the Bureau of Fisheries, the Bureau of the Census, and the Coast and Geodetic Survey already formed a hard core of seasoned research agencies in the Department of Commerce. In 1925, an executive order transferred both the Patent Office and the Bureau of Mines from the Department of the Interior, further strengthening Hoover's position as an administrator of research.

The secretary looked upon his establishment not simply as one carrying out scientific research but as an instrument to eliminate inefficiency from the American economy.[43] One of the ten points in his campaign against waste was the development "of pure and applied scientific research as the foundation of genuine labor-saving devices, better processes, and sounder methods." [44] The Bureau of Standards, an object of his personal attention, launched a coöperative program of simplified practices and commercial standardization. Part of the task of the new division that fostered these activities was to attend the many conferences of businessmen which Hoover called to Washington as part of his program of encouraging trade associations. The purpose was to stimulate requests from industry for simplified practices.[45]

The trade associations in turn supported research associates at the Bureau of Standards, the results of whose work were "published by the bureau" and made "available to the public at large." By 1924, some 29 associates represented 23 industrial organizations, among them the Portland Cement Association, which "has shown its appreciation of the value of fundamental research by employing a group of investigators stationed at the bureau to find out what Portland cement really is." [46]

The logic of Hoover's campaign against waste led him into fields already staked out by others, especially conservation. "A broad national policy is needed for the orderly development of all river and lake systems, that we may not suffer great losses through erratic development and failure in coördination to secure the maximum economic returns from each drainage basin." [47] Although the premises were different, his program strikingly paralleled those of the Geological Survey–Reclamation team in the Department of the Interior and of the Forest Service in the Department of Agriculture. The emphasis under Hoover was on water power, which the "progress of science and engineering" and "discoveries in transmission of electricity" now made more significant.[48] The result was a new center of conservation interest in the government and eventually, through Hoover's efforts in other capacities than as secretary of commerce, the building of Hoover Dam.[49]

The department was also pushed by expanding research technology into the regulation of both aviation and radio. These activities inherently required some kind of federal control in the same way that steamboats had back in the 1830's. Also like the explosions of a century earlier, the technical peculiarities of the new fields produced a need for more and better scientific data. The air commerce act of 1926 gave the Department of Commerce wide powers over aviation and led to the setting up of an Aeronautics Branch. While not supplanting the NACA in research, it led to an Aeronautical Division of the Bureau of Standards to carry out programs in which it had special interest.[50]

The babel caused by the broadcasting stations that sprang up after the war quickly rendered earlier laws ineffective. To distribute the necessarily limited number of frequencies available in the radio spectrum, Hoover early took action by calling for self-regulation.

But when an adverse court decision threatened the power of the secretary of commerce in this field, Congress passed legislation in 1927 setting up the Federal Radio Commission. This in turn led to the creation of a Radio Division in the Department of Commerce which became deeply enmeshed in the technical problems that emerged from attempts at regulation.[51]

The aviation and radio agencies went through several mutations before reaching a stable form. Although often involving research, they did not develop extensive laboratories of their own. Their position was analogous to that of the regulatory agencies in agriculture and public health which in part grew out of research and which existed in close conjunction with it.

In retrospect Hoover was proud of the fact that despite its increased activity the department grew little in either size or cost under his charge.[52] It was still, in 1928, peripheral to industrial research rather than its foundation. It hoped to encourage and stimulate projects rather than execute them as did the Department of Agriculture. Except in its regulatory activities it depended upon the voluntary coöperation of business to accomplish its research aims. Yet, given a series of strong successors and a favorable business climate, Hoover had laid the groundwork for a far-flung scientific organization. He himself was the personification of the engineer-administrator in an age that particularly admired technological progress. The year 1928 proved to be a peak, after which the department tended to lose its research functions to other agencies. During the middle 1920's, however, it was as strategic a place as there was in the government from which to assay the forces turned loose by the rise of industrial research and the fast pace of scientific discovery.

The National Research Fund

Late in 1925, Secretary Hoover began to make numerous public speeches on the subject of science, his interest running beyond the Department of Commerce program and even the confines of the government. Underlying his speeches was the beginning of a realization that the nation's scientific program was a single interrelated whole and that the burgeoning industrial research was creating a dangerous imbalance. Increasing national efficiency was laudable, but without a continuous supply of basic discoveries applied science could not maintain itself. In their final form Hoover's arguments followed a very definite logic. "Business and industry have realized the vivid values of the application of scientific discoveries." Federal and state governments also "support great laboratories, research departments and experimental stations, all devoted to applications of science to the many problems of industry and agriculture." Yet "the raw material for these laboratories comes alone from the ranks of pure science" supported from three sources — "the rest of the world, the universities, and the foundations." Estimating that industry and government in the United States spent $200,000,000 a year on applications, he set "the whole sum we have available to support pure science research at less than $10,000,000 a year, with probably less than 4000 men engaged in it, most of them dividing their time between it and teaching."

How, asked Hoover, "are we to secure the much wider and more liberal support to pure scientific research?" He saw three ways: — first, from the government; second, from business and industry; third, from benevolence. He recommended that the federal and state governments reverse the trend toward applied science and "accept the fact that the enlargement of our stock is no less an obligation of the state than its transmission," appropriating more money for basic research. He deplored the failure of benevolence to answer the recent calls of the Smithsonian, "the father of American science." But his main object was to get support for basic research from business and industry. "It is an appeal to self-interest, to insurance of every business and industry of its own future." With a national output of $50 billion annually in commodities "which could not be produced but for the discoveries of pure science," the nation "could well afford . . . to put back a hundredth of one per cent as an assurance of further progress." [53]

To do something about this idea of channeling industrial funds into basic research, Hoover went to New York in the first days of 1926 for conferences. Out of them emerged a kind of alliance between the secretary of commerce and the group that had been so prominent during World War I — John J. Carty, Gano Dunn, George Ellery Hale, and Robert A. Millikan. A committee formed under the National Academy of Sciences included, with Hoover as chairman, all these men plus such notables as Elihu Root, Owen D. Young, Andrew Mellon, and Charles Evans Hughes.[54] The National Research Fund, at first called the National Research Endowment, was to total $20,000,000, one-tenth subscribed each year by industry over a full decade. The money was not to be held as endowment but granted directly to investigators engaged in basic research.

On April 21, 1926, the National Research Fund was front-page news in the *New York Times*. The campaign was described as proceeding in "a unique manner. A few men of great prestige have made the appeal, approaching personally a few of the outstanding leaders of American industry." An effort was made to reassure corporations with a statement from Root, Hughes, and John W. Davis that contributions could be considered as investments in the interest of the stockholders. The article radiated confidence that the fund was well on the way. General Electric and American Telephone and Telegraph seemed to be in line, with Young and Carty on the committee, and many college presidents gave endorsements.[55]

The National Research Fund seemed to mark a real advance in the structure of science in America. Under the aegis of an important political leader, industry would support basic research in the universities with a large and continuous flow of money. Characteristic of its period, it stressed voluntary coöperation and eschewed any form of government control, relying on the enlightenment of business leaders. Only a few at the time expressed any reservations about the plan. J. M. Cattell, dubious about the inferiority of American basic research to Europe which the advocates of the Fund stressed so insistently, suggested that it would be desirable to spend a minute part of the money "that the board proposed to collect in determining whether the first statements that it makes are correct." [56] Hoover hinted gingerly that some corporations refused to give "because they have not grasped the essential differences" between applied science for immediate use and pure science as the basis of future inventions.[57] The pronouncement by the eminent lawyers of the legality of corporation gifts was a confession of uncertainty. Some statements indicated that the sights had been lowered to $10 million.[58]

None of these early shadows, however, can account for the complete and immediate failure of the National Research Fund. Only insignificant amounts for organizational purposes were deposited to it before 1930, when $379,660 in contributions was listed.[59] In 1934, "due to a variety of causes, most of which were influenced by the economic depression," $356,402.48 was returned to contributors. No mechanism was ever set up for making grants, and the records do not show that any money ever reached the hands of investigators. They indicate neither the reason for the long delay of four years in receiving actual contributions nor who had the responsibility for the detailed planning of the project. Hoover doubtless lost touch with it when he began his campaign for the Presidency. Perhaps the retirement and death of John J. Carty played a part. Certainly the depression would have curtailed the project even if it had been more active in the last years of prosperity. It is reasonable to assume that it was dormant long before the New Deal, which Hoover later blamed for its failure,[60] even came into existence.

The National Research Fund stands as a pioneer effort to redress the imbalance between basic and applied science. It was also the only large, new effort of the 1920's to deal with the whole pattern of science in the United States as a single unit. Its scale was sufficiently ample that its full operation would have made a measurable impression on American science. It might even have become a central organization for basic research. Whether it would eventually have achieved this promise is futile to ask, for it did not possess the stamina for the storms of the 1930's.

In the decade 1919–1929 government science accomplished its transition to peacetime and adapted its mechanism to a social structure in which business was dominant and industrial research a new colossus. The existing government scientific establishment held its ground and even expanded. Few real movements for coördinating science within the government arose. There was a beginning of a recognition that the estates of science — government, universities, foundations, and industry — were closely interrelated and that what one did affected the others. But the organizations that tried to deal with these interrelations — the NRC and the National Research Fund — shunned any connection with the government.

The decade of the 1920's was somewhat parallel to that of the 1870's. The old patterns of government science seemed to have re-emerged almost unchanged from their wartime jostling. But deep currents in American society itself made these progressively less well adapted to their functions, presaging a general reorientation. In the twentieth century, however, the wartime scars were deeper, and the depression of 1929 proved to be in itself profoundly disturbing to the structure of science in America.

XVIII

THE DEPRESSION AND THE NEW DEAL

1929–1939

THE depression that followed the stock-market crash of 1929 disrupted American society and with it the estates of science as had no other catastrophe except war. Earlier economic fluctuations had affected the federal budget somewhat and the short-run trend in appropriations for scientific bureaus had dipped, as in 1892 when a general cutback accompanied John Wesley Powell's fall from power. But by 1929 American science in the government, industry, and the universities was so interlaced both internally and with the whole economy that the serious disruptions of the great depression affected the whole research structure.

The Impact of the Great Depression

Since federal budgets are not made in a day, expenditures for research within the government continued upward at least through 1931.[1] By 1932 they began a precipitous decline which continued well into the Roosevelt administration. Estimates indicate that the 1932 high was not equaled until 1937. Most bureaus felt the impact of the depression by such deep cuts in funds that their programs were curtailed and personnel had to be dropped.

The cuts for science were politically nonpartisan, beginning under Hoover and continuing under the Democrats. Even the opponents of curtailment recognized this; witness the anguished cries of Congressman Fiorello H. La Guardia in December 1932. "Science knows no politics. Are we in this frenzy of economy, brought about by those who control the wealth of this country, seeking to put a barrier on science and research for the paltry sum of $39,113 out of an appropriation of $100,000,000?" He believed "the most humble research scientist in the Department of Agriculture is at this time contributing more to this country than the most useful member of Congress."[2] Even though Congress was ceasing to be a major opposer of scientific projects, this view did not prevail.

The Department of Agriculture carried a total research figure in its *Annual Reports* in these years that shows clearly the path of depression. In the fiscal year 1932 expenditures for research, including state payments, reached a peak of $21,500,000. This fell by $3,000,000 in 1933 and $2,000,000 more in 1934. The year 1935 showed only an insignificant increase, to $16,600,000. Not until 1937 did the total exceed that of 1932.[3] *Science* reported early in 1934 that 567 workers on scientific projects were dismissed from the department.[4] Especially hard hit were agencies such as the Biological Survey, which combined

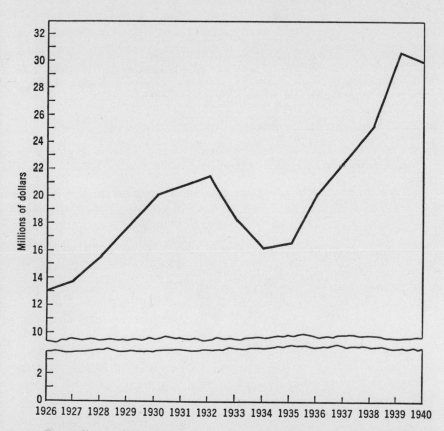

Fig. 5. Expenditures for research in the Department of Agriculture, 1926–1940, showing the effect of the depression. [Source: Secretary of Agriculture, *Annual Reports, 1926–1930* (Washington, 1927–1931); Director of Finance, Department of Agriculture, *Report, 1941* (Washington, 1941), 14.]

regulatory functions with research. One observer charged that "the policy appears to have been to continue the regulatory functions and to cut out completely the fact-finding activities on which all sound regulation policies must be based." [5]

For an agency untouched by the social sciences the cuts were even more drastic and the rebound slower. Total appropriation for the Bureau of Standards in fiscal 1931 was $3,904,000. By 1933 it had slipped to $2,308,000, and in 1934 it was only $1,755,000, less than half what it had been only two years before. From that trough it emerged only slowly, failing to regain the 1931 level before 1940. [6] For scientists in the government service, for administrators trying to cope with mounting problems and declining revenues, the depression was a major disruption, both officially and personally.

The other estates of science reacted very much as did the government. Industrial research reached a peak in 1931, hit bottom in 1933, and, rebounding somewhat earlier, was well on the way up by 1935. Universities equaled their previous high in 1936, as did total expenditures in all research in the United States. [7] Thus the Roosevelt administration found American science under pressure when it came into office. Its efforts to shape a research policy between 1933 and 1935 took place against a background of continuing disorganization and financial stringency.

In the closing days of the campaign of 1932, Herbert Hoover stated, "I . . . challenge the whole idea that we have ended the advance of America . . . What Governor Roosevelt has overlooked is the fact that we are yet but on the frontiers of development of science, and of invention." Progress in the last generation was "due to the scientific research, the opening of new invention, new flashes of light from the intelligence of our people." He was confident that "if we do not destroy this American system, if we continue to stimulate scientific research, if we continue to give it the impulse of initiative and enterprise, if we continue to build voluntary coöperative action instead of financial concentration . . . my children will enjoy the same opportunities that have come to me and to the whole 120,000,000 of my countrymen." [8]

This metaphor of science, the new frontier, is a common one, eloquently expressed by Franklin D. Roosevelt, among others. [9] But Hoover's use of it in the 1932 campaign had special connotations. He seemed to equate scientific research with the prosperity of the 1920's, the economic system then reigning, and the voluntary program he had developed as secretary of commerce. He was relying on research as a long-run answer to the fall in productivity that accompanied the depression.

The results of the election of 1932, in which Hoover was so thoroughly defeated, laid his line of reasoning open to the interpretation that research bore some responsibility for overproduction and hence for the depression. Many people, while perhaps admitting that research would eventually benefit the economy, were in no mood for long-range answers when the acute crisis demanded emergency action. Unlike a war, which stimulated science as well as disrupted it, the depression cast a cloud over the belief in its usefulness. Research programs could wait in a period of closed banks when the hungry unemployed trod the streets. Scattered statements appeared which even intimated that less research would be a good thing. One suggested "a slowing up of research in order that there may be time to discover, not new things, but the meaning of things already discovered." The physicist and chemist "seem to be traveling so fast as not to heed or care where or how or why they are going. Nor do they heed or care what misapplications are made of their discoveries." [10] Against this background the New Deal had to shape a scientific policy.

The large sums of money that the government began to spend during the first hundred days of the New Deal were designed to care for the unemployed and revive the economy, not to aid the hard-pressed scientific bureaus of the government nor to attack the depression by a long-range research program. The Democrats' promise in 1932 to make savings in government expenditures was a real policy for the regular scientific establishment, emergency funds lying tantalizingly out of its reach.

The plight of the Army Medical Library in these years illustrates the formidable barriers between the established research agencies and emergency money. Because the daughter of Dr. Harvey Cushing, the famous brain surgeon, had married James Roosevelt, the Army Medical Library had a spokesman with direct access to the President. In August 1933, Cushing wrote to Roosevelt, "You of course know all about the Surgeon-General's Library, for which John S. Billings was originally responsible. It is the only great medical library in the world, and the *Index Medicus* and the *Index-Catalogue* are probably more widely used throughout the world than any other medical books which have ever been published since the book of Isaiah." [11] The library needed a new building so badly that its collections were in physical danger from a leaky roof. Roosevelt, although admitting the value of the library, was hesitant. "The question naturally arises . . . as to the wisdom of asking for $2,000,000 for an expenditure of this kind at this time. If surplus monies were available, I would have no hesitancy in endorsing the request." [12]

Some time later the President faced squarely the possibility of using relief funds to build a library. "We are all tremendously keen about a new building for it. However, out of Public Works funds we must keep the District of Columbia somewhere within a reasonable ratio of expenditures compared with the population, remembering that these Public Works appropriations are primarily to relieve unemployment." To the President the most pressing needs which took up all available funds allotted to the District were: "a) One new building to take care of actual Government workers; b) A new sewage disposal plant, very much needed, as my nose on River trips testifies; c) A T. B. sanitarium . . . d) A stack room to take care of important current documents." [13] Those interested in government science could find small comfort in such priorities. To Cushing's protests Roosevelt replied, "I wish I were the dictator you assume me to be. I most assuredly do want to get the proper housing for the Surgeon General's Library started but it must be a monumental building and cannot be done out of Work Relief Funds: therefore, it will require an Act of Congress. We have had such demands for office space these two years that all special buildings of this type have been deferred." [14] The plight of the Army Medical Library was unchanged nearly twenty years later.[15]

The high official in the Roosevelt administration with the most extensive responsibility for science and the best background for understanding it was Henry A. Wallace, secretary of agriculture. He administered the largest single block of scientific agencies in the government. His family had been continuously active in the department since the days of Tama Jim Wilson. He could appreciate at first-hand the role that the science of genetics had played in the development of hybrid corn. In addition, the plight of the farmer and the glut of his products were major emergencies facing the government.

Wallace was emphatic that he did not wish a vacation from research. "We might just as well command the sun to stand still as to say that science should take a holiday. Science has turned scarcity into plenty. Merely because it has served us well is no reason why we should charge science with the responsibility of our failure to apportion production to need and to distribute the fruits of plenty equitably." [16] He, no less than Herbert Hoover, believed in the substitute frontier. The difference came in the kind of scientific research he envisaged and the object for which it was carried on. "Those who struggle beyond the new frontier will be those who know how to obey economic traffic lights, and drive social machines on the right-hand side of the road." [17]

The key to Wallace's concept of research lay in the raising of the social sciences to the same level as the natural sciences. "Actually, science has not given us the means of plenty until it has solved the economic and social as well as the technical difficulties involved." But he meant by social science more than statistical studies in the Bureau of Agricultural Economics. "The field of science is social as well as technical and includes human application as well as the discovery of scientific facts and principles." [18] Applied social science was really planning for a better life. He felt mankind could be freed from the grind of toil and the terrors of economic insecurity "only if the planning of the engineer and the scientist in their own field gives rise to comparable planning in our social world." [19] When this "better-controlled use of science and engineering" had achieved a higher percentage of leisure, more of man's energy would be "left over to enjoy the things which are nonmaterial and noneconomic, and I would include in this not only music, painting, literature and sport for sport's sake, but I would particularly include the idle curiosity of the scientist himself." [20] Thus through a process of social science and planning, research came the full circle to basic science again. However much this cycle was blurred in actual practice, the policy-makers for science in the New Deal period had to come to terms with this emphasis both on the social sciences and on social planning.

Wallace feared that the scientists themselves — the actual people who sat before him at the AAAS meeting in December 1933 — would be "a handicap rather than a help" in exploring beyond the new frontiers. "They have turned loose upon the world new productive power without regard to the social implications." In the past most "of the scientists and engineers were trained in *laissez-faire*, classical economics and in natural science based on the struggle for existence. They felt that competition was inherent in the very order of things, that 'dog eat dog' was almost a divine command." Wallace estimated that before 1933 three-fourths of them believed in orthodox economics. "Even today, I suspect that more than half of the engineers and scientists feel that the good old days will soon be back when a respectable engineer or scientist can be an orthodox stand-patter without having the slightest qualm of conscience." [21] Wallace was challenging American natural science to take some part in the social experiments of the New Deal, and the natural scientists, regardless of the secretary's unflattering opinion of their social and economic vision, firmly controlled the first great effort at participation.

The Science Advisory Board

Karl T. Compton, physicist and president of Massachusetts Institute of Technology, was on a ship from Boston headed for Bangor, Maine, in the summer of 1933 when his assistant sent him a radiogram: "Word received that you have been appointed chairman of committee to reorganize Federal Government." [22] The prospect of a scientist taking over to set the government right was not to be fulfilled, but the creation of a Science Advisory Board, which in fact was the impulse behind the inaccurate message, was a potentially important milestone in the path of central scientific organization for the government.

The inception of this new effort at coördination traced back to a request by Secretary Wallace that the National Research Council advise him on the reorganization of the Weather Bureau. [23] The chairman of the NRC, the geographer Isaiah Bowman, was evidently so imbued with the sense of the unusual crisis of the times that instead of handling the request routinely he made a counterproposal. Deprecating the fact that the NRC organized its work "according to the several fields of science rather than around the administrative and scientific problems of the Government," he suggested that it could work best through "*a general board* . . . which would then appoint committees to deal with specific problems as they arise one by one in the Departments of the Government." [24] Bowman's position was analogous to that of George Ellery Hale in 1915 — the leader of a group of scientists trying to reform the National Academy from within to secure vigorous action in an emergency. Wallace forwarded this proposal to President Roosevelt, who on July 31, 1933, issued an executive order creating "a Science Advisory Board with authority, acting through the machinery and under the jurisdiction of the National Academy of Sciences and the National Research Council, to appoint committees to deal with specific problems in the various departments." Besides Compton as chairman and Bowman, the list of nine members included W. W. Campbell, president of the Academy, and veterans of World War I science such as Millikan and Gano Dunn.[25]

As with the National Academy and the NRC, the government's interest and responsibility did not extend to financing the central organization itself. The Science Advisory Board begged its office space and a little expense money from the NRC. The Public Administration Clearing House plugged the gap with $2075 for September and October 1933. Then the Rockefeller Foundation solved the problem for a year by a grant of $50,000.[26] Significantly, the Division of Social Sciences of the Foundation was the source of this money. Thus the organization of the natural sciences had now become a problem for the social sciences, which were, however, unrepresented on the board itself. The members served without remuneration.

From its first meeting in August 1933, the Science Advisory Board embarked on an ambitious series of studies of government bureaus, looking forward to specific recommendations on policy and even personnel. This mission was the traditional one of the National Academy, but never in seventy years had that body made such a simultaneous attack on the problems of the government's scientific establishment. It inquired into the objectives, personnel, duplication, and coördination of scientific work.[27]

The membership of the subcommittees was drawn largely from university and industrial scientists. Only in rare cases did any government employee sit with a subcommittee, and none served on the main board, which considered their absence a distinct advantage over the National Academy or the NRC, where "many such scientists are distinguished and valuable members." [28] While in most cases waiting for requests from government officials, the board emphasized that "if the need for advice seems clear its possible usefulness has, in a number of cases, been pointed out to the head of a Department." [29] Assuming that it was a temporary organization, the board hoped that its experience "would determine whether some such agency should be continued, and, if so, in what form." [30] Thus the Science Advisory

Board had within it the embryo of an independent group of scientists not in the government acting as alert watchdogs of the federal research establishment.

Besides Wallace, the secretaries of commerce and interior and the director of the Bureau of the Budget called on the board for advice. The major agencies examined by subcommittees were the Weather Bureau, the Food and Drug Administration, the Bureau of Chemistry and Soils, the National Bureau of Standards, the Geological Survey, and the Bureau of Mines. Subcommittees also surveyed problems cutting across bureau lines — land use, mapping services, mineral policy, patent policy, and the land-grant colleges.[31] Some of the requests, as the director of the budget's inquiry "whether the . . . Government received adequate return from its contributions annually to the land-grant colleges," [32] seemed motivated by a desire to cut the scientific budget for the sake of economy. The Navy Department was more guarded, but a rather cumbersome double committee was set up as "a peacetime skeleton organization, so designed as to facilitate continued contact with the Department." A similar, less definite, arrangement was made with the War Department.[33]

The main body of conclusions tended to dramatize the plight that had already overtaken the bureaus and to plead eloquently for more support. The Weather Bureau should extend the air-mass analysis method over the whole United States.[34] The subcommittee for the National Bureau of Standards looked forward to "the happy day when appropriations may be increased." [35] The Bureau of Mines and the Geological Survey should not be recombined. In the latter agency "scientists who need books, collections, and maps about them for effective work are given desks in crowded offices where conditions for scientific work are intolerable. Nothing approaching these adverse conditions can be found in any other research organizations in the country, public or private." [36]

One subcommittee, in examining the Bureau of Chemistry and Soils in the Department of Agriculture, brushed against a perennial theme of government scientific organization. Beginning in 1880 the bureaus had pioneered in developing the problem approach and had organized themselves around it. Having proved its practicality, the problem approach now raised the question whether or not it was too practical. "Where the investigation is such that a given physical science is used merely as a tool the work should, the Committee believes, be carried out exclusively . . . in the bureau undertaking the main problem." This principle, however, now appeared not to apply to basic research. "Where the investigation involves the principle of a given physical science (e.g., chemistry) as such," in distinction to using these sciences as a tool, "the work . . . should be carried out in a bureau devoted exclusively to this field. To have small research laboratories devoted to fundamental problems in physical science scattered throughout several bureaus is wasteful and inefficient." [37] In form a revival of the principle on which the Department of Agriculture had been organized in the 1860's and 1870's, this observation was a token that basic research was becoming a more distinct entity in government science and that some people felt that it needed and deserved special handling.

The board, not content with reviving the role the National Academy had originally aspired to fill, set its sights early on the "great social objectives of science." [38] By September 15, 1933, Compton had already developed a "Recovery Program of Science Progress" which was in effect a New Deal for science. Submitted to Secretary of the Interior Harold L. Ickes, it proposed to enlist "the scientific and engineering groups in the country in a coöperative effort for the quick success of the National Industrial Recovery Program." Besides solving technical problems and advancing scientific knowledge along useful lines, the program would "provide employment and rebuild morale among the large body of scientists and engineers, together with still larger groups indirectly involved, such as mechanics, assistants, apparatus makers, and purveyors of equipment." [39]

The heart of the proposal was a fund of $16,000,000 to be expended during six years "in support of research in the natural sciences and their applications." The emphasis was on research as a background for the public-works program, for conservation, and as a basis for the creation of new industries. Even though this research would ostensibly strengthen the entire relief program, it had an additional objective because the unemployment "among scientifically and technically trained young men has been, and is, acute." [40] To provide research service for great public works by using unemployed scientists seemed to dovetail nicely. The program could be administered by the NRC, using university facilities as much as possible. The projects suggested as illustrations included meteorology, soil mechanics, sewage disposal, fog dissipation, cryogenic research, heavy hydrogen, long-distance transmission of electric power, mineral resources, and social problems of mechanization. Finally, part of the fund should be used for grants in aid of research in basic sciences.[41]

When Compton submitted this proposal, Public Works Administrator Harold Ickes "said that he was 99 per cent convinced that something of the sort should be done, but that there was unfortunately no provision under the law whereby public works funds could be expended for research but only for construction." [42] The program was dropped there, but in Compton's mind the idea itself was not dead.

During 1934, while the chairman attempted to remold his emergency program, the Science Advisory Board ran into stormy weather. In May a second executive order from President Roosevelt appointed six new members. No mention of the jurisdiction of the National Academy or nomination by its president appeared in the order, and two of the men named were not members of the Academy. One of them was Dr. Thomas Parran, state health commissioner of New York.[43] President W. W. Campbell and some of the members of the National Academy, who felt that their organization had been slighted,[44] formed a protest group aligned against Compton and Bowman. This division weakened the Science Advisory Board among those who might have been expected to give it the most wholehearted support. At the same time others criticized the board because it spoke only for scientists in private institutions, not for those in government service. An additional source of friction was the absence of the social scientists, who in other capacities were entering the Roosevelt administration in considerable numbers.

Meanwhile a group reasoning from entirely different premises was entering the field of the central organization of science. Planning as a concept in the government's operations was related to research in much the same way as regulation had been a generation earlier. To plan, one must have a body of data. To plan in the interest of the general welfare, one must be governed, as the early progressives had considered themselves to be, by scientific objectivity rather than by political pressure. Hence planning policy in the Roosevelt administration had a potential relation to research policy. During 1934 the two began to move closer together. A National Planning Board had existed for a year, from July 20, 1933 to June 30, 1934. The members were Frederic A. Delano, the President's uncle, Charles E. Merriam, and Wesley C. Mitchell. Delano had come to this position through an interest in city and park planning, while the other two were both distinguished social scientists. In April 1934, the National Planning Board approached the National Academy, which was by that time having doubts about Compton's group, requesting a report on the role of science in national planning.[45]

In the summer of 1934 a more adequate organization, called the National Resources Board, succeeded the first group with the same three men serving as an advisory committee and with Charles W. Eliot, 2nd, a landscape architect, as the executive officer.[46] But at the insistence of Harold Ickes a group of cabinet officers made up a majority of the new board.[47] With a small central staff, the National Resources Board worked through field offices in every part of the country. In addition, it appointed technical committees on the subjects of land, water, minerals, power, industry, and transportation. It was this board that eventually gained cognizance of Compton's plan for science.

In the fall of 1934, the Science Advisory Board got another chance for a new version of its program. Compton was now willing to speak out strongly against the "striking anomalies in our national policy" which led to the neglect of science and its power "to create new employment when this is desperately needed!"[48] He attacked both those who claimed that science required too much time and those who blamed it for unemployment. The government lagged far behind private business in providing for research in its bureaus, spending less than 0.5 per cent of its total budget. Yet "the displaced, trained scientists have been thrown on the unskilled labor market, where they frequently receive government pay in excess of that which they were earning in their professional work under the Civil Service!"[49] No other great nation was squandering its research activity in such a way. As a people, "and therefore as reflected in our national policies, we have been more lucky than intelligent." To utilize our resources more effectively "means scientific work on an increasing scale."[50]

Compton's new program was much more ambitious than the 1933 version. In the first place, he would allocate 0.5 per cent of Public Works emergency appropriations for scientific and engineering research. Second, he called for maintenance of the scientific bureaus of the government with adequate personnel and appropriations. Third, he called on the National Academy, the NRC, and a new Science Advisory Board to formulate programs of research both in and out of the government. But his main point was the appropriation of $5,000,000 annually for the support of scientific and engineering research outside the government. This venture in aid to science, "if put into effect, would be a new departure for our government, though in line with recent policy in foreign countries. If it were put into effect wisely, I believe it would yield returns of permanent value to the country exceeding those from almost any other comparable federal expenditure." [51] The total appropriation for the whole four points, embracing both government and nongovernment science, would come to $15,000,000 per year, or $75,000,000 over a trial five-year period.[52] At the Pittsburgh meeting of the AAAS in December 1934, Compton became president, and the scientists of the country seemed to swing solidly behind him.[53]

Roosevelt wrote to Compton that if "you will have the Science Advisory Board give consideration to the subject and submit a program with a budget, I will be glad to see that further attention is given to it." [54] The task of evaluating the program was delegated to Harold L. Ickes, who besides his many other duties was moving toward the coördination of research as chairman of the National Resources Board. He, in turn, delegated the review of the program to Delano, Merriam, Mitchell, and Eliot, the working members of that board. Since they were now entering the same field, their relation to the Science Advisory Board and the fate of the program for putting science to work were tied together.

F. A. Delano, in a reaction far from the stereotyped picture of the spendthrift habits of a New Dealer, told Compton that "your program is somewhat staggering in its size." He had sharp reservations about the mechanics. Congress had no way of committing itself in advance to $15,000,000 a year for five years. Where would this leave the scientific institutions at the end of the fifth period? Would grants taper off? Would they go to individuals or institutions? Should existing dispensing organizations have preference? "I feel, and I think I may safely say that my colleagues feel, that we cannot undertake the program for pure and applied science without considering the merits of similar but doubtless ambitious programs of the social sciences, of economics, and of education in general." He implied that he considered Compton a spokesman for the natural sciences alone when he said that lack of past progress in these other fields "is surely not an argument against their study." [55]

On January 21, 1935, the advisory committee of the National Resources Board — Delano, Merriam, Mitchell, and Eliot — met with Compton in New York, evidently endorsing Delano's point of view.[56] The advisory committee then shared in drafting a letter to Roosevelt for Ickes's signature which laid down basic policy both on the place of the Science Advisory Board and on the program for putting science to work.

Ickes's letter of January 31, 1935, considered Compton's effort as one of three programs, the others coming from the Social Science Research Council and from education. They all stressed the need for an advisory committee "to select and coördinate research projects in which the Federal Government is concerned. This is a planning function — planning for the full use of research resources of the country." The logic of this was to abolish the Science Advisory Board and replace it with a science committee under the National Resources Board which would coördinate not only natural science, but the social sciences and education as well.[57]

Compton's program for putting science to work fared no better. The advisory committee "does *not* recommend the appropriation or allocation of a large 'free' fund to any Science Research Committee for unspecified projects but does urge (1) adequate Federal appropriations to finance specific research projects which may be developed through a committee as outlined in the previous paragraph . . . and (2) generous support for scientific work carried on by the several branches of the Government." [58]

Roosevelt lost no time in approving Ickes's letter. He believed that the projects it mentioned "may be attained by allotment along definite lines from the proposed work fund." He then chose the National Resources Board to prepare a plan, "remembering always . . . that 90% of the amount expended must go to direct labor paid to persons taken from the relief rolls." [59] This memorandum sealed the doom of Compton's whole enterprise.

The Science Advisory Board did not give up immediately. In March, it proposed a modified form of its plan to put science to work calling for only $1,750,000 for two years, to be administered by NRC.[60] It also proposed that it be succeeded by a permanent Science

Advisory Board appointed by the President of the United States on nomination by the National Academy. Congress would appropriate $100,000 for the expense of the board, which would concentrate on providing scientific advisory service to the bureaus.[61] Neither of these plans prevailed.

In July 1935, President Roosevelt extended the life of the Science Advisory Board for six months to give the National Academy time to provide an agency to take over its work.[62] This action effectually returned the advisory service of scientists to the same position it had had before World War I, killing the Science Advisory Board and bypassing the NRC. The Academy dutifully set up a Government Relations and Science Advisory Committee which took over the records of the old Science Advisory Board. On December 26, 1935, President Roosevelt recognized this body by having a circular sent to all bureaus announcing its availability.[63] This was the end of the Science Advisory Board except for the forms of decent burial. The National Academy's committee carried on the advisory work for a few years on a diminishing scale. Then it quietly disappeared. "When government requests are received in the future they will be referred to specially appointed committees. This was the practice followed before the establishment of the dissolved committee." [64]

As Karl Compton looked back on his two years and a half of "extraordinary opportunity," he seemed to be one of those who, having hurried to Washington full of spirit and ambition in 1933, retired later, battered and disappointed.[65] The Science Advisory Board had tried to improve the federal scientific bureaus, to generate an emergency program to combat the depression, and to perpetuate itself as a permanent body. Its many successes in its first aim were swept away by its failure in the last two. Hurt by its narrow base and by quarrels within the National Academy, it nevertheless was an attempt to create a central scientific organization both for the government and for the country. Perhaps its most important contribution was to give a broadening experience to a generation that would have other chances before a decade had elapsed.

Research — A National Resource

The National Resources Board now had a clear field to set up a science committee of its own. Immediately after the President's memorandum of February 12, 1935, it invited the National Academy, the Social Science Research Council, and the American Council on Education to nominate members.[66] The principle of representation of existing organizations was an accepted one, but the inclusion of the social sciences and education was a new departure, or rather a return to the more inclusive organizational forms of the early republic. Natural science, instead of being the whole show, was actually outnumbered. The chairman through most of the life of the committee was Dr. Edwin B. Wilson, a statistician who combined as completely as anyone could the points of view of the social and the natural sciences.

At the first meeting, in March 1935, the main discussion centered around the channeling of relief funds into research. The committee expressed a "preference for the census or inventory type of projects, providing the collection of basic data that will be used by many different scientists, as contrasted with special studies in controversial fields." It also questioned the requirement that 90 per cent go to people from the relief rolls, suggesting 80 per cent instead.[67] Through the summer and fall of 1935, however, the interests of the science committee shifted away from the administration of relief.

In June 1935, the National Resources Board was abolished to be succeeded by the National Resources Committee, with identical personnel and functions but with more support.[68] As the entire agency became active in planning, the science committee turned more to studies in a sphere that it defined for itself in meetings of the following fall and winter. The theme revolved around the idea that national resources included human as well as natural resources.[69] In January 1936, at a dinner at the Cosmos Club, peace was declared among the National Academy, the NRC, Compton, and the new committee. That evening, some talk cropped up about "hard" and "soft" sciences and limiting the National Academy to the natural sciences alone. Behind such sentiments was the indication that the emphasis in the new group was to be on the social sciences.[70]

The first major studies concerned population problems and the social consequences of invention.[71] By following this line of investigation of human resources a step further, the committee arrived at the concept that research itself was a scarce human resource worthy of a careful study as a prelude to planning. In March 1937, the committee recommended a plan for studying the "interrelation of government and the intellectual life of the nation, whether in research, in education, or in technology." [72] This action led to a presidential letter of July 19, 1937, in which Roosevelt approved "a study of Federal Aids to Research and the place of research (including natural and social science) in the Federal Government." The President asserted that research "is one of the Nation's very greatest resources and the role of the Federal Government in supporting and stimulating it needs to be reëxamined." [73]

The resulting study, *Research — A National Resource*, was the most comprehensive examination of the federal research establishment ever made up to that time. Prepared by an *ad hoc* technical staff under the direction of Charles H. Judd, a psychologist from the University of Chicago, the study probed into the legal, social, and economic aspects of government science. Much of what they reported had existed before only in the unwritten lore of a few administrators. Perhaps its most significant accomplishment was showing government science against the larger background of the total research resources in America. Gifford Pinchot's group back in 1903 had treated government research as if it existed in a vacuum, the solution of its problems lying in reshuffling the bureaus into new patterns. The 1938 team, besides including the social sciences, looked into research in the other estates of science, the universities and industry. [74] The logic of the broadened view of research as a national resource led to a concept of the government's responsibility extending beyond its own establishment. The welfare of the total research establishment of the country was a question the federal government could not afford to ignore.

Despite its comprehensive view, the science committee did not achieve the status of a central scientific organization either inside or outside the government structure. It had no power to act nor even any organic sanction from Congress. Indeed, its value as a meeting place between the natural and social sciences might have been impaired by the possession of direct administrative power, since this would have raised the question of the relative weight to be given various branches.

The approach of war and its stern requirements quickly revealed that the science committee had neither the position nor the broad support from scientists necessary to organize an emergency war effort. Although one member proposed a study of mobilization problems, [75] the main interest of the committee seems to have been the protection of research in the face of disruptions.[76] The parent agency, which became the National Resources Planning Board in 1939, itself did not survive the war.

The Later New Deal

The predicament of science in the government in 1935 was strikingly analogous to its plight in 1830. An ambitious general program had failed and the executive in power was committed against it. Partisan passions were bitter in the wake of a change of administrations. Appropriations were down, the agencies tending to disorganization. Science had failed to make a secure theoretical place for itself in a new and popular political movement which was on the verge of fresh triumphs at the polls. Yet, as a century earlier, the government proved itself responsive to the needs for science and ingenious in providing practical answers to the problem of organizing its research establishment.

The year 1935 marks the end of the emergency phase of the New Deal and the beginning of a considered effort to find more basic solutions to the social and economic problems facing the nation. Research in the natural sciences as well as the social sciences had a better chance in the new climate. As the first emergency agencies gave way to more permanent bureaus, research began to make a better showing in programs popularly considered typical of the Roosevelt administration.

The Works Projects Administration, which emerged from the general reorganization of relief in 1935, had as its goal providing jobs on which people could use their skill.[77] One writer estimated that of 6,000,000 workers tabulated on relief 80,000 were professional and technical people.[78] This group, less some 15,000 musicians and many others not scientists, was the pool with which the WPA research program had to work.

The WPA developed a series of projects that "provided professional, technical, and, on occasion, manual workers to assist in scientific and technological research and experimentation in tax-supported universities and colleges." Nearly every state university had such projects, which were under the supervision of faculty members.[79] In a single short period in 1939–1940, approximately 50 reports in the fields of mathematics, biology, pathology and therapy, and scientific technology were registered. The fact that these reports were in the form of articles published in scholarly journals indicated that their quality met accepted standards.[80] To the charge of "boondoggling," Administrator Harry Hopkins replied that he was proud that "we have used skilled engineers, architects, and others in research work, in connection with universities, and I have no objection if they want to call that 'boondoggling.' " [81]

Some WPA funds went to other agencies of the government. Although these were not usually used for research, they helped some bureaus — the Forest Service, the National Park Service, the Bureau of Entomology and Plant Quarantine — that had scientific interests.[82] Significantly, the National Resources Committee drew its administrative expenses through the WPA. A great scientific bureau, however, did not develop. The WPA's lease on life was too tenuous to allow planning of projects more than a year ahead.[83] The restriction on payment of funds to persons not on the relief rolls continued to limit scientific projects much more severely than they did general relief. This had been apparent during the deliberations on Compton's program in 1935. An allowance of under 10 per cent [84] might be sufficient for building roads, but in research the creative individual is an essential element, and even in the depths of the depression he was not usually found on the relief rolls.

By an entirely different route the Tennessee Valley Authority became a kind of research agency. Its point of departure, of course, was the government's nitrate plant at Muscle Shoals, left over from World War I. But as a unified program for water control and resource development in the Tennessee Valley the authority faced all the engineering and conservation problems for which the government had earlier taken responsibility. Its multipurpose approach also raised economic and social questions which could be answered only by research. However, because of the questionable wisdom of building a multipurpose government research center on a regional basis, the TVA turned instead to existing federal, state, and local agencies. The Department of Agriculture, the Geological Survey, and the Bureau of the Census were the major federal agencies called upon, while among state and local units the land-grant colleges played an important role. The bulk of the program came under the heads of fertilizer and munitions, agricultural resources, forest resources, minerals, fish and game, topographic mapping, and health problems.[85] In a very loose sense the TVA was a gigantic experiment in applied science. Perhaps this accounts for the frequent appearance in TVA literature of terms such as "unified development," "a controlled river," and "Valley resources," and the relative scarcity of references to research as a separate activity.

The destiny of science in the later New Deal essentially depended not on the few new agencies, but upon the health of the regular establishment. Beginning in fiscal 1936, total expenditures for research turned sharply upward. By 1938, they stood well clear of the depression trough, and, at an estimated $75,000,000,[86] easily topped all previous highs. They even gained on the total budget, reaching slightly more than 1 per cent. While this percentage was less than in the Progressive Era before the first World War,[87] the total budget had grown so enormously that the two periods could now be considered as comparable in their emphasis on research.

Besides the financial recovery of the federal scientific establishment after 1935, important qualitative shifts in emphasis and procedure appeared which reflect the theme of "research — a national resource." The conservation movement, which was the intellectual seedbed of this apt slogan, became again a fighting faith for the first time since the days of Pinchot. Harold Ickes, a political if not a scientific heir of the Progressives, conceived of his Department of the Interior as a Department of Conservation and attempted persistently to get the name changed by Congress. The Department of Agriculture, which with some reason felt that this move would mean the transfer of the Forest Service, had no intention of losing its role as a great center of conservation in the government.[88]

By 1935, the shape of a new conservation policy began to emerge. Ickes, divining that the movement would now cover not only the products of the public domain but also the soil itself, had started a Soil Erosion Service in his department. However, with the great dust storms of the mid-thirties and the collapse of the first Agricultural Adjustment Administration, the Department of Agriculture took the soil problem away from him to embark on a great conservation program.[89]

The Soil Conservation Service, established in 1935 under Hugh H. Bennett, embraced crop control and extension service as well as a certain amount of research. Organized around local conservation districts, its emphasis was heavily on practical application of science to check erosion and rehabilitate farms.[90] In spite of some initial suspicion from the older bureaus, Bennett soon made a place for his agency in the Department of Agriculture.[91] Bennett was a zealot in the tradition of Harvey Wiley. As one observer put it, the department had had an erosion program, "but was content to rest on research. Mr. Bennett, however, was a crusader fired with an enthusiasm that could not be dampened by the skepticism of some of his research-minded associates." [92] The arrangement was to leave much of the research to the established bureaus.

By 1939, Secretary Wallace was stressing the interrelated character of the conservation program. Instead of separate problems in forestry, wildlife, grazing, soil, and crop adjustment, "there is one unified land use problem" of which the others are merely aspects. "This problem involves the whole pattern of soil, climate, topography, and social institutions; it has to do with social and economic conditions, as well as with the physical problems of crop, livestock, and timber production, and of soil and water conservation." [93]

Thus conservation research in general and soil research in particular thrived throughout the great domain of the Department of Agriculture and its state experiment stations. Such a remote field as soil microbiology benefited from the interest in the new conservation, even if few dreamed that this line of research was on the verge of brilliant achievements in the discovery of antibiotics.[94]

The renewed emphasis on conservation and other qualitative changes began to be reflected in the structure of the department during the second New Deal. Two basic trends which marked significant modifications in the classic mold had become evident by 1935. The first was a reaccentuation of basic research after several decades of shift in the other direction. The second was a drive for more coördination of research than either the department's bureaus or the state experiment stations provided.

The Bankhead-Jones Act of 1935 provided for appropriations for scientific, technical, economic, and other research into laws and principles underlying basic problems in agriculture. Beginning with $1,000,000 for the first year, these funds were for work in addition to the regular program, not substitutes for it. "In thus appropriating funds for basic research, in addition to funds for highly specific problems, Congress recognized that fundamental research may often be more practical than short-cut research." [95] To carry out the Bankhead-Jones program, the department followed the pattern of the Forest Service and the Bureau of Mines in establishing regional laboratories located solely because of technical requirements. Each of the laboratories, which eventually numbered nine, concentrated on a problem particularly appropriate to its region — soy beans, vegetable breeding, grass breeding, for example. Facilities used were normally at a land-grant college, with other institutions in the area entering into memoranda of understanding.[96]

The regional pattern was emphasized and the state pattern further modified when the Agricultural Adjustment Act of 1938, concerned with disposing of huge surpluses, provided $4,000,000 for establishing laboratories to conduct research on industrial utilization of agricultural products and $4,000,000 annually afterward. Out of this program came the four regional research laboratories at New Orleans, at Peoria, Illinois, at Albany, California, and at Wyndmoor, Pennsylvania.[97] The Northern Regional Research Laboratory at Peoria was soon to gain fame for its role in developing mass production of penicillin.

Besides the superimposition of two regional systems on the older state-by-state research, the department, beginning with a memorandum of the secretary in 1934,[98] built up a great central research installation at Beltsville, Maryland, near Washington. By providing extensive field and laboratory facilities for all the scientific bureaus, Beltsville became the hub of actual research as the old department had never been. In achieving a kind of centralization of knowledge, and in stressing the interwoven nature of its research activities, the department in the late 1930's seemed well on the way to an over-all coördination of its research empire, a trend that was to culminate in the Agricultural Research Administration of 1942.

Other agencies besides the Department of Agriculture responded to the trend toward basic research and more comprehensive programs. The Public Health Service had begun to show a broader outlook as far back as 1930, when it moved its Hygienic Laboratory to Bethesda, Maryland, and renamed it the National Institutes of Health. The Social Security Act of 1935 pumped large funds into state and local health agencies through the Public Health Service, making allotments based on population, special health problems, and the financial need of the various states.[99] Then in 1937 Congress authorized the National Cancer Institute, a step that began a new era for research in the Public Health Service.

In the first place, the National Cancer Institute was a clear departure into the field of noninfectious diseases and a full recognition that health itself as a part of the public welfare was a proper subject of study by the federal government. In the second place, the National Cancer Institute made no attempt to conduct all its research within its own walls on the federal payroll. Its major activities were advanced training for specialists, fellowships within the service, and the distribution of grants-in-aid. Thus close ties were established with the medical schools and private investigators working on cancer. Between 1938 and 1940 the Institute made thirty-three grants totaling $220,000.[100]

When research was considered as a national resource, it could not be walled off in the separate compartments of government, the universities, the foundations, and industry. With its fellowships and grants, the National Cancer Institute was experimenting with means of getting over the barriers between the estates of science. The technique was old, having been tried by the National Board of Health back in 1880, but now there were both a new urgency and a new hope of successful results.

Other agencies, confronted with the same situation, also experimented with methods for crossing the barriers between federal and nonfederal research. The National Advisory Committee for Aeronautics, even while building large research facilities of its own, had maintained close relations with the engineering departments of several universities. In its new and relatively well-defined field the NACA was able to formulate some of its research program in such a way that universities were willing to sign contracts to undertake specific projects. The fact that the committee was composed largely of scientists aided in finding common ground for the contract provisions. In 1939, NACA had contracts for twelve special investigations at ten universities.[101]

Cumulatively, the qualitative changes in government science during the later New Deal presaged a new era even if war had not intervened. The research responsibilities of the government were now so large, so important to its major functions, and so interwoven with one another that important decisions of policy could not be postponed long. For instance, the old assumption that the universities could be trusted to handle all basic research unaided was open to increasing question. Some new move for a central scientific organization appeared called for which would not only coördinate the federal research establishment, but also adjust the total program of the nation in all the estates of science. The prosperity of the federal establishment in 1938 resembled that of 1916, but the shift to applied science had now been at least partially reversed and the interrelatedness of the whole research effort was rapidly coming to the fore as the most pressing problem.

The Shadow of War

The accomplishments of government science during the later New Deal, geared to recovery from an economic depression, gave little sign that they occurred in a period of heightening international tension. Even more ominous, the dictatorships threatening democracy and the freedom of science benefited from the claim that a free government could not efficiently organize its research for war. In the face of this challenge the government reacted only very slowly before 1939. The radar programs of both the Navy and the Army Signal Corps came to fruition in operational equipment by the later thirties,[102] but these achievements neither cost much money nor indicated activity in other areas.

A better measure of the military research program is the over-all experience of the Army. In 1932, little research was going on, with the average expenditure from 1924 to 1933 only $4,000,000. In 1934, a board headed by former Secretary of War Newton D. Baker recommended that "more definite and continuing appropriations should be available for research and development programs." [103] The usual budget for research and development then went up to $9,000,000, divided into $5,000,000 for the Air Corps and $4,000,000 for all other branches — a slight gain which reflected the heightening international tension. By 1936, storm clouds were more obvious, forcing the Army General Staff to take a new look at research. But its conclusion in the face of danger was that "the Army needs large quantities of excellent equipment that has already been developed. The amount of funds allocated to Research and Development in former years is in excess of the proper proportion for the item in consideration of the rearmament program." [104] Hence, the research outlay should be cut to the $5,000,000–$7,000,000 bracket. This "1936 paradox" was not unlike the early reaction to the depression; the crisis was much too serious to wait for research. This understandable response reflected clearly the assumption that the coming war would not be dominated by technological change based on scientific research.

As war broke in 1939, the military research program proved too small to bear the burden when the realization dawned that the course of the war would be affected and even determined by weapons as yet completely undeveloped. The depression-oriented observers who coined the phrase "research — a national resource" now found that their slogan had a deadly new meaning. Research had become a key military resource involving the nation's total research establishment, not just the few laboratories within the armed forces. The question facing the country in 1939 was not how to expand the Army and Navy research facilities, but how to bring the nation's whole scientific ability to bear on research applicable to the war.

The decade 1929–1939, viewed for itself and not as a prelude to what was to follow, was one of increasing use of research in the government. Much energy during the decade went to combating the effects of the depression. But this effort itself combined with the cascading scientific discoveries to give the government a more stable and more flexible establishment.

The Roosevelt administration, like nearly all its predecessors, developed no over-all policy for or against science. The essentials of the New Deal rested on other bases than research and its results. Yet, on a pragmatic level, the government in the New Deal years threw off the blight of the depression and raised the scientific establishment to unprecedented opulence. The more consistent inclusion of the social sciences added a new dimension to the government's research effort, while the coupling of science with planning increased its effectiveness.

Viewed as the last stepping stone to World War II, the decade of the depression and the New Deal made definite contributions. The concept of research as a national resource served well in the crisis. The techniques such as the use of contracts for joining federal and nonfederal research were valuable in setting up an emergency organization.

Even the failure to erect an adequate central scientific organization was a bequest to the future. A whole generation of leaders had toiled and worried over the problem, whether in the National Academy, the National Research Council, the Science Advisory Board, or the National Resources Committee. No earlier period had seen so much questioning of the relation of science to the government and to society. The generation that wrestled with its problems unsuccessfully in the 1930's was to have within another decade unprecedented opportunity and responsibility.

XIX

PROSPECT AND RETROSPECT
AT THE BEGINNING OF A NEW ERA

1940

THE year 1940 marked the beginning of a new era in the relations of the federal government and science. So far as a line can be drawn across the continuous path of history, this date separates the first century and a half of American experience in the field from what has come after. As the scale of operations changed completly, science moved dramatically to the center of the stage. By the time the bombs fell on Hiroshima and Nagasaki, the entire country was aware that science was a political, economic, and social force of the first magnitude. Most thoughtful people also realized, although vaguely and with foreboding, that democracy had to reconcile itself with this new power, whose seemingly sudden intrusion into the structure of the state was a challenge to the resiliency of our institutions. This new realization was clear by the end of the war and dominated the deliberations of the decade after 1945.

World War II and the Postwar Readjustment

The men who put together the wartime research empire were conscious that they were operating in huge new ways.[1] They tended to stress the inadequacy of older institutions, rejecting all thought of getting along solely with the National Academy or the National Research Council. Yet even while making new departures, they used more of the accumulated experience of the nation than they perhaps were aware.

The establishment of the National Defense Research Committee in 1940 had many overtones from past ventures in organizing science. It emerged from conferences between President Roosevelt and four key scientists. Of these, Vannevar Bush as president of the Carnegie

Institution of Washington and James B. Conant as president of Harvard represented the foundations and the universities. Frank B. Jewett was president of the National Academy and a leader of industrial research. Karl T. Compton, president of Massachusetts Institute of Technology, was no stranger to the councils of the government science. These four made a *de facto* committee which stood for all the great estates of science. Their aim was to bring the whole scientific resource of the country to bear on weapons research.

Although the new NDRC bypassed the National Academy and the National Research Council, it relied for its authority on the old Council of National Defense of 1916. Of the eight men on the NDRC, the president of the National Academy and the commissioner of patents served *ex officio*, two additional members being appointed by the secretaries of war and the Navy. The rest were appointed without respect to other offices. Once set up as a government agency, the NDRC lost no time calling on the National Academy and the NRC for coöperation, thus getting full use of the services they were prepared to render.

An important precedent for the NDRC was the National Advisory Committee for Aeronautics, of which Bush was at this time both chairman and an admirer. The committee form, the military representatives, and the predominance of independent members, were all NACA characteristics. More significant, the field of aeronautics was explicitly excluded from the jurisdiction of the NDRC, an admission that the NACA was the one research outfit in the government already organized for the emergency.

While taking the bulk of its projects from lists submitted by the armed services, the NDRC insisted on the right of independent judgment concerning what to undertake as well as how to carry it on. Deciding at the outset not to build laboratories or to engage directly in research of its own, it developed contracts by which universities and industrial firms could take on the work. It placed contracts with the institution best fitted for each project without regard to any state or geographic pattern. While adopting the rule that no one should make a profit out of research, it equally undertook to pay its own way, including indirect costs.

These decisions taken together revolutionized the relation of the federal government to the other estates of science. Yet each decision taken singly was the culmination of a long line of experimentation in governmental forms. The right of independent judgment had been an ideal since the days of the National Institute. The procedure of contract research had been gradually taking shape for a long time, and the abandonment of the geographic pattern had already begun even in its stronghold, the Department of Agriculture. The principle of no profit from research was enshrined in the organic act of the National Academy, while the NRC and the Science Advisory Board had each in its day moved a step in the direction of paying full compensation for research services. Even the exclusion of the evaluation of inventions from the research set-up reflected a judgment on the government's experience in both the Civil War and World War I.

Although the NDRC moved ahead rapidly during its first year, it soon proved too narrow an organization for the emergency because its field was limited to weapons research. In June 1941, President Roosevelt by an executive order established the Office of Scientific Research and Development in response to shortcomings that the scientists, especially Vannevar Bush, had observed. In the first place, a wide gap existed between the weapons produced by NDRC research and the battlefield. The omitted step, which corresponded to engineering development, was emphasized in the change of title. Research and development were here coupled in a union that was to become standard in government terminology. In the second place, the NDRC had little machinery for correlating its own research with that carried on directly by the Army, the Navy, and the NACA. In the third place, some satisfactory arrangement had to be made to bring military medicine into the general organization. The NDRC continued as a branch of the OSRD.

One of the provisions in the executive order creating the OSRD defined its relation to American science as a whole. It was to "serve as a center for mobilization of the scientific personnel and resources of the Nation in order to assure maximum utilization of such personnel and resources in developing and applying the results of scientific research to defense purposes." In the degree that the coming war was to engulf the whole life of the nation, the OSRD became a central organization for all the estates of science. Another provision empowered it to "coördinate, aid, where desirable, supplement the experimental and other scientific and medical research activities relating to national defense carried on by the Departments of War and Navy and other departments and agencies of the Federal Government."[2]

Again insofar as the war was to become total, the OSRD became a central organization for the entire government.

Abandoning the committee form for the agency itself, Bush became director, responsible to the President. This line of authority avoided the difficulty that had appeared at least as early as the National Institute and the Smithsonian of entrusting public money to a private or semiprivate organization.

Even in a field as spectacularly new as nuclear fission, the building blocks of the emergency organization had a long history.[3] President Roosevelt in appointing the first Advisory Committee on Uranium in 1939 called on the director of the National Bureau of Standards, Lyman J. Briggs, to be its chairman. In early 1941, after this group had been reconstituted under NDRC, Briggs felt the need of an impartial review of the problem, which resulted in the appointment of a committee by the National Academy. Thus, the ancient government advisory organization, whose shortcomings had been the point of origin for so many of its successors, still had a role to play on this newest of weapons. The National Academy committee, under the chairmanship of A. H. Compton, submitted a report in December 1941, which figured in the decision to go into the uranium program on a larger scale.

The pressing problem of secrecy arose in the early days of the uranium program because of the possibility that the basic data of physics might bear directly on a weapon of crucial military importance. After the attempt to stop publication by international voluntary agreement had failed, the Division of Physical Sciences of the National Research Council was the body that took up the matter. It eventually established a reference committee to control publication in all fields of military interest.

This voluntary program, unfortunately, only scratched the surface of the problem of secrecy and security, for it did not cover the war work itself or the research done under direct government auspices. As the OSRD operations proliferated, scientists came up against all the inconveniences that went with secrecy — classified documents, loyalty checks, and compartmentalization of research. In this difficult field alone — the balancing of the military need for secrecy with science's need for a free flow of information — was past experience in the government lacking. Despite the oath question at the time of the Civil War and certain restrictions during the first World War, experience could not help because it did not exist. In contrast to the problems of central organization, neither government officials nor scientists had given the conflict of security and science any serious thought. Indeed, the past policy of the government had always been at one with the traditions of science in fostering wide publication and free exchange of information. The troublous history of the security problem during the war and after is a measure of its novelty in the relation of government to science.

Rising to a great climax in 1945, the war-research effort exceeded by many fold both the World War I program and the level of activity of the immediate prewar period. In terms of money, expenditures reached a high of $1.6 billion as compared to only $100 million in 1940.[4] So ready were the President and Congress to support military research that money was not the limiting factor on the size of the research effort. Long before the point had been reached at which the government ceased to be willing to spend, the shortage of trained manpower began to govern the size of the programs. This confidence, this willingness to gamble the critical resources of the country on the prognoses of scientists, is one of the measures of Franklin D. Roosevelt's stature as commander-in-chief, and the success of radar and nuclear fission and many other programs were impressive vindication of his judgment.

Many of the characteristics of the wartime research effort were in fact permanent changes in the government's relation to science, more so than even the leaders at the time expected. Expenditures of the order of $1 billion or more became established. The predominance of weapons research and wide resort to the contract were also prominent features which not only changed the shape of government science but also deeply affected the universities as well. As all the estates of science were drawn into a single great effort of applied science, the interests of basic science suffered not only in the government, but in their accustomed home in the private institutions. In the inheritance of these patterns, postwar science was more akin to the war period than to any previous era of peace.

Despite its great accomplishments, the OSRD was not an answer to the perennial questions of central scientific organization in the government and the country. Vannevar Bush, the director, never meant it to be a permanent structure. In this he differed from George Ellery Hale, who, in 1918, saw his wartime NRC as a permanent agency temporarily diverted from its long-range program. As the end of the war approached, Bush advanced compelling reasons against the indefinite continuance of OSRD. The scale was much too large for peacetime military research. OSRD had tended toward development as the war proceeded, approaching the programs carried on within the military services. At the end of the war with Germany many trained men would be needed on reconversion problems, making it impossible for OSRD to hold its staff. By the summer of 1944, the OSRD leaders were thinking actively about the demobilization of their agency.

Along one line the dismantlement of the OSRD meant a shift of weapons research from the emergency agency to the armed services.[5] Projects were divided into those which should be terminated and those which were of continuing interest to the military. At the same time the War and Navy Departments began to seek a means of administering the research they would receive. Unlike their reaction after World War I, the services recognized that research would remain an important factor and attempted to adjust their organization to it. Even before the establishment of the Department of Defense in 1947, a Research and Development Board undertook to coördinate the military research and to adjust its relations with civilian scientists. This, in turn, gave way in 1953 to the office of Assistant Secretary of Defense (Research and Development). The Office of Naval Research, established in 1946, not only provided for its own service but became an agency that gave great emphasis to basic research.

The atomic-energy program, important to the military but possessing many other aspects as well, followed a different course. Having outgrown the OSRD by 1943, it had completed the war as a part of the Corps of Engineers of the Army. But in 1946, the Congress, picking up the tradition of civilian control that had developed after the Civil War, created the Atomic Energy Commission. Combining research and production, the AEC relied heavily on the use of contracts. Among its new departures were the National Laboratories, created by contracts with corporations formed specifically to carry on the research. Almost immediately the AEC's expenditures made it the largest civilian scientific outfit in the government, overshadowing the size of the Department of Agriculture by more than three times.

A third line leading from the OSRD was the federal government's concern for basic research. This need, which had clearly appeared before the war, was of deep concern to Bush, who believed that OSRD's programs in basic research should be turned over to permanent organizations while the war was still on. Out of this desire came President Roosevelt's letter of November 17, 1944, which set in motion the investigation leading to the report, *Science, The Endless Frontier*, with its suggestion of a National Research Foundation. The discussion of this idea and the attempt to draft and pass a suitable law took five years. Finally, in 1950, the National Science Foundation took its place among the agencies of the government.

Although the changes brought by these successors of the OSRD have been so profound that much attention has been focused on them in the postwar period, the government's scientific establishment which had taken shape before 1916 was still a living part of the government. Indeed, some of the older agencies have had great bursts of splendor. In 1938, the Public Health Service spent only a little more for research than the Bureau of Entomology and Plant Quarantine. In 1953, its $56,000,000 almost precisely matched the expenditures of the whole Department of Agriculture. As had occurred so often in the past, the effect of the new efforts was superimposed on the existing structures. The old agencies still had their usefulness, and the old problems still had to be solved. The mighty edifice of government science dominated the scene in the middle of the twentieth century as a Gothic cathedral dominated a thirteenth-century landscape. The work of many hands over many years, it universally inspired admiration, wonder, and fear.

The First Century and a Half

A look backward over the republic's first 150 years of experience with science shows a coherent pattern on two distinct levels. On the pragmatic plane of science responding to the needs of society, the story is one of accomplishment. On the higher plane of the attempt to create a comprehensive organization of science as a fundamental institution within the state, the record is fraught with yearning.

From the beginning the government proved able and willing to use the science of the day not only for its own internal needs but as a boon to its citizens. The Lewis and Clark expedition and the Coast Survey were early examples of what became after 1830 the systematic use of science in all phases of exploration, both of the westward reaches of the continent and of the sea lanes that carried American commerce around the world. During the Civil War, glimpses of other uses of science widened its scope, and in the two generations following 1865 a great scientific establishment arose within the government, developing its own typical organization around the problems it faced, recruiting a specialized personnel, and contributing greatly to the scientific resources of the country. With World War I, the establishment had to shift into the field of weapons research on a wide scale for the first time. By the 1920's and 1930's, science was so important a part of society that government science not only became thoroughly interlaced with the other research institutions of the country but received major shocks from the social and economic upheaval of the great depression.

The history of the pragmatic response of government science is not without its shadows. Some of its agencies failed to survive, for instance, the National Board of Health. Indifference and occasional outright hostility made the way harder in Congress. To reconcile the actualities of civil service with the needs of science was difficult both before and after the introduction of the merit system. As the other estates of science arose — research in the universities, in the foundations, and in industry — the government often suffered by comparison. Because the public interest seemed to demand a practical use assured in advance for all its work, it sometimes had to endure the scorn of those with a discriminating taste for pure knowledge.

Over the long sweep of a century and a half, however, the shadows are only minor. When the needs cried out, science answered with measurable results. Many of the nation's most illustrious names appear on the roll of those who responded. Many of the more obscure ones brought science to the government in spite of personal sacrifice and inadequate reward. Some of the institutions they built were among the greatest of their kind in the world. The practice of whole industries, the knowledge of whole branches of science, and the well-being of the entire population of the country would be markedly impaired without the cumulative results of government science. And since the results of science during most of the 150 years flowed freely over national boundaries, the government's contributions entered the general reserve of knowledge available for the use of all the world. Whatever the United States borrowed from European sources impoverished no one, while what she put back enriched both the practice and the knowledge of others.

No generation has been satisfied that pragmatic response to needs was the whole of the relation between the government and science. In the first place, the necessity always appeared of arranging the piecemeal responses of science into a coherent pattern. In the second place, the conviction persisted that the government had a responsibility for science independent of its practical usefulness. This duty emphasized the basic discoveries that made science an integral part of civilization.

The welding of these two basic attitudes produced a long series of experiments with central scientific organization. No period in the history of the republic is altogether devoid of specific proposals to meet this combined aspiration. The American Philosophical Society, the American Academy of Arts and Sciences, and a National University were the candidates in the first era. Then the National Institute, the Smithsonian Institution, and finally the National Academy of Sciences arose from a generation of lively debate after 1830. A department of science in the executive branch of the government loomed briefly in the 1880's. The National Research Council emerged from World War I. The peculiar character of the decade of the 1920's was reflected in the National Research Fund. The Science Advisory Board and the science committee of the National Resources Committee were products of the depression and the New Deal. The OSRD was the crowning effort of World War II. With all their variation in approach, support, and membership, these organizations aimed to coördinate the government establishment and at the same time to stand for science as an ideal and knowledge as an aspect of civilization.

Along the way, other more specialized institutions edged into the stream of central scientific organization. The Coast Survey dominated the pre-Civil War era. The Geological Survey had immense influence in the 1880's. The Forest Service grew up in the Progressive Era to stand for the whole conservation point of view. The Carnegie Institution of Washington was the private foundations' effort to make the capital city a true scientific metropolis. In the 1940's, weapons research under the armed forces seemed to many to become the whole of government science, basic as well as applied. In the long run, however, these agencies were only seasonal mushrooms, depending on special circumstances and often on unusually gifted individual administrators. The names of Bache, Powell, and Pinchot come first to mind. The increasing scarcity of the phenomenon after 1900 indicates that only cataclysmic events such as World War II could give a specialized group any color of a central organization.

None of the candidates for the position has successfully achieved a truly dominant position either in the country or in government science. All have sooner or later missed their mark. In the course of this history of frustration some general phenomena have emerged. In the first place, an institution that had to depend on endowment had too little elasticity to fill the role, witness the Smithsonian's experience during and after the Civil War. In the second place, the public functions inherent in the role made its fulfilment impossible for a purely private institution. The National Institute, the Smithsonian, the National Academy, and the National Research Council never surmounted the difficulty that arose when they asked Congress to turn over public money. Could Congress give over control to a private body, even one made up of experts or one in which congressmen as individuals sat? That this question never received a clear-cut affirmative is evident in the Smithsonian's inability to keep the NACA and in the lack of federal support of the National Academy, the peacetime NRC, and the Science Advisory Board.

In the third place, the executive branch of the government never saw its way clear to yielding its right to reject the advice of an outside body. The National Academy never gained the right to name scientists for appointment or to limit the President's selection to a panel named by them. By the executive orders creating the National Research Council, the Science Advisory Board, and the OSRD, the President asserted his general authority in the field. Thus, a channel of responsibility between the organization and the federal executive emerges as a requirement for success.

Finally, an institution had to have not only a legal authority within the government, but a moral authority with all the estates of science in the country and a position of honor among the great scientific societies of the world. This broader dimension was a concern of the earliest organizations as well as the later ones. But in the twentieth century, government research became so colossal that by its use of funds and personnel it could control the dynamics of the other estates of science. With this dominant position, the approbation of all science became an absolute necessity. To be truly representative of the varied interests of the professional natural scientists, engineers, and social scientists who demand a voice, implies a certain amount of independence in the face of the government's interests. The need for reconciliation of the government's legitimate demand for responsibility and the scientists' essential stake in independence is one way of stating the unsolved dilemma of all attempts at central scientific organization.

The failure of these many groups to become a central organization does not mean that they strove entirely in vain. Although they did not successfully solve the basic problem, most of them not only survived but made illustrious records in which their friends and members take justifiable pride. The Smithsonian has entered its second century still holding to the ideal of pure research and the universality of its scope that Joseph Henry set for it. The National Academy still had a role to fill as an administrative framework for certain types of advice even when the age of nuclear fission dawned. The National Research Council has proved highly useful to both the OSRD and the National Science Foundation, even though they were its successors as central organizations. Hence the very efforts to solve the problem have, while failing to reach their goal, immeasurably enriched the scientific life of the nation.

Science in a Democracy

From the beginning of the republic, men had two attitudes toward science. On the one hand, seeing its freedom and the withering of superstition and blind authority under its examination, they concluded that it was cut of the same pattern as democracy. On the other hand, observing the inaccessibility of its lore to the untutored masses and the support it traditionally received from aristocratic sources, they feared that inconstant democracy could never have the discrimination to foster science on the same plane with authoritarian states. In the twentieth century, this second mistrustful attitude took two forms. One fear saw dictatorships using science with horribly efficient purpose to destroy a free world that had the research but could not mobilize it. The other fear saw science itself becoming so powerful and so complex that the people could not understand it and their representative institutions could not control it. Against these extremes stands the long experience of American democracy.

In a narrowly partisan sense, science has seldom been a political issue. No party has ever been consistently pro-science and the other anti-science. From the day of Thomas Jefferson and John Adams to that of Herbert Hoover and Franklin D. Roosevelt, a friendliness to science has been common ground for those who differed on other issues and whose parties professed to stand at opposite poles. The turning points in policy have seldom coincided with changes in party control of the government. The scientists themselves have held every shade of opinion. Some ages have condemned them as dangerous radicals; others have mistrusted them as standpatters.

In a broader way, however, politics has been a factor in government science from the time the Constitutional Convention puzzled about the powers of the federal government over internal improvements. The shift of political center of gravity in the Union that accompanied the Civil War was necessary to the opening of new possibilities. In the Progressive Era, the power to gather data and the power to regulate moved close together. Beginning with the first World War, science became a recognized part of the military — the most direct power available to the state. By the 1930's, research was recognized as a national resource; during World War II it proved its right to the title.

The Congress as the representatives of the people and the controller of the purse had to come to a working relation with science from the time of its First Session when patent applications began to arrive. On occasion it tried to administer science directly through its own committees, as in the case of the Wilkes expedition collections. But by the Civil War, it had retired from this pretension. Throughout most of the nineteenth century, Congress was a major check on the growth of government science. Only by such techniques as the organic act buried in an appropriation bill was the federal establishment built. However, in the twentieth century, the Congress, in spite of misgivings about its role in providing for research, has ceased to be a negatively limiting factor on the use of science in the government. Given adequate preparation and presentation of programs, Congress has shown itself able and willing to support research.[6] As popular indifference to science has given way to admiration for its more spectacular accomplishments, Congress has reflected the change in its attitude toward such activities as public health.

The executive control of science was in the first years centered largely in the President himself. Jefferson had marked interest in science, but his whole civil service scarcely contained another to equal him. After sifting down through the cabinet, effective control of science reached the level of the bureau chief during the Civil War, notably in Alexander Dallas Bache, Charles Henry Davis, and Joseph Henry. As the establishment grew, the bureau chief became more firmly entrenched, although after the Powell period the incumbents seemed a little less like statesmen. During the early twentieth century the effective administration of research remained low in the pyramid of authority, even though Gifford Pinchot momentarily carried conservation issues into the upper reaches of Theodore Roosevelt's administration, only to metamorphose himself into a politician. By the time of the great depression, science began to press decisions on higher officials, pushing up to the President for the crucial choices concerning the manufacturing of an atomic bomb. Although the machinery of advice by which decisions are made in higher echelons is a major problem, the executive has a firm base for control and an extensive body of experience on the lower level.

How the government can keep science in bounds is only one side of the relation. The controlling structure itself is the product of an experience profoundly affected by science. Democracy as measured by twentieth century practice in the United States has had the benefit of a long partnership with science, not a long record of hostility. The conservationists of the Progressive Era profoundly believed that only a government informed by science could rule justly in the public interest. If their belief was too simple, the experience of the nation has nevertheless been in their favor. A democracy that has in fact enjoyed the results of science has been more tolerable, more humane, and more able to fulfill its responsibilities to its people. After the industry of the country and the military forces of the world came to draw their power from research technology, a government without considerable scientific competence could not have governed at all. During a century and a half, science has not only contributed to the power of the government but to the ability of the people to maintain their freedom.

CHRONOLOGY

1787   Constitutional Convention
1789   Organization of the government
1790   First patent law
      First census
1792   The Mint
1798   Medical care for merchant seamen
1800   Library of Congress
1802   Army Corps of Engineers
      United States Military Academy, West Point, New York
1803   Lewis and Clark expedition
1807   Coast Survey act
1813   Federal law establishing vaccine agent
1816   Columbian Institute
1818   Discontinuance of the Coast Survey
      Army Medical Department
1819   Long expedition to the Rockies
1822   Repeal of vaccine law
1824   Survey act and increased activity of the Corps of Topographical
        Engineers
1825   John Quincy Adams's first annual message
1830   Authorization of work on weights and measures
      Navy Depot of Charts and Instruments
1832   Reëstablishment of the Coast Survey
1836   Reorganization of the Patent Office
1838   Federal law regulating steamboats
      United States Exploring Expedition
1839   First use of Patent Office fund for agricultural studies
1840   National Institute for the Promotion of Science
1842   Act in support of Morse's telegraph
      United States Botanic Garden
      Naval Observatory
      Frémont's first expedition to the Rockies
1845   United States Naval Academy
1846   Mexican War: Emory's reconnaissance from Fort Leavenworth to
        San Diego
      Smithsonian Institution
1847   David Dale Owen's and others' geological surveys of federally
        owned lands
      Lynch's Dead Sea Expedition

1848   United States and Mexican Boundary Survey
       American Association for the Advancement of Science
1849   Grant to Charles G. Page for electric locomotive
       Nautical Almanac Office
       Gilliss's U. S. Naval Astronomical Expedition to Chile
1851   Herndon and Gibbon's explorations of the valley of the Amazon
       Bache's presidential address before the AAAS
1852   Kane's expedition to the Arctic
       Perry's Japan expedition
1853   Pacific railroad surveys
       Ringgold and Rodgers's U. S. North Pacific Exploring Expedition
1859   President Buchanan vetoes land-grant college bill
1861   Government Printing Office
       Outbreak of the Civil War
1862   Act creating the Department of Agriculture
       Homestead Act
       Morrill Act for land-grant colleges
       Reorganization of the Navy Department
1863   Navy Permanent Commission
       National Academy of Sciences
       Army Signal Corps
1866   Navy Hydrographic Office separated from Naval Observatory
       Disbanding of the Corps of Topographical Engineers
1867   Clarence King's Geological Survey of the Fortieth Parallel
       Army Medical Museum
1868   Army Medical Library
1869   Lt. G. M. Wheeler's Geographical Surveys West of the Hundredth
          Meridian
1870   Meteorological work begins in Army Signal Corps
       J. W. Powell's Geographical and Topographical Survey of the
          Colorado River of the West
       Reorganization of the Marine Hospital Service
1871   Fish Commission
1873   F. V. Hayden's Geological and Geographical Survey of the
          Territories
1879   U. S. Geological Survey
       National Board of Health
1880   Bureau of American Ethnology, Smithsonian Institution
1881   Founding of the magazine *Science*
       Reorganization of Division of Entomology, Department of Agri-
          culture

1883   Pendleton Civil Service Act; U. S. Civil Service Commission established

1884   Bureau of Animal Industry, Department of Agriculture
Allison Commission

1886   Division of Economic Ornithology and Mammalogy, Department of Agriculture, forerunner of Biological Survey

1887   Hygienic Laboratory, Marine Hospital Service
Hatch Act for Agricultural Experiment Stations

1888   Powell's irrigation survey
Office of Experiment Stations, Department of Agriculture

1889   Elevation of head of Department of Agriculture to cabinet rank

1890   Act transferring meteorological service from the Army and creating Weather Bureau, Department of Agriculture

1891   Astrophysical Observatory, Smithsonian Institution

1893   Army Medical School

1896   National Academy Committee on Forestry

1898   Spanish-American War

1900   Yellow-fever board in Cuba
Lacey Act to control shipment of game birds in interstate commerce

1901   Bureau of Chemistry, Department of Agriculture
Bureau of Plant Industry, Department of Agriculture
Bureau of Soils, Department of Agriculture
National Bureau of Standards

1902   Newlands Act and establishment of Reclamation Service
Reorganization of Marine Hospital and Public Health Service
Bureau of the Census

1903   Philippine Bureau of Science
Committee on Organization of Scientific Work

1905   Transfer of forest reserves from Department of Interior to Department of Agriculture; change of name of Division of Forestry to Forest Service

1906   Pure Food and Drug Act

1908   Conference of governors on conservation

1910   Bureau of Mines

1912   Public Health Service

1914   Smith-Lever Act authorizing extension work, Department of Agriculture

1915   National Advisory Committee for Aeronautics
Naval Consulting Board

1916   National Park Service
National Research Council

1917 Entry into World War I
1918 Chemical Warfare Service
1921 Bureau of the Budget
1922 Bureau of Agricultural Economics, Department of Agriculture
1923 Naval Research Laboratory
      Bureau of Home Economics, Department of Agriculture
1926 National Research Fund
      Aeronautics Branch established in Commerce Department
1927 Radio Division of Department of Commerce
1930 National Institute of Health
1933 Science Advisory Board
      Tennessee Valley Authority
1934 Agricultural Research Center
1935 Bankhead-Jones Act for Agricultural Research
      National Resources Committee
      Soil Conservation Service
1937 National Cancer Institute
1938 *Research — A National Resource*
      Agricultural Adjustment Act
1940 National Defense Research Committee
1941 Office of Scientific Research and Development
      Entry into World War II
1942 Agricultural Research Administration
1946 Atomic Energy Commission
      Office of Naval Research
1950 National Science Foundation

BIBLIOGRAPHIC NOTE

Few people have written directly on the subject of this book, yet so many have touched it that the literature from which it is drawn is immense. Because the path runs at a sharp angle to the established lines of bibliography, many works of history, science, and government refer to it for a moment, then go off on their divergent courses. Although the notes cite but a portion of the material consulted, they serve as the detailed bibliography which the organization of the text arranges automatically by subject in a roughly chronological pattern. The following remarks contain only a few general reflections on the sources.

This study's one true predecessor is the series of papers by George Brown Goode, most conveniently available in *A Memorial of George Brown Goode* (Smithsonian Institution, *Annual Report for 1897*, Part II, Washington, 1901). Besides being a scientist and a distinguished public servant, Goode had the gift of historical understanding and a sense for the structure of living institutions. Studies have become more elaborate since his day, notably *Research — A National Resource* (Washington, 1938–1940), Vannevar Bush, *Science, the Endless Frontier* (Washington, 1945), and the President's Scientific Research Board, *Science and Public Policy* (5 vols., Washington, 1947). But these uneven cross sections only hint at the long lines of historical development.

The period after 1940 is already better served than any other period. James P. Baxter, 3rd, *Scientists Against Time* (Boston, 1946); Henry D. Smyth, *Atomic Energy for Military Purposes* (Princeton, 1945); and Irvin Stewart, *Organizing Scientific Research for War* (Office of Scientific Research and Development, *Science in World War II*, Boston, 1948), cover important aspects of the relation of the government and science in World War II. Don K. Price, *Government and Science* (New York, 1954), emphasizes postwar developments. The presence of this body of literature played a part in the selection of 1940 as the terminal date for the present study.

BIBLIOGRAPHIES

A recent list bearing directly on this study is Kathrine O. Murra and Helen D. Jones, "The Administration of Research: A Selective Bibliography," in President's Scientific Research Board, *Science and Public Policy* (Washington, 1947), III, 253–324. For a segment of science before 1865, Max Meisel, *A Bibliography of American Natural History: The Pioneer Century, 1769–1865* (3 vols., Brooklyn, 1924–1929), is a masterwork of the bibliographer's art.

Entry into the world of government documents may be made through Anne M. Boyd and Rae E. Rips, *United States Government Publications* (3rd ed., New York, 1949), and Laurence F. Schmeckebier, *Government Publications and Their Use* (Washington, 1939).

On general American historical literature, Oscar Handlin, *et al.*, *Harvard Guide to American History* (Cambridge, Mass., 1954), has been available for the latter part of this study. Also useful are H. P. Beers, *Bibliographies in American History: Guide to Materials for Research* (New York, 1942), and *Writings on American History* (1902–   ). The relative paucity of items bearing on science in these splendid works is a commentary on the backward state of this field of historical study. For the history of science, the "Critical Bibliographies" appearing in *Isis* are indispensable.

The greatest body of bibliographic information is scattered through hundreds of works primarily devoted to something else. For example, the Institute for Government Research, *Service Monographs of the United States Government* (62 vols., Baltimore and Washington, 1918–1930), contain bibliographies on the history of most of the government's bureaus.

MANUSCRIPTS

Since this study is largely based on printed sources, it can only indirectly serve as a guide to subjects that would benefit from information derived from manuscripts. The greatest collection of unpublished material bearing on this subject is in the National Archives, which stands as a challenge to the present generation of historians. National Archives, *Guide to the Records in the National Archives* (Washington, 1948), provides a general orientation, while specifically useful is Nathan Reingold, "The National Archives and the History of Science in America," *Isis*, 46:22–28 (1955). For particular subjects the National Archives, *Preliminary Inventories*, are valuable.

For certain topics of great importance on which printed material failed, manuscripts in the extensive and little-used private papers of American scientists have been consulted on an occasional basis. The most important of these were the Benjamin Peirce papers in the Harvard University Archives, the historic letter file of the Gray Herbarium of Harvard University, and the Pinchot papers at the Library of Congress.

PRINTED PRIMARY SOURCES

Government Documents

Official papers play a prominent part in this study. *American State Papers: Documents, Legislative and Executive, 1789–1837* (38 vols.,

Washington, 1832–1861) is a collection arranged by subject matter for which there is no counterpart in later periods. The congressional series may be entered by the *Check List of United States Public Documents 1789–1909* (Washington, 1911), and *Tables of and Annotated Index to the Congressional Series of United States Public Documents 1817–1893* (Washington, 1902). Proceedings of Congress appear in *Annals of Congress 1789–1824* (42 vols., Washington, 1834–1856), *Congressional Debates 1825–1837* (29 vols., Washington, 1825–1837), *Congressional Globe* (109 vols., Washington, 1834–1873), and *Congressional Record* (Washington, 1873–). Presidential papers appear in J. D. Richardson, comp., *Compilation of the Messages and Papers of the Presidents, 1789–1897* (10 vols., Nashville, 1905).

In the twentieth century, the flood of government publications has been so great that the congressional series has been substantially outgrown. *Annual Reports*, both of the cabinet secretaries and of bureau chiefs, are an extensive source, uneven in quality but often very significant.

Other Printed Primary Sources

The publications of some nongovernmental institutions have been important to this study, especially the *Annual Reports* and *Proceedings* of the National Academy of Sciences. The Smithsonian Institution has a literature of its own. Besides the *Annual Reports*, the early history is covered by three great collections of documents edited by W. J. Rhees.

Collected papers of scientists seldom include much material that sheds light on the organization of science, although L. H. Butterfield, ed., *Letters of Benjamin Rush* (2 vols., Princeton, 1951), is an exception. Among the papers of the few major statesmen actively concerned with the administration of science, occasional relevant material appears. Some of the more useful are Roy P. Basler, ed., *The Collected Works of Abraham Lincoln* (8 vols., New Brunswick, N. J., 1953); Paul L. Ford, ed., *The Works of Thomas Jefferson* (12 vols., New York, 1904); and Elting E. Morison, ed., *The Letters of Theodore Roosevelt* (8 vols., Cambridge, Mass., 1951–1954).

A few of the standard published document collections of American history contain material of interest. Among them are Max Farrand, ed., *The Records of the Federal Convention of 1787* (4 vols., New Haven, 1911–1937); and A. P. Nasatir, ed., *Before Lewis and Clark* (Saint Louis, 1952).

Periodicals are often primary sources, although many scientific journals reveal almost nothing of the relations of science to the government. For the middle nineteenth century the *American Journal of Science and Arts* (1818–) had wide coverage and general interest. Since 1883, the magazine *Science* forms a continuous series of scientific news items, editorial comment, and articles which often provide sources of both fact and opinion.

AUTOBIOGRAPHIES AND BIOGRAPHIES

Autobiographies and biographies have played such an important role that they deserve a separate listing. The overemphasis on the individual hero in science may account in part for this situation. But more fundamental is the fact that biographers have stalked in whatever fields their quarry has led them into. Hence, they have often penetrated deep into the brambles of the relation of science to the government during the same years that the historians concerned with impersonal trends ignored the subject.

Autobiographies

Autobiographies have sometimes provided the most specific accounts of the inner workings of government science. Among these may be mentioned especially Charles F. Adams, ed., *Memoirs of John Quincy Adams* (12 vols., Philadelphia, 1875–1876); Bailey K. Ashford, *A Soldier in Science* (New York, 1934); David Fairchild, *The World Was My Garden* (New York, 1938); Herbert Hoover, *Memoirs* (3 vols., New York, 1951–1952); Leland O. Howard, *Fighting the Insects* (New York, 1933); Robert A. Millikan, *Autobiography* (New York, 1950); Simon Newcomb, *The Reminiscences of an Astronomer* (Boston, 1903); Gifford Pinchot, *Breaking New Ground* (New York, 1947); Harvey W. Wiley, *An Autobiography* (Indianapolis, 1930).

Biographies

Among collective biographical works, Allen Johnson and Dumas Malone, eds., *Dictionary of American Biography* (22 vols., New York, 1928–1944), is preëminent. For this study the National Academy of Sciences, *Biographical Memoirs* (1877–   ) is also important.

Biographies of political leaders sometimes brush the relation of science to the government, for example, Irving Brant, *James Madison* (4 vols., Indianapolis, 1941–1950); and Dumas Malone, *Jefferson and His Time* (2 vols. to date, Boston, 1948–1952).

The most directly relevant of the biographies of scientists and administrators are Florian Cajori, *The Chequered Career of Ferdinand Rudolph Hassler, First Superintendent of the United States Coast Survey* (Boston, 1929); Thomas Coulson, *Joseph Henry: His Life and Work* (Princeton, 1950); William H. Dall, *Spencer Fullerton Baird* (Philadelphia, 1915); William C. Darrah, *Powell of the Colorado* (Princeton,

1951); Charles H. Davis, *Life of Charles Henry Davis, Rear Admiral, 1807–1877* (Boston, 1899); Merle M. Odgers, *Alexander Dallas Bache: Scientist and Educator, 1806–1867* (Philadlphia, 1947); Charles Schuchert and Clara M. Le Vene, *O. C. Marsh: Pioneer in Paleontology* (New Haven, 1940); Wallace Stegner, *Beyond the Hundredth Meridian: John Wesley Powell and the Second Opening of the West* (Boston, 1954).

Others, though less useful or less directly applicable because their subjects were not so closely associated with the government, contain much valuable material. Among them are Frank Cameron, *Cottrell: Samaritan of Science* (Garden City, New York, 1952); John F. Fulton, *Harvey Cushing, A Biography* (Springfield, Ill., 1946); Fielding H. Garrison, *John Shaw Billings: A Memoir* (New York, 1915); Charles L. Lewis, *Matthew Fontaine Maury: The Pathfinder of the Seas* (Annapolis, 1927); James P. Munro, *A Life of Francis Amasa Walker* (New York, 1923); Jesse S. Myer, *Life and Letters of Dr. William Beaumont* (Saint Louis, 1939); Andrew D. Rogers, III, *Erwin Frink Smith: A Story of North American Plant Pathology* (Philadelphia, 1952); Andrew D. Rodgers, III, *John Torrey: A Story of North American Botany* (Princeton, 1942); Charles C. Sellers, *Charles Willson Peale* (2 vols., Philadelphia, 1947); Robert Shankland, *Steve Mather of the National Parks* (New York, 1951).

OTHER SECONDARY WORKS

In a perfectly ordered world the author of a synthesis such as this one would eschew primary sources entirely. Monographs of uniform quality stacked neatly along the path of his story would serve him well enough. Unfortunately, the primary sources listed above are testimony that no such simple situation exists in this field. While secondary works have been used as much as possible, their varying quality and chaotic arrangement have left many gaps unexplored and many important questions unanswered. Prepared by men with a wide range of qualifications for the task and with many different aims, these secondary materials neveretheless provide much relevant information. For convenience, they may be divided roughly into four groups.

The first category contains those publications produced by the government's efforts to understand itself. Some government documents are really monographs. Among these are Esther C. Brunauer, *International Councils of Scientific Unions, Brussels and Cambridge* (Department of State, Publication 2413, Washington, 1945); Department of Agriculture, *Yearbook for 1899* (Washington, 1900); Lloyd N. Scott, *Naval Consulting Board of the United States* (Washington, 1920); A. C. True, *A History of Agricultural Experimentation and Research in the United*

States, 1607–1925 (Department of Agriculture, *Miscellaneous Publication No. 251*, Washington, 1937); A. C. True, *A History of Agricultural Education in the United States, 1785–1925* (Department of Agriculture, *Miscellaneous Publication No. 36*, Washington, 1929); Mark S. Watson, *Chief of Staff: Prewar Plans and Preparations (United States Army in World War II. The War Department*, Washington, 1950).

The second major category contains books written by scientists and other active participants. Among these are I. W. Bailey and H. A. Spoehr, *The Role of Research in the Development of Forestry in America* (New York, 1929); Vannevar Bush, *Modern Arms and Free Men* (New York, 1949); Grosvenor B. Clarkson, *Industrial America in the World War: The Strategy Behind the Line, 1917–1918* (Boston, 1923); A. A. Fries and C. J. West, *Chemical Warfare* (New York, 1921); George W. Gray, *Frontiers of Flight: The Story of NACA Research* (New York, 1948); T. Swann Harding, *Two Blades of Grass: A History of Scientific Development in the U. S. Department of Agriculture* (Norman, Okla., 1947); Edgar E. Hume, *Victories of Army Medicine* (Philadelphia, 1943); George P. Merrill, *Contributions to the History of American Geology* (National Museum, *Annual Report for 1904*, Washington, 1906); Paul H. Oehser, *Sons of Science: The Story of the Smithsonian Institution and Its Leaders* (New York, 1949); Frederick W. True, *A History of the First Half-Century of the National Academy of Sciences, 1863–1913* (Washington, 1913); R. C. Williams, *The United States Public Health Service, 1798–1950* (Washington, 1951); R. M. Yerkes, ed., *The New World of Science: Its Development During the War* (New York, 1920).

The third category contains books primarily directed to the problems of public administration. The Institute for Government Research, *Service Monographs of the United States Government* (62 vols., Baltimore and Washington, 1918–1930), is the core of this literature and a source of great importance. Other studies stimulated by professional interest in public administration are Richard G. Axt, *The Federal Government and Financing Higher Education* (New York, 1952); Carleton R. Ball, *Federal, State, and Local Administrative Relationships in Agriculture* (2 vols., Berkeley, 1938); Robert H. Connery, *Government Problems in Wildlife Conservation* (New York, 1935); John M. Gaus and Leon O. Wolcott, *Public Administration and the United States Department of Agriculture* (Chicago, 1940); Richard Hofstadter and C. D. Hardy, *The Development and Scope of Higher Education in the United States* (New York, 1952); V. O. Key, Jr., *The Administration of Federal Grants to States* (Chicago, 1937); R. D. Leigh, *Federal Health Administration in the United States* (New York, 1927); A. W. MacMahon and J. D. Millett, *Federal Administrators: A Biographical Approach to the Problem of Departmental Management* (New York, 1939); Lloyd M. Short, *The Development of National Administrative Organization in the United States* (Baltimore, 1923). Belonging to the public administration group but having a broader historical scope as well are Leonard D. White's three volumes, *The Federalists* (New York, 1948), *The Jeffersonians* (New York, 1951), and *The Jacksonians* (New York, 1954).

The fourth category contains the works of authors whose primary interest is historical. Some of these books center around science, such as George W. Adams, *Doctors in Blue: The Medical History of the Union Army in the Civil War* (New York, 1952); Ralph S. Bates, *Scientific Societies in the United States* (New York, 1945); Richard H. Shryock, *American Medical Research, Past and Present* (New York, 1947); Dirk J. Struik, *Yankee Science in the Making* (Boston, 1948). Some general works on a whole period are useful, notably George Dangerfield, *The Era of Good Feelings* (New York, 1952); Frederic L. Paxson, *American Democracy and the World War* (3 vols., Boston, 1936–1948). Some of the newer studies of technology by economic historians are relevant. Among them are O. E. Anderson, Jr., *Refrigeration in America: A History of a New Technology and Its Impact* (Princeton, 1953); Louis C. Hunter, *Steamboats on the Western Rivers: An Economic and Technological History* (Cambridge, Mass., 1949). Special studies occasionally have had larger implications than their own limited subjects. Among these key monographs are James P. Baxter, 3rd, *The Introduction of the Iron Clad Warship* (Cambridge, Mass., 1933); I. B. Holley, Jr., *Ideas and Weapons: Exploitation of the Aerial Weapon by the United States during World War I: A Study in the Relationship of Technological Advance, Military Doctrine, and the Development of Weapons* (New Haven, 1953); Gerstle Mack, *The Land Divided: A History of the Panama Canal and Other Isthmian Canal Projects* (New York, 1944). Monographs on subjects other than science are often useful. Examples of the most significant are Sidney Forman, *West Point: A History of the United States Military Academy* (New York, 1950); F. Stansbury Haydon, *Aeronautics in the Union and Confederate Armies* (Baltimore, 1941); W. Turrentine Jackson, *Wagon Roads West* (Berkeley, 1952); Jeannette Mirsky, *To the Arctic! The Story of Northern Exploration from Earliest Times to the Present* (New York, 1948); F. Paul Prucha, *Broadax and Bayonet: The Role of the United States Army in the Development of the Northwest, 1815–1860* (Madison, Wis., 1953); F. A. Shannon, *The Organization and Administration of the Union Army, 1861–1865* (2 vols., Cleveland, 1928); Richard J. Storr, *The Beginnings of Graduate Education in America* (Chicago, 1953).

Of the four categories, the second and third can have only a limited expansion in covering the period 1789–1940. Direct participants and students of public administration will naturally tend to focus upon the current scene. Only by a great increase in the quality and quantity of historical studies can the true significance of the first century and a half of the government's experience with science gradually emerge.

REFERENCES

Chapter I. First Attempts to Form a Policy

1. Max Farrand, *The Records of the Federal Convention of 1787* (New Haven, 1911–1937), II, 325.
2. *Ibid.*
3. *Ibid.*, 615.
4. Dr. Benjamin Rush was at this time publicly advocating such an institution. L. H. Butterfield, ed., *Letters of Benjamin Rush* (Princeton, 1951), I, 491–495.
5. *Annals of Congress*, May 3, 1790, 1 Cong., 2 Sess., 1551.
6. W. S. Holt, *The Bureau of the Census: Its History, Activities and Organization* (Institute for Government Research, *Service Monograph No. 53*, Washington, 1929), 2–4.
7. Florian Cajori, *The Early Mathematical Sciences in North and South America* (Boston, 1928), 80.
8. Butterfield, *Letters of Rush*, I, 312.
9. E. G. Inlow, *The Patent Grant* (Baltimore, 1950), 44, 45; Merrill Jensen, *The New Nation: A History of the United States During the Confederation, 1781–1789* (New York, 1950), 286.
10. E. F. Smith, *Priestley in America, 1794–1804* (Philadelphia, 1920), 50.
11. *Annals of Congress*, April 15, 1789, 1 Cong., 1 Sess., 143.
12. *Annals of Congress*, December 30, 1791, 2 Cong., 1 Sess., 299.
13. *Ibid.*, 314.
14. *Ibid.*, 313.
15. *Ibid.*, 431.
16. *National Cyclopedia of American Biography* (New York, 1893–1919), IX, 287.
17. John Churchman, *The Magnetic Atlas, or Various Charts of the Whole Terraqueous Globe* (London, 1794), 71–72.
18. *Annals of Congress*, March 4, 1790, 1 Cong., 2 Sess., II, 1413.
19. For example, Thomas Jefferson, Monticello, to Oliver Evans, January 16, 1814, in A. A. Lipscomb and A. E. Bergh, eds., *Writings of Thomas Jefferson* (Memorial Edition, Washington, 1905), XIV, 64; P. J. Federico, "Operation of the Patent Act of 1790," *Journal of the Patent Office Society*, 18:241–242 (1936); Dumas Malone, *Jefferson and His Time, II, Jefferson and the Rights of Man* (Boston, 1951), 282–285.
20. *American State Papers, Misc.*, I, 45.
21. Federico, *Journal of the Patent Office Society*, 18:251 (1936).
22. Thomas Jefferson to Hugh Williamson, April 1, 1792, *Report of Inventions of the U.S. Patent Office*, 62 Cong., 3 Sess., H. R. Doc. 1110, 215.
23. L. D. White, *The Jeffersonians: A Study in Administrative History, 1801–1829* (New York, 1951), 205.
24. G. A. Weber, *The Patent Office: Its History, Activities, and Organization* (Institute for Government Research, *Service Monograph No. 31*, Baltimore, 1924), 8.
25. *American State Papers, Misc.*, II, 188.

26. *Annals of Congress*, January 17, 1792; March 22, 1792, 2 Cong., 1 Sess., 74, 112; February 3, 1796, 4 Cong., 1 Sess., 288.

27. *American State Papers, Misc.*, I, 140.

28. *Annals of Congress*, February 3, 1796, 4 Cong., 1 Sess., 288.

29. Thomas Jefferson, Monticello, to George Washington, February 23, 1795, *Writings of Jefferson* (Memorial Edition), XIX, 108–109; George Washington, Philadelphia, to Thomas Jefferson, March 15, 1795, in J. C. Fitzpatrick, ed., *Writings of George Washington* (Washington, 1931–1941), XXXIV, 147.

30. Thomas Jefferson, "The Anas," in P. L. Ford, ed., *The Works of Thomas Jefferson* (Federal Edition, New York, 1904), I, 330.

31. *American State Papers, Misc.*, I, 153.

32. *Annals of Congress*, December 12, 21, 26, 1796, 4 Cong., 2 Sess., 1600–1601, 1694–1710; Irving Brant, *James Madison* (Indianapolis, 1941–1950), III, 447–448.

33. *American State Papers, Misc.*, I, 154.

34. *Ibid.*

35. J. A. Tobey, *The National Government and Public Health* (Baltimore, 1926), 21.

36. L. D. White, *The Federalists: A Study in Administrative History* (New York, 1948), 121.

37. R. D. Leigh, *Federal Health Administration in the United States* (New York, 1927), 81.

38. *Ibid.*, 82–84.

39. Rush to Jefferson, June 15, 1805, in Butterfield, *Letters of Rush*, II, 896.

40. J. P. Watson, *The Bureau of the Mint* (Institute for Government Research, *Service Monograph No. 37*, Washington, 1926), 4.

41. White, *Federalists*, 142; Edward Ford, *David Rittenhouse, Astronomer-Patriot, 1732–1796* (Philadelphia, 1946), 183–185.

42. White, *Federalists*, 140.

43. Butterfield, *Letters of Rush*, II, 897.

44. Watson, *Bureau of the Mint*, 17.

45. C. D. Hellman, "Jefferson's Efforts towards the Decimalization of United States Weights and Measures," *Isis*, 16:289–297 (1931).

46. *American State Papers, Misc.*, I, 149.

47. *Annals of Congress*, May 14, 1796, 4 Cong., 1 Sess., 1381–1383.

48. C. A. Browne, *Thomas Jefferson and the Scientific Trends of His Time* (Waltham, Mass., 1943), 387.

49. *American State Papers, Misc.*, I, 202.

50. White, *Federalists*, 303.

Chapter II. Theory and Action in the Jeffersonian Era

1. Thomas Moore, quoted in George Dangerfield, *The Era of Good Feelings* (New York, 1952), 157.

2. L. H. Butterfield, ed., *Letters of Benjamin Rush* (Princeton, 1951), I, 779.

3. E. T. Martin, *Thomas Jefferson: Scientist* (New York, 1952), 51.

4. C. A. Browne, *Thomas Jefferson and the Scientific Trends of His Time* (Waltham, Mass., 1943), 380.

5. William Cobbett, quoted in Martin, *Jefferson*, 218.

6. C. C. Sellers, *Charles Willson Peale* (Philadelphia, 1947), II, 133.

7. Florian Cajori, *The Early Mathematical Sciences in North and South America* (Boston, 1928), 79.

8. Browne, *Jefferson and Scientific Trends*, 405.

9. Sellers, *Peale*, II, 150.

10. *Ibid.*

11. J. D. Richardson, comp., *Compilation of the Messages and Papers of the Presidents, 1789–1897* (Nashville, 1905), I, 379.

12. *Ibid.*, 409–410.

13. Joel Barlow, "Prospectus of a National Institution to be Established in the United States," in *A Memorial of George Brown Goode* (Smithsonian Institution, *Annual Report for 1897*, Part II, Washington, 1901), 63–81.

14. Thomas Jefferson to Joel Barlow, February 24, 1806, in Paul L. Ford, ed., *The Works of Thomas Jefferson* (Federal Edition, New York, 1904), VIII, 232–233.

15. Thomas Jefferson to Joel Barlow, December 10, 1807, *ibid.*, 529–530.

16. S. L. Mitchill, *A Discourse on the Character and Services of Thomas Jefferson, More Especially as a Promoter of Natural and Physical Sciences* (New York, 1826), 33–34.

17. R. G. Thwaites, *A Brief History of Rocky Mountain Exploration with Especial Reference to the Expedition of Lewis and Clark* (New York, 1904), 67.

18. *Ibid.*, 73–77; Bernard DeVoto, *The Course of Empire* (Boston, 1952), 345–348.

19. A. P. Nasatir, *Before Lewis and Clark* (Saint Louis, 1952), II, 712.

20. Thwaites, *Rocky Mountain Exploration*, 93–94.

21. *Ibid.*, 95.

22. Thomas Jefferson to Caspar Wistar, February 28, 1803, in Ford, *Works of Jefferson*, IX, 422.

23. Thwaites, *Rocky Mountain Exploration*, 99; DeVoto, *Course of Empire*, 426; Butterfield, *Letters of Rush*, II, 868.

24. For the various publications stemming from the expedition, see Max Meisel, *A Bibliography of American Natural History* (Brooklyn, 1924), II, 88–103.

25. Joseph Ewan, "Frederick Pursh, 1774–1820, and his Botanical Associates," *Proceedings of the American Philosophical Society*, 96:610 (1952).

26. DeVoto, *Course of Empire*, 424–426.

27. *American State Papers, Misc.*, I, 390–391.

28. *Ibid.*, II, 85–88.

29. *Ibid.*, II, 117–119.

30. Thomas Jefferson to Henry Dearborn, June 22, 1807, in A. A. Lipscomb and A. E. Bergh, eds., *Writings of Jefferson* (Memorial Edition, Washington, 1905), XI, 252.

31. Sidney Forman, *West Point: A History of the United States Military Academy* (New York, 1950), 14.

32. W. S. Holt, *Office of Chief of Engineers of the Army: Its Non-Military History, Activities, and Organization* (Institute for Government Research, Service Monograph No. 27, Baltimore, 1923), 2.

33. Forman, *West Point*, 20–35.

34. D. E. Smith and Jekuthiel Ginsburg, *A History of Mathematics in America before 1900* (Carus Mathematical Monographs B, Chicago, 1934), 77.

35. Cajori, *Early Mathematical Sciences*, 70.

36. *Annals of Congress*, December 15, 1806, 9 Cong., 2 Sess., 151.

37. National Archives, *Guide to the Material in the National Archives* (Washington, 1940), 90.

38. Florian Cajori, *The Chequered Career of Ferdinand Rudolph Hassler, First Superintendent of the United States Coast Survey: A Chapter in the History of Science in America* (Boston, 1929), 37–44.

39. *Annals of Congress*, December 15, 1806, January 6, 1807, February 5, 1807, February 10, 1807, 9 Cong., 2 Sess., 151–152, 259–260, 456, 1252–1253.

40. Cajori, *Hassler*, 191.

41. *Ibid.*, 47–49.

42. *Ibid.*, 77–78.

43. *Ibid.*

44. *Ibid.*, 80–81.

45. *Ibid.*, 89–92.

46. *Ibid.*, 118–119.

47. *American State Papers, Misc.*, II, 403.

48. Asbury Dickins to Dr. William Darlington, December 16, 1819, quoted in M. E. Pickard, "Government and Science in the United States: Historical Backgrounds III. The Smithsonian Institution," *Journal of the History of Medicine and Allied Sciences*, 1:259–260 (1946).

49. *Ibid.*, 257.

50. Richard Rathbun, "The Columbian Institute for the Promotion of Arts and Sciences . . . ," U. S. National Museum, *Bulletin No. 101* (Washington, 1917), 19.

51. Pickard, *Journal of the History of Medicine*, 1:258, 260 (1946).

52. Rathbun, U. S. National Museum, *Bulletin No. 101*, 48–51.

53. *American State Papers, Misc.*, II, 753.

54. C. O. Paullin, "Early Movements for a National Observatory, 1802–1842," *Records of the Columbia Historical Society*, 26:42–43 (1923).

55. Rathbun, U. S. National Museum, *Bulletin No. 101*, 64.

56. Mitchill, *Discourse on Jefferson*, 7.

57. C. M. Wiltse, *John C. Calhoun, Nationalist, 1782–1828* (Indianapolis and New York, 1944), 167–168, 182–185.

58. Edwin James, *Account of an Expedition from Pittsburgh to the Rocky Mountains . . . under the Command of Major Stephen H. Long . . .* (Philadelphia, 1823), I, 3–4, 423–425.

59. A. D. Rodgers, III, *John Torrey: A Story of North American Botany* (Princeton, 1942), 48–53.

60. Holt, *Army Engineers*, 5.

61. *Ibid.*, 6–7.

62. G. F. Fiebeger, "Sylvanus Thayer," *Dictionary of American Biography*, XVIII, 410–411; Forman, *West Point*, 36–60.

63. H. D. Dale, "Stephen Harriman Long," *DAB*, XI, 380; Dirk J. Struik, *Yankee Science in the Making* (Boston, 1948), 245.

64. J. A. Tobey, *Medical Department of the Army* (Institute for Government Research, *Service Monograph No. 45*, Baltimore, 1927), 10–11; Leonard D. White, *The Jeffersonians: A Study in Administrative History, 1801–1829* (New York, 1951), 247–248.

65. J. S. Myer, *Life and Letters of Dr. William Beaumont* (Saint Louis, 1939), 118–120, 156–167, 251.

66. *Statutes at Large*, II, 806.

67. *American State Papers, Misc.*, II, 565.

68. *Annals of Congress*, January 30, February 22, March 28, and April 13, 1822, 17 Cong., 1 Sess., 851–852, 1134, 1382, 1545.

69. *American State Papers, Misc.*, II, 921.

70. Dangerfield, *Era of Good Feelings*, 160.

71. Richardson, *Messages*, II, 311.

72. *Ibid.*, 312.

73. Dangerfield, *Era of Good Feelings*, 351.

74. Richardson, *Messages*, II, 316.

75. *Ibid.*

76. *Ibid.*, 312, 216.

77. C. F. Adams, ed., *The Memoirs of John Quincy Adams Comprising Portions of his Diary from 1795 to 1848* (Philadelphia, 1875–1876), VII, 62–63.

78. *Annals of Congress*, December 13, 1825, 19 Cong., 1 Sess., 802.

79. Dangerfield, *Era of Good Feelings*, 352.

80. Adams, *Memoirs*, VII, 145.

81. Secretary of the Navy, *Report*, 20 Cong., 2 Sess., November 27, 1828, Sen. Doc. 1 (Ser. 181), 130.

82. *American State Papers, Naval Affairs*, III, 547.

83. Committee on Naval Affairs, *Report*, 20 Cong., 2 Sess., February 23, 1829, Sen. Doc. 94 (Ser. 182), 10.

Chapter III. Practical Achievements in the Age of the Common Man

1. G. P. Merrill, *Contributions to a History of American State Geological and Natural History Surveys* (U. S. National Museum, *Bulletin 109*, Washington, 1920), 150–153.

2. G. A. Weber, *The Patent Office: Its History, Activities and Organization* (Institute for Government Research, *Service Monograph No. 31*, Baltimore, 1924), 9; among early employees of the office were Charles G. Page and Titian R. Peale.

3. Commissioner of Patents, *Report*, January 17, 1838, 25 Cong., 2 Sess., Sen. Doc. 105 (Ser. 315), 1.

4. Weber, *Patent Office*, 9.

5. Commissioner of Patents, *Report*, January 17, 1838, 4–6.

6. U. S. Bureau of the Census, *Historical Statistics of the United States, 1789–1945* (Washington, 1949), 313.

7. Carleton Mabee, *The American Leonardo: A Life of Samuel F. B. Morse* (New York, 1943), 195, 206.

8. *Ibid.*, 282.

9. *Ibid.*, 285.

10. *Ibid.*, 258.

11. Charles G. Page, Washington, to the Secretary of the Navy, April 16, 1849; July 5, 1849; October 2, 1849; November 28, 1851, Miscellaneous Letters of the Secretary of the Navy, National Archives.

12. C. W. Mitman, "Charles G. Page," *DAB*, XIV, 135–136; *Congressional Globe*, September 23, 1850, 31 Cong., 1 Sess., 1924; Malcolm MacLaren, *The Rise of the Electrical Industry During the Nineteenth Century* (Princeton, 1943), 87–88.

13. L. C. Hunter, *Steamboats on the Western Rivers: An Economic and Technological History* (Cambridge, Mass., 1949), 175.

14. Secretary of the Treasury, *Report*, March 3, 1831, 21 Cong., 2 Sess., H. R. Ex. Doc. 131 (Ser. 209), 1; S. L. Wright, *The Story of the Franklin Institute* (Philadelphia, 1938), 12.

15. Franklin Institute, *Report*, March 1, 1836, H. R. Doc. 162 (Ser. 289), 3–9.

16. Hunter, *Steamboats*, 525–526.

17. Committee to Test Inventions, *Report*, February 7, 1839, 25 Cong., 3 Sess., H. R. Ex. Doc. 170 (Ser. 347), 2; Secretary of the Navy, *Report*, June 24, 1842, 27 Cong., 2 Sess., Sen. Doc. 336 (Ser. 399), 1.

18. *Appleton's Cyclopaedia of American Biography* (New York, 1898–1899), III, 451; A. C. Fieldner, A. W. Gauger, and G. R. Yohe, "Gas and Fuel Chemistry," *Industrial and Engineering Chemistry*, 43:1039 (1951).

19. Hunter, *Steamboats*, 463; R. D. Leigh, *Federal Health Administration in the United States* (New York, 1927), 85–88.

20. Committee on Foreign Affairs, *Report*, January 20, 1832, 22 Cong., 1 Sess., H. R. Doc. 226 (Ser. 225), 1–2.

21. G. A. Weber, *The Bureau of Standards* (Institute for Government Research, *Service Monograph No. 35*, Baltimore, 1925), 35; W. Hallock and H. T. Wade, *Outline of the Evolution of Weights and Measures and the Metric System* (Boston, 1906), 110.

22. G. A. Weber, *The Coast and Geodetic Survey: Its History, Activities and Organization* (Institute for Government Research, *Service Monograph No. 16*, Baltimore, 1923), 3.

23. *Statutes at Large*, July 10, 1832, IV, 571.

24. Florian Cajori, *The Chequered Career of Ferdinand Rudolph Hassler, First Superintendent of the United States Coast Survey: A Chapter in the History of Science in America* (Boston, 1929), 189.

25. *Ibid.*, 192, 198.

26. *Ibid.*, 201.

27. *Ibid.*, 203.

28. *Ibid.*, 178.

29. F. R. Hassler, *A Report . . . Showing the Progress of the Coast Survey*, January 3, 1842, 27 Cong., 2 Sess., H. R. Ex. Doc. 28 (Ser. 402), 5.

30. Martin Van Buren, "First Annual Message," December 5, 1837, in J. D. Richardson, comp., *Compilation of the Messages and Papers of the Presidents, 1789–1897* (Nashville, 1905), III, 393.

31. *Congressional Globe*, June 22, 1841, 27 Cong., 1 Sess., 88; December 16, 1842, 27 Cong., 3 Sess., 57.

32. Cajori, *Hassler*, 217.

33. *Congressional Globe*, December 16, 1842, 27 Cong., 3 Sess., 58.

34. C. F. Adams, ed., *Memoirs of John Quincy Adams* (Philadelphia, 1875–1876), XI, 335.

35. *Congressional Globe*, February 25, 1843, 27 Cong., 3 Sess., 351.

36. See R. F. Almy, "J. N. Reynolds: A Brief Biography with Particular Reference to Poe and Symmes," *The Colophon*, n.s., 2:227–245 (1937); Committee on Naval Affairs, *Report*, March 25, 1828, 20 Cong., 1 Sess., H. R. Doc. 209 (Ser. 178), 4–15.

37. Committee on Commerce, *Report*, February 7, 1835, 23 Cong., 2 Sess., H. R. Report 94 (Ser. 276).

38. J. N. Reynolds, *Address on the Subject of a Surveying and Exploring Expedition to the Pacific Ocean and South Seas* (New York, 1836), *passim*.

39. D. C. Haskell, *The United States Exploring Expedition, 1838–1842, and its Publications, 1844–1874: A Bibliography* (New York, 1942), 2.

40. Charles Wilkes, *Narrative of the United States Exploring Expedition During the Years 1838, 1839, 1840, 1841, 1842* (Philadelphia, 1845), I, xiii.

41. Letters in Historic Letter File, Gray Herbarium, Harvard University; for instance, Charles Pickering, Philadelphia, to Asa Gray, September 23, 1836.

42. For a picture of the Navy in the late 1830's by an officer who was at one time scheduled to go on the expedition, see M. F. Maury, "Our Navy: Scraps from the Lucky Bag," *Southern Literary Messenger*, 6:233–240 (1840); 11:83–91 (1845).

43. G. S. Bryan, "The Purpose, Equipment and Personnel of the Wilkes Expedition," *Proceedings of the American Philosophical Society*, 82:553 (1940).

44. Asa Gray, New York, to N. W. Folwell, April 4, 1837, Cornell University Library.

45. *Exploring Expedition: Correspondence between J. N. Reynolds and the Hon. Mahlon Dickerson* [New York, 1838?], *passim*.

46. M. F. Maury, "Lucky Bag," *Southern Literary Messenger*, 6:797 (1840).

47. See [Joel R. Poinsett], "The Exploring Expedition," *North American Review*, 56:259–264 (1843).

48. Wilkes, *Narrative*, I, 353.

49. *Ibid.*, 352, 353.

50. Charles Pickering, Washington, to Asa Gray, April 27, 1838, Gray Herbarium, Harvard University.

51. Haskell, *Bibliography*, 18–19.

52. S. E. Morison, *History of United States Naval Operations in World War II*, vol. 7, *Aleutians, Gilberts, and Marshalls* (Boston, 1951), 76, 153.

53. See pp. 70–76.

54. *U. S. Government Organization Manual 1953–54* (Washington, 1953), 32–33.

55. For full details on this phase of the expedition, see Haskell, *Bibliography*.

56. *Ibid.*, 16.

57. G. A. Weber, *The Hydrographic Office* (Institute for Government Research, *Service Monograph No. 42*, Baltimore, 1926), 12.

58. *Ibid.*, 8–11, 23; C. O. Paullin, "Early Movements for a National Observatory," *Records of the Columbia Historical Society*, 25:48–50 (1923).

59. Weber, *Hydrographic Office*, 14, 16.

60. Paullin, *Records of the Columbia Historical Society*, 25, 48.

61. Max Meisel, *A Bibliography of American Natural History* (Brooklyn, 1924–1929), II, 531–532.

62. D. D. Owen, *Report*, June 6, 1840, 26 Cong., 1 Sess., H. R. Ex. Doc. 239 (Ser. 368), 10–161; Meisel, *American Natural History*, II, 677–678.

63. See F. P. Prucha, *Broadax and Bayonet: The Role of the United States Army in the Development of the Northwest, 1815–1860* (Madison, Wis., 1953), 189.

64. See L. K. Eaton, "Military Medicine on the Louisiana Frontier: A Letter of Melines Conklin Leavenworth to Dr. Eli Todd," *Bulletin of the History of Medicine*, 24:247–253 (1950).

65. G. P. Merrill, *Contributions to the History of American Geology* (Smithsonian Institution, *Annual Report. Report of the U. S. National Museum*, Washington, 1904), 303.

66. F. G. Hill, "Government Engineering Aid to Railroads before the Civil War," *Journal of Economic History*, 11:240–242 (1951).

67. W. S. Holt, *Office of Chief of Engineers of the Army: Its Non-Military History, Activities, and Organization* (Institute for Government Research, *Service Monograph No. 27*, Baltimore, 1923), 8.

68. W. T. Jackson, *Wagon Roads West* (Berkeley and Los Angeles, 1952), 2.

69. Meisel, *American Natural History*, II, 618; J. N. Nicollet, *Report*, January 11, 1845, 28 Cong., 2 Sess., H. R. Doc. 52 (Ser. 464), 3–5.

Chapter IV. The Fulfillment of Smithson's Will

1. The literature on the Smithsonian Institution is voluminous. For documents, see W. J. Rhees, ed., *The Smithsonian Institution: Documents Relative to its Origin and History* (Smithsonian Institution, *Miscellaneous Collections*, XVII, Washington, 1879), hereafter cited as Rhees, *Documents*; W. J. Rhees, ed., *The Smithsonian Institution: Journals of the Board of Regents* (Smithsonian Institution, *Miscellaneous Collections*, XVIII, Washington, 1879), hereafter cited as Rhees, *Journals*; W. J. Rhees, ed., *The Smithsonian Institution: Documents Relative to its Origin and History, 1835–1899* (Smithsonian Institution, *Miscellaneous Collections*, XLII–XLIII, Washington, 1901), hereafter cited as Rhees, *Origin*, I and II. For the early period see G. B. Goode, ed., *The Smithsonian Institution: 1846–1896: The History of the First Half-Century* (Washington, 1897); for the most recent general history, P. H. Oehser, *Sons of Science: The Story of the Smithsonian Institution and its Leaders* (New York, 1949); for the most detailed analysis of the early period, M. E. Pickard, "Government and Science in the United States: Historical Backgrounds III. The Smithsonian Institution," *Journal of the History of Medicine and Allied Sciences*, 1:254–280, 446–481 (1946).

2. Rhees, *Origin*, I, 6.

3. Pickard, *Journal of the History of Medicine*, 1:447 (1946).

4. C. F. Adams, ed., *Memoirs of John Quincy Adams Comprising Portions of his Diary from 1795 to 1848* (Philadelphia, 1875–1876), IX, 244.

5. Rhees, *Origin*, I, 140.

6. *Ibid.*, 138.

7. *Ibid.*, 128.

8. *Ibid.*, 140.

9. G. B. Goode, "The Genesis of the United States National Museum" in *A Memorial of George Brown Goode* (Smithsonian Institution, *Annual Report for 1897*, Part II, Washington, 1901), 85.

10. J. D. Richardson, comp., *Compilation of the Messages and Papers of the Presidents, 1789–1897* (Nashville, 1905), III, 506.

11. President of the United States, *Message . . . upon . . . the Bequest of James Smithson . . .*, December 10, 1838, 25 Cong., 3 Sess., H. R. Ex. Doc. 11 (Ser. 345), 2.

12. *Ibid.*, 838–839.

13. *Ibid.*, 856, 860–862.

14. Rhees, *Origin*, I, 155.

15. *Ibid.*, 163–181.
16. Rhees, *Documents*, 849.
17. Rhees, *Origin*, I, 146.
18. Adams, *Memoirs*, XI, 65.
19. Rhees, *Documents*, 842.
20. Rhees, *Origin*, I, 184.
21. Adams, *Memoirs*, XI, 336.
22. *Ibid.*, X, 57.
23. *Ibid.*, 23.
24. *Ibid.*, 57.
25. *Ibid.*, 25.
26. *Ibid.*, 191.
27. *Ibid.*, 139; XI, 112.
28. For general accounts see Goode, *Memorial*, 97–109; Pickard, *Journal of the History of Medicine*, 1:254, 280 (1946).
29. Goode, *Memorial*, 99.
30. *Constitution and By-Laws of the National Institute for the Promotion of Science* (Washington, 1840), 11.
31. Quoted in Pickard, *Journal of the History of Medicine*, 1:267 (1946).
32. Goode, *Memorial*, 86–88.
33. Pickard, *Journal of the History of Medicine*, 1:266 (1946); Rhees, *Documents*, 240.
34. J. J. Abert, quoted in Goode, *Memorial*, 129.
35. *Congressional Globe*, July 7, 1842, 27 Cong., 2 Sess., 729.
36. Goode, *Memorial*, 157–163; Pickard, *Journal of the History of Medicine*, 1:271 (1946).
37. *Ibid.*, 271; Goode, *Memorial*, 116–118.
38. D. C. Haskell, *The United States Exploring Expedition, 1838–1842, and its Publications, 1844–1874: A Bibliography* (New York, 1942), 7.
39. J. R. Poinsett, *Calendar of Poinsett Papers in the Henry D. Gilpin Collection* . . . (Philadelphia, 1941), 165, 168.
40. Haskell, *Bibliography*, 150.
41. Goode, *Memorial*, 130–131.
42. *Ibid.*, 124; Pickard, *Journal of the History of Medicine*, 1:278 (1946); *Reply of Colonel Abert and Mr. Markoe to the Hon. Mr. Tappan, of the United States Senate* (Washington, 1843).
43. National Institute for the Promotion of Science, *Third Bulletin of the Proceedings* (Washington, 1845), 392–415.
44. Goode, "The First National Scientific Congress," *Memorial*, 469–477; Pickard, *Journal of the History of Medicine*, 1:279 (1946).
45. Goode, *Memorial*, 133; Pickard, *Journal of the History of Medicine*, 1:275–276, 284 (1946).
46. Goode, *Memorial*, 469–477.
47. Rhees, *Origin*, I, 266.
48. *Congressional Globe*, January 21, 1845, 28 Cong., 2 Sess., 163.
49. Rhees, *Origin*, I, 281.
50. *Ibid.*, 385.
51. Pickard, *Journal of the History of Medicine*, 1:470–471 (1946).
52. Adams, *Memoirs*, XII, 195, 219.
53. See R. W. Leopold, *Robert Dale Owen: A Biography* (Cambridge, Mass., 1940), 221–227.

54. Rhees, *Origin*, I, 349.
55. Leopold, *Owen*, 225.
56. Rhees, *Origin*, I, 391.
57. *Ibid.*, 411.
58. *Ibid.*, 350.
59. *Ibid.*, 434.
60. *Ibid.*, 433.
61. *Ibid.*, 422, 428.
62. M. M. Odgers, "Bache as an Educator," *Proceedings of the American Philosophical Society*, 84:125–186 (1941).
63. Rhees, *Origin*, II, 11; Thomas Coulson, *Joseph Henry: His Life and Work* (Princeton, 1950), 178–179.
64. *Ibid.*, 179.
65. Rhees, *Origin*, II, 12.
66. Oehser, *Sons of Science*, 40.
67. Rhees, *Documents*, 944.
68. *Ibid.*, 945.
69. *Ibid.*, 955.
70. *Ibid.*, 945–946.
71. *Ibid.*, 946–947.
72. Rhees, *Journals*, 26.
73. Rhees, *Documents*, 961–994.
74. Rhees, *Origin*, I, 451.
75. *Ibid.*, 605, 623, 625.
76. Leopold, *Owen*, 243.
77. W. H. Dall, *Spencer Fullerton Baird: A Biography, Including Selections from his Correspondence with Audubon, Agassiz, Dana, and Others* (Philadelphia, 1915), 156.
78. Coulson, *Henry*, 183.
79. *Ibid.*, 209.
80. Oehser, *Sons of Science*, 43.
81. See Oehser, *Sons of Science*, 44–45; Rhees, *Origin*, I, 512–600; Rhees, *Journals*, 113; Coulson, *Henry*, 213.
82. This terminology is from Goode, *Memorial*, 113.
83. Pickard, *Journal of the History of Medicine*, 1:428 (1946).
84. Smithsonian Institution, *Annual Report for 1857* (Washington, 1858), 14.
85. Smithsonian Institution, *Annual Report for 1858* (Washington, 1859), 14.
86. Dall, *Baird*, 389.
87. See Goode, *Memorial*, 147.
88. Goode, *Smithsonian Institution*, 576.
89. Smithsonian Institution, *Annual Report for 1853* (Washington, 1854), 19–21; Smithsonian Institution, *Annual Report for 1859* (Washington, 1860), 385–395.
90. Goode, *Smithsonian Institution*, 612.
91. Smithsonian Institution, *Annual Report for 1855* (Washington, 1856), 23.
92. Oehser, *Sons of Science*, 51.
93. Smithsonian Institution, *Annual Report for 1856* (Washington, 1857), 34.

94. See Marcus Benjamin, "Meteorology," in Goode, *Smithsonian Institution*, 647-678; Oehser, *Sons of Science*, 47.

95. Coulson, *Henry*, 246.

96. *Annual Report for 1856*, 221.

97. Rhees, *Journals*, 139-150.

98. Coulson, *Henry*, 235.

99. Joseph Henry, Washington, to Asa Gray, June 2, 1855, Gray Herbarium of Harvard University.

100. *Annual Report for 1853*, 22.

101. *Annual Report for 1859*, 17.

Chapter V. The Great Explorations and Surveys

1. Max Meisel, *A Bibliography of American Natural History* (Brooklyn, 1924-1929), II, III, contains extensive systematic information on the literature of the expeditions of this period.

2. G. P. Merrill, *Contributions to the History of American Geology* (Smithsonian Institution, *Annual Report. Report of the U. S. National Museum*, Washington, 1904), 411-413, hereafter cited as Merrill, *History of American Geology*.

3. Meisel, *American Natural History*, III, 106.

4. Merrill, *History of American Geology*, 415.

5. *Ibid.*, 417-422.

6. Meisel, *American Natural History*, III, x-xi.

7. *Ibid.*, II, 713.

8. S. F. Baird, Washington, to Captain George B. McClellan, November 6, 1852, in W. H. Dall, *Spencer Fullerton Baird* (Philadelphia, 1915), 283.

9. Smithsonian Institution, *Annual Report for 1877* (Washington, 1878), 105-109.

10. Meisel, *American Natural History*, III, 10, 15-16.

11. See W. T. Jackson, *Wagon Roads West* (Berkeley and Los Angeles, 1952), *passim*.

12. The expeditions under Lt. John Pope (1855), Lt. J. C. Ives (1857), and Capt. Randolph B. Marcy (1854); Meisel, *American Natural History*, III, 236, 241, 271.

13. Bureau of Explorations and Surveys, War Department, *Report*, November 30, 1857, 35 Cong., 1 Sess., Sen. Ex. Doc. 11 (Ser. 920), 38-39.

14. Secretary of War, *Report*, December 6, 1858, 35 Cong., 2 Sess., Sen. Ex. Doc. 1 (Ser. 975), 620.

15. Meisel, *American Natural History*, III, 265-267.

16. *Ibid.*, 99-105, 275-278.

17. For a map see Merrill, *History of American Geology*, 450.

18. Meisel, *American Natural History*, III, 190.

19. Jackson, *Wagon Roads West*, 243.

20. Gerstle Mack, *The Land Divided: A History of the Panama Canal and Other Isthmian Canal Projects* (New York, 1944), 111.

21. *Ibid.*, 242; Meisel, *American Natural History*, III, 273-275.

22. Meisel, *American Natural History*, III, 313-314; P. J. Scheips, "Buchanan and the Chiriqui Naval Station Sites," *Military Affairs* (Summer 1954), 64-80.

23. Mack, *Land Divided*, 112.

24. Meisel, *American Natural History*, III, 42; W. F. Lynch, *Commerce and the Holy Land* (Philadelphia, 1860).

25. Secretary of the Navy, *Report*, December 6, 1853, 33 Cong., 1 Sess., Sen. Ex. Doc. 1 (Ser. 692), 329–389.

26. W. L. Herndon and Lardner Gibbon, *Exploration of the Valley of the Amazon*, 1853, 32 Cong., 2 Sess., Sen. Ex. Doc. 36, I, 63.

27. *Ibid.*, 341; see also D. M. Dozer, "Matthew Fontaine Maury's Letter of Instruction to William Lewis Herndon," *The Hispanic American Historical Review*, 28:212–228 (1949).

28. Meisel, *American Natural History*, III, 220.

29. Lieutenant J. M. Gilliss, *The U. S. Naval Astronomical Expedition* (Washington, 1855), 33 Cong., 1 Sess., H. R. Ex. Doc. 121 (Ser. 728).

30. R. S. Dugan, "James Melville Gilliss," *Dictionary of American Biography*, VII, 292–293; B. A. Gould, "James Melville Gilliss," National Academy of Sciences, *Biographical Memoirs*, 1:165 (1877).

31. M. C. Perry, ed., *Narrative of the Expedition of an American Squadron to the China Seas and Japan, Performed in the Years 1852, 1853, and 1854 . . .* (Washington, 1856), I, 78–79.

32. See A. H. Dupree, "Science vs. the Military: Dr. James Morrow and the Perry Expedition," *Pacific Historical Review*, 22:29–37 (1953).

33. S. F. Baird, Washington, to George P. Marsh, March 8, 1853, in Dall, *Baird*, 298.

34. A. B. Cole, *Yankee Surveyors in the Shogun's Seas: Records of the U. S. Surveying Expedition in the North Pacific Ocean, 1853–1856* (Princeton, 1947), 5–7.

35. *Ibid.*, 24.

36. Asa Gray, "Diagnostic Characters of New Species of Phaenogamous Plants, collected in Japan by Charles Wright, Botanist of the U. S. North Pacific Exploring Expedition . . . with Observations upon the Relations of the Japanese Flora to That of North America and of Other Parts of the North American Temperate Zone," *Memoirs of the American Academy of Arts and Sciences* [N. S.], 6:377–452 (1858).

37. See Jeannette Mirsky, *To the Arctic! The Story of Northern Exploration from Earliest Times to the Present* (New York, 1948), 137–160.

38. Zachary Taylor, "Special Message to the Senate and House of Representatives of the United States," January 4, 1850, in J. D. Richardson, comp., *Compilation of the Messages and Papers of the Presidents, 1789–1897* (Nashville, 1905), V, 25–26.

39. See J. E. Nourse, *American Explorations in the Ice Zones: The Expeditions of De Haven, Kane, Rodgers, Hayes, Hall, Schwatka, and De Long . . . with a Brief Notice of the Antarctic Cruise under Lieut. Wilkes, 1840 . . .* (Boston, 1884), 44–49.

40. Smithsonian Institution, *Annual Report for 1858* (Washington, 1859), 18; Nourse, *Ice Zones*, 65–89.

41. Superintendent of the Coast Survey, [*Annual*] *Report*, 1848, 30 Cong., 2 Sess., Sen. Ex. Doc. 1 (Ser. 529), 4–5.

42. *Ibid.*, 5.

43. A. D. Bache, "On the Progress of the Survey of the Coast of the United States," *Proceedings of the American Association for the Advancement of Science*, 2:164 (1849).

44. Superintendent of the Coast Survey, [*Annual*] *Report*, 1848, 3.

45. R. D. Owen, Washington, to A. D. Bache, December 2, 1845, Coast Survey Records, National Archives.

46. See M. M. Odgers, *Alexander Dallas Bache: Scientist and Educator, 1806–1867* (Washington, 1947), 149–150.

47. Superintendent of the Coast Survey, *Annual Report*, 1851, 32 Cong., 1 Sess., Sen. Ex. Doc. 3 (Ser. 615), 9.

48. *Congressional Globe*, December 31, 1850, 31 Cong., 2 Sess., 150.

49. Bache, *Proceedings of the American Association for the Advancement of Science*, 2:166 (1849).

50. Superintendent of the Coast Survey, *[Annual] Report*, 1850, 31 Cong., 2 Sess., Sen. Ex. Doc. 7 (Ser. 588), 1–2.

51. *Congressional Globe*, September 22, 27, 1850, 31 Cong., 1 Sess., 1630–1632, 1987–1988.

52. Superintendent of the Coast Survey, *Annual Report*, 1860, 36 Cong., 2 Sess., H. R. Ex. Doc. 14 (Ser. 1098), 25.

53. See "Commemoration of the Life and Work of Alexander Dallas Bache and Symposium on Geomagnetism," *Proceedings of the American Philosophical Society*, 84:125–351 (1941).

54. Superintendent of the Coast Survey, *Annual Report*, 1858, 35 Cong., 2 Sess., Sen. Ex. Doc. 14 (Ser. 990), 35.

55. "Commemoration of Bache," *Proceedings of the American Philosophical Society*, 84:274 (1941).

56. Superintendent of the Coast Survey, *Annual Report*, 1860, 29.

57. S. C. Walker, "Application of the Galvanic Circuit to the Astronomical Clock and Telegraph Register in Determining Local Differences of Longitude, and in Astronomical Observations Generally," *American Journal of Science and the Arts* [2d Ser.], 7:206 (1849).

58. Superintendent of the Coast Survey, *Annual Report*, 1851, 541; 1858, 39.

59. Superintendent of the Coast Survey, *[Annual] Report*, 1855, 34 Cong., 1 Sess., Sen. Ex. Doc. 22 (Ser. 826), 7.

60. Superintendent of the Coast Survey, *Annual Report*, 1848, 38; 1860, 28.

61. Superintendent of the Coast Survey, *[Annual] Report*, 1847, 30 Cong., 1 Sess., Sen. Ex. Doc. 6 (Ser. 505), 25–26; Superintendent of the Coast Survey, *[Annual] Report*, 1849, 31 Cong., 1 Sess., Sen. Ex. Doc. 5 (Ser. 553), 28.

62. E. C. Agassiz, ed., *Louis Agassiz: His Life and Correspondence* (Boston, 1885), II, 480–487.

63. Secretary of the Treasury, *Report*, December 16, 1858, 33 Cong., 2 Sess., Sen. Ex. Doc. 6 (Ser. 980), 118.

64. See G. A. Weber, *The Naval Observatory: Its History, Activities, and Organization* (Institute for Government Research, *Service Monograph No. 39*, Baltimore, 1926), 17–28; C. L. Lewis, *Matthew Fontaine Maury: The Pathfinder of the Seas* (Annapolis, Md., 1927).

65. *Ibid.*, 51–60.

66. Millard Fillmore, Second Annual Message, December 2, 1851, in Richardson, *Messages*, V, 134.

67. Lewis, *Maury*, 89.

68. See B. A. Gould, "Gilliss," National Academy, *Biographical Memoirs*, 1:177 (1877).

69. "Notice of M. F. Maury, *Physical Geography of the Sea*," *American Journal of Science* [2d Ser.], 19:449 (1855).

70. J. D. Dana, New Haven, Connecticut, to A. D. Bache, May 15, 1858, Bache Papers, Library of Congress.

71. Weber, *Naval Observatory*, 27.

72. See C. H. Davis, *Life of Charles Henry Davis, Rear Admiral, 1807-1877* (Boston, 1899).

73. Secretary of the Navy, *Report*, June 21, 1852, 23 Cong., 1 Sess., Sen. Ex. Doc. 88 (Ser. 620), 1-16.

74. Lewis, *Maury*, 108-112.

75. Committee on Naval Affairs, *Report*, February 25, 1856, 34 Cong., 1 Sess., Sen. Rep. Com. 8 (Ser. 836), 17.

76. *Congressional Globe*, July 10, 1856, 34 Cong., 1 Sess., 1594.

77. Davis, *Life of Davis*, 99-100.

78. E. R. Miller, "The Evolution of Meteorological Institutions in the United States," *Monthly Weather Review*, 59:1-6 (1931).

79. *Ibid.*, 2.

80. Lewis, *Maury*, 80.

81. Miller, *Monthly Weather Review*, 59:3 (1931).

82. Secretary of the Treasury, *Letter . . . upon . . . introducing into the United States Foreign Trees and Plants . . .*, April 6, 1832, 22 Cong., 1 Sess., H. R. Doc. 198 (Ser. 220); Secretary of the Treasury, *Letter . . . in Relation to the Growth and Manufacture of Silk . . .*, February 7, 1828, 20 Cong., 1 Sess., H. R. Doc. 158 (Ser. 172), 3-124.

83. See K. A. Ryerson, "History and Significance of the Foreign Plant Introduction Work of the United States Department of Agriculture," *Agricultural History*, 7:111-112 (1933).

84. W. R. Maxon, "Henry Perrine," *DAB*, XIV, 480-481; Committee on Agriculture, *Report*, April 26, 1832, 22 Cong., 1 Sess., H. R. Doc. 454 (Ser. 226), 3-4.

85. Commissioner of Patents, *Report*, January 17, 1838, 25 Cong., 2 Sess., Sen. Doc. 105 (Ser. 315), 1.

86. A. C. True, *A History of Agricultural Experimentation and Research in the United States, 1607-1925* (U. S. Department of Agriculture, *Miscelaneous Publications 251*, Washington, 1927), 24.

87. *Ibid.*, 26.

88. See *Liebig and After Liebig: A Century of Progress in Agricultural Chemistry* (Washington, 1942); P. W. Bidwell and J. I. Falconer, *History of Agriculture in Northern United States, 1620-1860* (Washington, 1925), 319; A. L. Demaree, *The American Agricultural Press, 1819-1860* (New York, 1941), 65-69.

89. Adams, *Memoirs*, XII, 188-189, 197.

90. True, *Agricultural Research*, 34.

91. *Ibid.*, 27-33.

92. Edward Wiest, *Agricultural Organization in the United States* (University of Kentucky, *Studies in Economics and Sociology*, II, Lexington, Ky., 1923), 293-295.

93. Richardson, *Messages*, V, 18, 86, 127-128.

94. Lyman Carrier, "The United States Agricultural Society, 1852-1860; its Relation to the Origin of the United States Department of Agriculture and Land Grant Colleges," *Agricultural History*, 9:284-285 (1937).

95. See E. D. Ross, "The 'Father' of the Land-Grant College," *Agricultural History*, 12:151-186 (1938); A. C. True, *A History of Agricultural Education*

in the United States, 1785–1925 (U. S. Department of Agriculture, *Miscellaneous Publications No. 36*, Washington, 1929), 89–99; J. G. Randall, *Civil War and Reconstruction* (New York, 1937), 379–380.

Chapter VI. Bache and the Quest for a Central Scientific Organization

1. H. L. Fairchild, "The History of the American Association for the Advancement of Science," *Science* [N. S.], 59:366–368, 410 (1924).
2. *Proceedings of the American Association for the Advancement of Science*, 2:66 (1849).
3. *Ibid.*, 5:v–viii (1851).
4. A. D. Bache, "Address," *Proceedings of the American Association for the Advancement of Science*, 6:xli–lx (1851).
5. T. Romeyn Beck, Albany, N. Y., to Joseph Henry, November 29, 1847, in W. J. Rhees, ed., *The Smithsonian Institution: Documents Relative to its Origin and History* (Smithsonian Institution, *Miscellaneous Collections*, XVII, Washington, 1879), 961–962.
6. In the preparation of this section I have benefited from information furnished by Edward Lurie.
7. R. J. Storr, *The Beginnings of Graduate Education in America* (Chicago, 1953), 82–93.

Chapter VII. The Civil War

1. H. S. Commager, ed., *The Blue and the Gray* (Indianapolis, 1950), I, xv.
2. See Thomas Weber, *The Northern Railroads in the Civil War, 1861–1865* (New York, 1952); for a general analysis of the Civil War and science, see I. B. Cohen, "Science and the Civil War," *Technology Review*, 48:167–170, 192–193 (1946).
3. D. C. Rhodes, "Daniel Craig McCallum," *Dictionary of American Biography*, XI, 565–566.
4. Weber, *Northern Railroads*, 230.
5. Comte de Paris, *History of the Civil War in America* (Philadelphia, 1875–1888), I, 300.
6. See J. P. Baxter, 3rd, *The Introduction of the Ironclad Warship* (Cambridge, Mass., 1933), 3; Allan Westcott, ed., *American Sea Power Since 1775* (Philadelphia, 1947), 99.
7. Westcott, *American Sea Power*, 99–102.
8. Baxter, *Ironclad*, 22–25.
9. *Ibid.*, 48–49, 211–218.
10. D. S. Peterson, *Admiral John A. Dahlgren, Father of U. S. Naval Ordnance* (New York, 1945), 57–58.
11. M. V. Dahlgren, *Memoir of John A. Dahlgren, Rear-Admiral United States Navy* (Boston, 1882), 234–235.
12. *Ibid.*, 227–228.
13. *Ibid.*, 376, 274.
14. See A. D. Chandler, Jr., "Du Pont, Dahlgren, and the Civil War Nitre Shortage," *Military Affairs*, 13:144 (1949).
15. C. O. Paullin, "A Half-Century of Naval Administration in America in 1861–1911," *United States Naval Institute Proceedings*, 38:1309–1336 (1912), 39:165–195 (1913).

16. Dahlgren, *Memoir of Dahlgren*, 384, 391.

17. *Ibid.*, 372.

18. This story may be pieced together from Lincoln's papers as published in R. P. Basler, ed., *The Collected Works of Abraham Lincoln* (New Brunswick, N. J., 1953), V, 354; VI, 3–5, 163, 342, 367, 559–561.

19. Secretary of the Navy, *Report . . . December, 1864* (Washington, 1864), xxix.

20. F. A. Taylor, "Benjamin Franklin Isherwood," *DAB*, IX, 515–516.

21. Secretary of the Navy, *Report . . . December, 1863* (Washington, 1863), xix.

22. Baxter, *Ironclad*, 248–257; Paullin, *United States Naval Institute Proceedings*, 39:173 (1913).

23. *Ibid.*, 38:1325 (1912).

24. See for example, Select Committee, *Report [on] a National Foundry . . .* March 3, 1835, 23 Cong., 2 Sess., H. R. Doc. 141 (Ser. 276), 1–37.

25. A. P. Van Gelder and Hugo Schlatter, *History of the Explosives Industry in America* (New York, 1927), 24; C. D. Rhodes, "Thomas Jackson Rodman," *DAB*, XVI, 80–81.

26. C. W. Mitman, "Robert Parker Parrott," *DAB*, XIV, 260–261.

27. F. A. Shannon, *The Organization and Administration of the Union Army, 1861–1865* (Cleveland, 1928), I, 127–147.

28. Brigadier General A. B. Dyer, Chief of Ordnance, to Secretary of War Stanton, December 5, 1864, in *The War of the Rebellion: A Compilation of the Official Records of the Union and Confederate Armies* (Washington, 1880–1901), ser. iii, IV, 971, hereafter cited as *Army Official Records*.

29. This conclusion and comparison are both indicated by Shannon, *The Union Army*, I, 147–148.

30. The most complete account is F. S. Haydon, *Aeronautics in the Union and Confederate Armies* (Baltimore, 1941).

31. *Ibid.*, 25–27.

32. *Ibid.*, 180–181; *Army Official Records*, ser. iii, I, 283.

33. Haydon, *Aeronautics*, 267.

34. *Ibid.*, 280–292.

35. See R. H. Shryock, *American Medical Research: Past and Present* (New York, 1947), chap. 2; G. W. Adams, *Doctors in Blue: The Medical History of the Union Army in the Civil War* (New York, 1952).

36. *Ibid.*, 4–5, 43.

37. *Army Official Records*, ser. iii, I, 258–259; Adams, *Doctors in Blue*, 8.

38. R. S. Dugan, "Benjamin Apthorp Gould," *DAB*, VII, 447–448.

39. H. E. Brown, *The Medical Department of the United States Army from 1775 to 1873* (Washington, 1873), 256–257.

40. E. E. Hume, "The Army Medical Library of Washington, the Largest Medical Library That Has Ever Existed," *Isis*, 26:423–432 (1937).

41. C. W. Burr and A. H. Quinn, "Silas Weir Mitchell," *DAB*, XIII, 62–63.

42. Thomas Coulson, *Joseph Henry: His Life and Work* (Princeton, 1950), 237–238.

43. *Ibid.*, 242; Smithsonian Institution, *Annual Report for 1862* (Washington, 1863), 43–44.

44. Smithsonian Institution, *Annual Report for 1862*, 13–14.

45. *Ibid.*, 33.

46. Smithsonian Institution, *Annual Report for 1861* (Washington, 1862), 13.

47. Smithsonian Institution, *Annual Report for 1865* (Washington, 1866), 60–61.

48. Commissioner of Patents, *Report [on] Results of Meteorological Observations Made under the Direction of the United States Patent Office and the Smithsonian Institution from the Years 1854–1859* . . . 36 Cong., 1 Sess., H. R. Ex. Doc. 55 (Ser. 1053), 1864, v–vii.

49. Smithsonian Institution, *Annual Report for 1862*, 28–29.

50. *Ibid.*, 15.

51. Coulson, *Henry*, 244, 308–309.

52. *Ibid.*, 239, 252–253.

53. See p. 137.

54. C. H. Davis, Washington, to Mrs. C. H. Davis, June 14, 1861, in C. H. Davis, *Life of Charles Henry Davis, Rear Admiral 1807–1877* (Boston, 1899), 124.

55. Secretary of the Navy, *Report*, December 1863, xxxii.

56. Davis, *Life of Davis*, 121.

57. Gideon Welles, Washington, to S. F. Du Pont and others, June 25, 1861, in Welles, "Admiral Farragut and New Orleans," *Galaxy*, 12:672 (1871).

58. See Paullin, *United States Naval Institute Proceedings*, 38:1324 (1912).

59. Davis, *Life of Davis*, 283–284; H. P. Beers, "The Bureau of Navigation, 1862–1942," *American Archivist*, 6:212–252 (1943).

60. Davis, *Life of Davis*, 284–285.

61. Chief of Engineers, "Annual Report for Year Ending June 30, 1865," *Army Official Records*, ser. iii, V, 169.

62. Engineers Bureau, "Report," in Secretary of War, *Annual Report*, December 1861, 37 Cong., 2 Sess., Sen. Ex. Doc. 1 (Ser. 1118), 94; *ibid.*, December 1862, 37 Cong., 3 Sess., H. R. Ex. Doc. 1 (Ser. 1159), 15.

63. *Army Official Records*, i, V, 25.

64. The correspondence between Bache and Peirce in the Peirce Papers, Harvard University Archives, indicates the range and nature of Lazzaroni activity.

65. B. Peirce [Cambridge], Mass., to A. D. Bache, January 5, 1863, Harvard University Archives.

66. B. Peirce [Cambridge], to A. D. Bache, January 30, 1863, Harvard University Archives.

67. The most detailed version of his reasons, here quoted, Henry gave in a long letter to Louis Agassiz, August 13, 1864, Peirce Papers, Harvard University Archives, which outlined the whole history of his connection with the National Academy. Several shorter accounts appear in letters written in the spring of 1863.

68. Davis, *Life of Davis*, 290.

69. Gideon Welles, Washington, to Charles Henry Davis, February 11, 1863, in F. W. True, *A History of the First Half-Century of the National Academy of Sciences, 1863–1913* (Washington, 1913), 1–2.

70. Joseph Henry, Washington, to Asa Gray, April 15, 1863, Gray Herbarium of Harvard University.

71. Joseph Henry, Washington, to Asa Gray, January 22, 1864, Gray Herbarium.

72. Paullin, *United States Naval Institute Proceedings*, 38:1324–1325 (1912); Reports of the Permanent Commission, 1863, 1864, 1865, National Archives.

73. B. Peirce [Cambridge], to A. D. Bache, February 5, 1863, Harvard University Archives.

74. *Ibid.*

75. E. C. Agassiz, ed., *Louis Agassiz: His Life and Correspondence* (Boston, 1885), II, 569–570.

76. John Torrey places Agassiz's arrival on February 20, but Davis's testimony and Wilson's ability to act on the 20th indicate the 19th for his arrival.

77. John Torrey, New York, to Asa Gray, March 9, 1863, Gray Herbarium.

78. C. H. Davis to Mrs. Davis, March 7 [1863], in Davis, *Life of Davis*, 202.

79. E. S. Holden, *Memorials of William Cranch Bond, Director of the Harvard College Observatory 1840–1859, and of his Son, George Phillips Bond, Director of the Harvard College Observatory 1859–1865* (San Francisco, 1897), 75–76, 162, 163.

80. Henry to Agassiz, August 13, 1864.

81. True, *National Academy*, 5.

82. *Ibid.*, 352.

83. Torrey to Gray, March 9, 1863.

84. Henry to Agassiz, August 13, 1864.

85. C. H. Davis to Mrs. Davis, February 27, 1863, in Davis, *Life of Davis*, 291.

86. *Congressional Globe*, March 3, 1863, 37 Cong., 3 Sess., 1500–1501.

87. True, *National Academy*, 6.

88. Torrey to Gray, March 9, 1863.

89. Davis to Mrs. Davis, March 7 [1863], in Davis, *Life of Davis*, 292.

90. Donald Fleming, *John William Draper and the Religion of Science* (Philadelphia, 1950), 110.

91. Torrey to Gray, March 9, 1863.

92. Asa Gray, Cambridge, to Joseph Henry, April 18, 1863, Gray Herbarium.

93. For an explicit reference to this point of view, see Asa Gray, Cambridge, to Joseph Henry, April 11, 1874, Gray Herbarium.

94. B. Peirce, Cambridge, to A. D. Bache, March 27, 1863, Harvard University Archives.

95. Joseph Henry, Washington, to Asa Gray, April 15, 1863, Gray Herbarium.

96. True, *National Academy*, 23.

97. *Ibid.*, 25–26.

98. B. Peirce, Cambridge, to A. D. Bache, March 6, 1863, Harvard University Archives.

99. Quoted in True, *National Academy*, 21.

100. *Ibid.*, 19.

101. M. L. Ames, *Life and Letters of Peter and Susan Lesley* (New York, 1909), I, 419.

102. True, *National Academy*, 62.

103. W. B. Rogers, Boston, to A. D. Bache, May 31, 1863, Rogers Papers, M. I. T. Library.

104. B. Peirce [Cambridge], to A. D. Bache, May 27, 1863, Harvard University Archives.

105. Asa Gray, Cambridge, to George Engelmann, September 2, 1863, typed copy at Gray Herbarium.

106. Peirce to Bache, October 18, 1863, Harvard University Archives.

107. B. Peirce [Cambridge], to A. D. Bache, October 26, 1863, Harvard University Archives.

108. Three papers by Agassiz were the only ones on natural history. Bache and Peirce gave two papers each. F. A. P. Barnard, "On the Force of Fired Gun-Powder, and the Pressure to which Heavy Guns are Actually Subjected in Firing," and Joseph Henry, "On Materials of Combustion for Lamps in Light-Houses," were the only papers related to the war or the government. *Annual of the National Academy of Sciences for 1863–1864* (Washington, 1865), 62–63.

109. Joseph Henry, Washington, to Asa Gray, January 22, 1864, Gray Herbarium.

110. True, *National Academy*, 206–213.

111. *Ibid.*, 213–215.

112. *Ibid.*, 215–217.

113. *Ibid.*, 219.

114. *Ibid.*, 219–225.

115. *Ibid.*, 226.

116. M. M. Odgers, *Alexander Dallas Bache: Scientist and Educator, 1806–1867* (Philadelphia, 1947), 210.

117. Henry to Agassiz, August 13, 1864.

118. True, *National Academy*, 29.

119. *Ibid.*, 34.

120. See, for example, Asa Gray, Cambridge, to George Engelmann, July 20, 1866 and February 27, 1867, typed copies in the Gray Herbarium.

121. Joseph Henry, Washington, to Asa Gray, July 8, 1868, Gray Herbarium.

122. True, *National Academy*, 13–15.

123. Joseph Henry, Washington, to Asa Gray, March 28, 1870, Gray Herbarium.

124. Henry considered as rivals a National Institute, chartered in New York, and the American Union Academy of Arts, Literature, and Sciences, of Washington, whose president was John W. Draper. Joseph Henry, Washington, to Asa Gray, March 22, 1870, Gray Herbarium.

Chapter VIII. The Evolution of Research in Agriculture

1. See pp. 113–114.

2. E. D. Ross, "The Civil War Agricultural New Deal," *Social Forces*, 15:99 (1936); T. S. Harding, *Two Blades of Grass: A History of Scientific Development in the U. S. Department of Agriculture* (Norman, Okla., 1947), 19–23.

3. See R. M. Robbins, *Our Landed Heritage: The Public Domain, 1776–1936* (Princeton, 1942), 203–204.

4. E. D. Ross, "Lincoln and Agriculture," *Agricultural History*, 3:54 (1929).

5. E. D. Ross, "The United States Department of Agriculture during the Commissionership," *Agricultural History*, 20:132–133 (1946).

6. C. H. Greathouse, *Historical Sketch of the U. S. Department of Agri-*

culture (Washington, 1898), 41–52, contains a collection of basic laws before 1900.

7. *Ibid.*, 47.

8. See A. C. True, *A History of Agricultural Education in the United States* (U. S. Department of Agriculture, *Miscellaneous Publications No. 36*, Washington, 1929), 99; E. D. Ross, "The 'Father' of the Land-Grant College," *Agricultural History*, 12:164 (1938).

9. *Ibid.*, 174.

10. True, *Agricultural Education*, 108.

11. *Ibid.*, 111.

12. See R. G. Axt, *The Federal Government and Financing Higher Education* (New York, 1952), 44–50.

13. For a discussion of the Constitutional position of the Department of Agriculture, see F. G. Caffey, "A Brief Statutory History of the United States Department of Agriculture," *Case and Comment*, 22:22 (1916).

14. See J. M. Swank, *The Department of Agriculture: Its History and Objects* (Washington, 1872), 23; for a tolerant view of Newton, see Harding, *Two Blades*, 24–27; for an unfavorable interpretation see Ross, *Agricultural History*, 20:133 (1946).

15. Commissioner of Agriculture, *Report for 1862* (Washington, 1863), 20.

16. C. A. Browne, "Charles M. Wetherill," *Dictionary of American Biography*, XX, 22–23.

17. Commissioner of Agriculture, *Report for 1864* (Washington, 1865), 540.

18. Commissioner of Agriculture, *Report for 1866* (Washington, 1867), 45.

19. W. A. Taylor, "William Saunders," *DAB*, XVI, 383–384.

20. Ross, *Agricultural History*, 3:61 (1929).

21. Harding, *Two Blades*, 30.

22. *Ibid.*, 148.

23. Commissioner of Agriculture, *Report on the Diseases of Cattle in the United States* (Washington, 1871), 118.

24. *Ibid.*, 156–170, 171–174.

25. See R. J. Dubos, *Louis Pasteur, Free Lance of Science* (Boston, 1950), 19.

26. A. C. True, *A History of Agricultural Experimentation and Research in the United States 1607–1925* (U. S. Department of Agriculture, *Miscellaneous Publications No. 251*, Washington, 1937), 48–49.

27. C. C. Parry, Washington, to Asa Gray, September 19, 1871, Gray Herbarium.

28. F. Watts, Washington, to C. C. Parry, September 27, 1871, "Correspondence Relating to the Dismissal of the Late Botanist to the Department of Agriculture, at Washington," *American Naturalist*, 6:40 (1872).

29. F. Watts, Washington, to Asa Gray, November [December] 8, 1871, *ibid.*, 42.

30. True, *Agricultural Research*, 183.

31. Asa Gray, Cambridge, Mass., to Joseph D. Hooker, June 3, 1866, Library of the Royal Botanic Garden, Kew.

32. For an annual summary, see Chief of the Division of Accounts and Disbursements, "Report," *Annual Reports of the Department of Agriculture for the Year Ended June 30, 1915* (Washington, 1916), 251.

33. "Review of the Commissioner of Agriculture, *Reports for 1881 and 1882*," *Science*, 1:143 (1883).

34. L. D. White, *The Jacksonians: A Study in Administrative History* (New York, 1954), 534-537.

35. J. S. Mill, *Considerations on Representative Government* (New York, 1882), 304.

36. See L. O. Howard, *A History of Applied Entomology* (Smithsonian Institution, *Miscellaneous Collection No. 84*, Washington, 1931); L. O. Howard, *Fighting the Insects: The Story of an Entomologist, Telling of the Life and Experience of the Writer* (New York, 1933); G. A. Weber, *The Bureau of Entomology* (Institute for Government Research, *Service Monograph No. 60*, Washington, 1930).

37. Howard, *Applied Entomology*, 79-81.

38. *Ibid.*, 82.

39. *Ibid.*, 86.

40. Weber, *Bureau of Entomology*, 7.

41. Howard, *Applied Entomology*, 167.

42. Howard, *Fighting the Insects*, 65-66.

43. Weber, *Bureau of Entomology*, 9.

44. *Ibid.*, 18.

45. *Ibid.*, 20-21.

46. See Howard, *Fighting the Insects*, 144-154.

47. Weber, *Bureau of Entomology*, 10.

48. Howard, *Applied Entomology*, 117.

49. *Ibid.*, 134-138.

50. Weber, *Bureau of Entomology*, 16-17.

51. Howard, *Applied Entomology*, 167.

52. Howard, *Fighting the Insects*, 69-70.

53. Quoted in F. W. Powell, *The Bureau of Animal Industry* (Institute for Government Research, *Service Monograph No. 41*, Baltimore, 1927), 5-6.

54. *Ibid.*, 6.

55. *Ibid.*, 10.

56. Harding, *Two Blades*, 149.

57. Powell, *Bureau of Animal Industry*, 122.

58. G. F. Thompson, "Administrative Work of the Federal Government in Relation to the Animal Industry," Department of Agriculture, *Yearbook*, *1899* (Washington, 1900), 442-445.

59. Theobald Smith and F. L. Kilborne, "Investigations into the Nature, Causation, and Prevention of Southern Cattle Fever," Bureau of Animal Industry, *Report for 1891* (Washington, 1893), 178; D. E. Salmon, "Some Examples of the Development of Knowledge Concerning Animal Diseases," Department of Agriculture, *Yearbook*, *1899*, 124; Hans Zinsser, "Theobald Smith, 1859-1928," National Academy of Sciences, *Biographical Memoirs*, 17:261-303 (1937).

60. S. H. Gage, "Theobald Smith, 1859-1934," *Cornell Veterinarian*, 25:210-211 (1935).

61. Smith and Kilborne, Bureau of Animal Industry, *Report for 1891*, 179.

62. Quoted in Gage, *Cornell Veterinarian*, 25:212 (1935).

63. C. R. Ball, *Federal, State, and Local Administrative Relationships in Agriculture* (Berkeley, 1938), I, 395-396.

64. Powell, *Bureau of Animal Industry*, 152-153.

65. Powell, *Bureau of Animal Industry*, 17–18; for a discussion of the category "line agency" as those responsible for the substantive activities of the department, see J. M. Gaus and L. O. Wolcott, *Public Administration and the United States Department of Agriculture* (Chicago, 1940), 262–265.

66. Harding, *Two Blades*, 152–154; R. R. Henley, "Marion Dorset," *DAB*, XXI, 258.

67. Harding, *Two Blades*, 83.

68. See A. D. Rodgers, III, *Erwin Frink Smith: A Study of North American Plant Pathology* (Philadelphia, 1952).

69. F. W. Powell, *The Bureau of Plant Industry: Its History, Activities and Organization* (Institute for Government Research, *Service Monograph No. 47*, Baltimore, 1927), 8; Harding, *Two Blades*, 84.

70. Secretary of Agriculture, *Report for 1895* (Washington, 1896), 57.

71. Powell, *Bureau of Plant Industry*, 101.

72. *Ibid.*, 12.

73. *Ibid.*, 8–9.

74. See David Fairchild, *The World Was My Garden* (New York, 1938).

75. Harding, *Two Blades*, 102–103.

76. See p. 176.

77. See Ball, *Administrative Relationships*.

78. True, *Agricultural Education*, 126–129.

79. *Ibid.*, 194–200.

80. Quoted, *ibid.*, 202.

81. *Ibid.*, 210–211.

82. Quoted in Milton Conover, *Office of Experiment Stations* (Institute for Government Research, *Service Monograph No. 32*, Baltimore, 1924), 55.

83. Benjamin Harrow, "Wilbur Olin Atwater," *DAB*, I, 417–418.

84. Quoted in Conover, *Office of Experiment Stations*, 55.

85. Quoted, *ibid.*, 53–54.

86. For a general discussion see V. O. Key, *The Administration of Federal Grants to States* (Chicago, 1937), 42–50.

87. *Science*, 4:508 (1884).

88. E. W. Hilgard, "Letter to the Editor," *Science*, 5:23 (1885).

89. *Science*, 5:21 (1885).

90. Conover, *Experiment Stations*, 57.

91. *Ibid.*, 59.

92. *Ibid.*, 59 n.

93. Key, *Federal Grants to States*, 8. In 1936 the federal government contributed 30.8 per cent.

94. J. M. Rusk, "The Duty of the Hour," *North American Review*, 152:430 (1891).

95. Gaus and Wolcott, *Department of Agriculture*, 15–17.

96. Ball, *Administrative Relationships*, II, 749–750.

97. A. W. MacMahon and J. D. Millett, *Federal Administrators: A Biographical Approach to the Problem of Departmental Management* (New York, 1939), 212–213.

98. Howard, *Applied Entomology*, 92.

99. *Who Was Who* (1943–1950), II, 142.

100. MacMahon and Millett, *Federal Administrators*, 215–216.

101. Howard, *Fighting the Insects*, 35–36.

102. *Ibid.*, 36.

103. C. W. Dabney, Jr., "Civil Service in the Department of Agriculture," Office of Experiment Stations, *Circular 33* (Washington, 1897), 2–3.

104. *Ibid.*, 4.

105. *Ibid.*, 10.

106. See MacMahon and Millett, *Federal Administrators*, 318–336, for a detailed account of service records of bureau chiefs in the Department of Agriculture who have attained office by promotion.

107. Dabney, "Civil Service," *Circular 33*, 4.

108. Secretary of Agriculture, *Report for 1913* (Washington, 1914), 9.

109. J. M. Phalen, "Theobald Smith," *DAB*, XXI, 665–667.

110. E. D. Merrill, "Autobiographical: Early Years, the Philippines, California," *Asa Gray Bulletin* [N. S.], 2:343 and *passim* (1953).

111. L. B. Schmidt, "James Wilson," *DAB*, XX, 330–331.

112. True, *Agricultural Research*, 187–188.

113. H. W. Wiley, "The Relation of Chemistry to the Progress of Agriculture," U. S. Department of Agriculture, *Yearbook, 1899* (Washington, 1900), 239; Harding, *Two Blades*, 30.

114. "The Laboratory in Modern Science," *Science*, 3:172 (1884).

115. Ball, *Administrative Relationships*, I, 273.

116. *Ibid.*, 275.

117. Gifford Pinchot, *Breaking New Ground* (New York, 1947), 311.

118. See H. W. Wiley, *An Autobiography* (Indianapolis, 1930), 168–169, 177–178; Wiley, "Relation of Chemistry to Agriculture," U. S. Department of Agriculture, *Yearbook, 1899*, 241.

119. For example, C. W. Dabney to H. W. Wiley, January 16, 1894, Letters of the Assistant Secretary of Agriculture, National Archives.

120. Ball, *Administrative Relationships*, I, 175, 181–182.

121. Gaus and Wolcott, *Department of Agriculture*, 20–21, 23.

122. Quoted in Harding, *Two Blades*, 32.

123. G. A. Weber, *The Food, Drug, and Insecticide Administration: Its History, Activities and Organization* (Institute for Government Research, Service Monograph No. 50, Baltimore, 1928), 2.

124. *Ibid.*, 2, 5.

125. *Ibid.*, 8.

126. Wiley, *Autobiography*, 215–220.

127. Commissioner of Agriculture, *Report for 1888* (Washington, 1889), 221.

128. True, *Agricultural Research*, 190.

129. Weber, *Food, Drug, and Insecticide Administration*, 18.

130. Wiley, *Autobiography*, 232–233.

131. *Ibid.*, 190–191, 231.

132. *Ibid.*, 238–239; Weber, *Food, Drug, and Insecticide Administration*, 15.

133. Wiley, *Autobiography*, 241.

134. Theodore Roosevelt, Washington, to H. H. Rusby, January 7, 1909, in E. E. Morison, ed., *The Letters of Theodore Roosevelt* (Cambridge, Mass., 1952), VI, 1467–1468.

135. Theodore Roosevelt, Washington, to Ira Remsen, January 16, 1908, in Morison, ed., *Roosevelt Letters*, VI, 908–909.

136. *Ibid.*, 908 n.

137. See O. E. Anderson, *Refrigeration in America* (Princeton, 1953), 122, 132, 140, 152, 173, 183.

138. C. R. Barnett, "Seaman Asahel Knapp," *DAB*, X, 452–453; R. B. Fosdick, *The Story of the Rockefeller Foundation* (New York, 1952), 181–182.

139. True, *Agricultural Education*, 288–289.

140. Ball, *Administrative Relationships*, I, 34–35.

141. Amount disbursed, $24,044,657.08, Secretary of Agriculture, *Report for 1915* (Washington, 1916), 251.

142. Secretary of Agriculture, *Report for 1913* (Washington, 1914), 9.

143. Secretary of Agriculture, *Report for 1914* (Washington, 1915), 42–43.

Chapter IX. The Decline of Science in the Military Services

1. See G. A. Weber, *The Naval Observatory* (Institute for Government Research, *Service Monograph No. 39*, Baltimore, 1926), 28–45; Simon Newcomb, *The Reminiscences of an Astronomer* (Boston, 1903), Chapter IV.

2. R. S. Dugan, "Edward Singleton Holden," *Dictionary of American Biography*, IX, 136–137.

3. H. B. Lemon, "Albert Abraham Michelson," *DAB*, XII, 593–596.

4. Newcomb, *Reminiscences*, 212–213.

5. E. W. Brown, "George W. Hill," *DAB*, IX, 32–33.

6. Newcomb, *Reminiscences*, 220–221.

7. See, for example, "The National Observatory," *Science*, 2:415 (1883).

8. Quoted in *Science*, 7:178–179 (1886).

9. See Newcomb, *Reminiscences*, 128–136.

10. Weber, *Naval Observatory*, 31.

11. For example: 1880, $70,415; 1900, $83,620; 1915, $85,630. Weber, *Naval Observatory*, 87–88; "The United States Naval Observatory," *Science* [N. S.], 7:111–113 (1898).

12. Gerstle Mack, *The Land Divided: A History of the Panama Canal and Other Isthmian Canal Projects* (New York, 1944), 168.

13. *Ibid.*, 169.

14. Interoceanic Canal Commission, *Report*, April 18, 1879, 46 Cong., 1 Sess., Sen. Ex. Doc. 15 (Ser. 1869).

15. G. A. Weber, *The Hydrographic Office* (Institute for Government Research, *Service Monograph No. 42*, Baltimore, 1926), 26–40.

16. [Lewis J. Darter, Jr.], *List of Climatological Records in the National Archives* (Washington, 1942), xi.

17. Smithsonian Institution, *Report for 1869* (Washington, 1871), 51.

18. E. R. Miller, "New Light on the Beginnings of the Weather Bureau from the Papers of Increase A. Lapham," *Monthly Weather Review*, 59:65–70 (1931).

19. *Ibid.*, 69.

20. [Darter], *List*, xv–xvi.

21. H. E. Paine, quoted in Miller, *Monthly Weather Review*, 59:68 (1931).

22. G. A. Weber, *The Weather Bureau: Its History, Activities and Organization* (Institute for Government Research, *Service Monograph No. 9*, New York, 1922), 4.

23. E. R. Miller, "The Evolution of Meteorological Institutions in the United States," *Monthly Weather Review*, 59:4 (1931).

24. See, for example, C. G. Rossby, "The Scientific Basis of Modern Me-

teorology," *Climate and Man, Yearbook of Agriculture* (Washington, 1941), 652.

25. See Cleveland Abbe, "The Meteorological Work of the U. S. Signal Service 1870 to 1891," in Oliver L. Fassig, ed., *Report of the International Meteorological Congress . . . 1893*, U. S. Department of Agriculture, Weather Bureau, *Bulletin 11* (Washington, 1895), Part 2, 232; Weber, *Weather Bureau*, 16.

26. W. J. Humphreys, "Cleveland Abbe," *DAB*, I, 1-2.

27. See Abbe, "Meteorological Work," Weather Bureau, *Bulletin 11*, 241–242; [Darter], *List*, xxviii; "The U. S. Signal Service, II," *Science*, 2:387 (1883).

28. Joint Commission to Consider the Present Organization of the Signal Service [etc.] . . . , *Testimony*, March 16, 1886, 49 Cong., 1 Sess., Sen. Misc. Doc. 82 (Ser. 2345), 19*-20*, hereafter cited as Allison Commission, *Testimony*.

29. Abbe, "Meteorological Work," Weather Bureau, *Bulletin 11*, 737-738.

30. Allison Commission, *Testimony*, 209-230.

31. *Ibid.*, 127.

32. *Ibid.*, 11*.

33. *Science*, 6:397 (1885).

34. Quoted in [Darter], *List*, xxxii.

35. See Chapter XI.

36. See Allison Commission, *Testimony*.

37. *Ibid.*, 302.

38. *Ibid.*, 265.

39. *Ibid.*, 261.

40. *Ibid.*, 10*.

41. *Ibid.*, 795-796.

42. [Darter], *List*, xxii.

43. Allison Commission, *Testimony*, 82-86.

44. *Ibid.*, 1002.

45. Abbe, "Meteorological Work," Weather Bureau, *Bulletin 11*, 241.

46. [Darter], *List*, xix-xx.

47. *Ibid.*, xxxvi-xxxvii.

48. Miller, *Monthly Weather Review*, 59:4 (1931).

49. Weber, *Weather Bureau*, 70.

50. See A. W. MacMahon and J. D. Millett, *Federal Administrators: A Biographical Approach to the Problem of Departmental Management* (New York, 1939), 384-387.

51. [Darter], *List*, xxxviii.

52. Cleveland Abbe, "The Needs of Meteorology," *Science* [N. S.], 1:181–182 (1895).

53. For an account of the international circumpolar stations, with emphasis on Greely, see Jeannette Mirsky, *To the Arctic! The Story of Northern Exploration from Earliest Times to the Present* (New York, 1948), 185-195; J. E. Caswell, *The Utilization of the Scientific Reports of the United States Arctic Expedition, 1850-1909* (Stanford, Calif., 1951).

54. W. S. Holt, *Office of Chief of Engineers of the Army: Its Non-Military History, Activities and Organization* (Institute for Government Research, Service Monograph No. 27, Baltimore, 1923), 11.

55. *Ibid.*, 24, 31.

Chapter X. The Geological Survey

1. For a brief account see G. P. Merrill, *Contributions to the History of American Geology* (U. S. National Museum, *Report, 1904*), 607–616; for the reports, see Clarence King, *Report of the Geological Exploration of the Fortieth Parallel* (Washington, 1870–1880), 7 vols. and atlas; in the preparation of this and the following chapter I have had the privilege of consulting an early draft of two chapters of Thomas G. Manning's forthcoming history of the Geological Survey.

2. Merrill, *American Geology*, 608.

3. John Wesley Powell, Washington, to Carl Schurz, November 1, 1878, in National Academy of Sciences, *A Report on the Surveys of the Territories*, December 3, 1878, 45 Cong., 3 Sess., H. R. Misc. Doc. 5 (Ser. 1861), 22, hereafter cited as National Academy, *Surveys of the Territories*.

4. Merrill, *American Geology*, 616.

5. W. C. Darrah, *Powell of the Colorado* (Princeton, 1951), 243.

6. *The U. S. Geological Survey: Its History, Activities, and Organization* (Institute for Government Research, *Service Monograph No. 1*, New York, 1918), 5, hereafter cited as *Geological Survey*.

7. G. M. Wheeler, *Annual Report upon the Geographical Surveys West of the One Hundredth Meridian* . . . (Appendix JJ of the Chief of Engineers, *Annual Report for 1876*, Washington, 1876), iii.

8. *Ibid.*, iv.

9. National Academy, *Surveys of the Territories*, 22.

10. *Geological Survey*, 5–6.

11. See pp. 91–92 for the account of the work of D. D. Owen, Charles T. Jackson, and J. D. Whitney in the Mississippi Valley and Michigan.

12. J. D. Whitney, Boston, to George W. Julian, January 12, 1865, in E. T. Brewster, ed., *Life and Letters of Josiah Dwight Whitney* (Boston, 1909), 243.

13. See G. P. Merrill, *American Geology*, 592–607.

14. *Ibid.*, 595–596.

15. *Ibid.*, 596.

16. J. L. Gray, *Letters of Asa Gray* (Boston, 1893), II, 669–676.

17. Asa Gray and J. D. Hooker, "The Vegetation of the Rocky Mountain Region and a Comparison with that of Other Parts of the World," United States Geological and Geographical Survey of the Territories, *Bulletin*, VI (1881), 1–77.

18. *Geological Survey*, 4.

19. National Academy, *Surveys of the Territories*, 22.

20. Since Powell reappeared in American historiography in W. P. Webb, *The Great Plains* (Boston, 1931), he has become one of the most studied figures in the annals of American science. W. C. Darrah, *Powell*, and Wallace Stegner, *Beyond the Hundredth Meridian: John Wesley Powell and the Second Opening of the West* (Boston, 1954) are the accounts largely used here. T. G. Manning's forthcoming history of the Geological Survey will make the coverage still more complete.

21. Darrah, *Powell*, 92.

22. *Ibid.*, 92–93.

23. Quoted in Stegner, *Powell*, 387–388.

24. See W. M. Davis, "John Wesley Powell, 1834-1902," National Academy of Sciences, *Biographical Memoirs*, VIII (1919), 32-34.

25. Darrah, *Powell*, 205.

26. See Stegner, *Powell*, 116-117.

27. *Geological Survey*, 4.

28. Stegner, *Powell*, 146-174.

29. J. W. Powell to the Secretary of the Interior, April 24, 1874, in *Geographical and Geological Surveys West of the Mississippi*, 43 Cong., 1 Sess. (May 26, 1874), H. R. Report 612 (Ser. 1626), 10, hereafter cited as *Surveys West of the Mississippi*.

30. Washington, 1878.

31. G. A. Weber, *The Coast and Geodetic Survey: Its History, Activities and Organization* (Institute for Government Research, *Service Monograph No. 16*, Baltimore, 1923), 8; for an account of early work in the West, see Superintendent of the Coast Survey, *Annual Report for 1872* (Washington, 1875), 39-40.

32. Weber, *Coast and Geodetic Survey*, 9.

33. Joint Commission to Consider the Present Organization of the Signal Service [etc.] . . . , *Testimony*, March 16, 1886, 49 Cong., 1 Sess., Sen. Misc. Doc. 82 (Ser. 2345), 61, hereafter cited as Allison Commission, *Testimony*.

34. *Surveys West of the Mississippi*, 32-33.

35. See L. A. Du Bridge, "Science and National Security," *Science*, 120: 1082 (1954).

36. *Surveys West of the Mississippi*, 1.

37. *Ibid., passim*; Darrah, *Powell*, 207-211.

38. Stegner, *Powell*, 207-208.

39. Charles Schuchert and C. M. Le Vene, *O. C. Marsh: Pioneer in Paleontology* (New Haven, 1940), 249.

40. See Allan Nevins, *Abram S. Hewitt: With Some Account of Peter Cooper* (New York, 1935).

41. Stegner, *Powell*, 232-233; H. N. Smith, "King, Powell, and the Establishment of the United States Geological Survey," *Mississippi Valley Historical Review*, 34:42 (1947).

42. Schuchert and Le Vene, *Marsh*, 252.

43. National Academy, *Surveys of the Territories*, 11.

44. *Ibid.*, 16.

45. *Ibid.*, 17.

46. *Ibid.*, 21.

47. *Ibid.*, 23.

48. *Ibid.*, 24-25.

49. *Ibid.*, 26-27.

50. Stegner, *Powell*, 235.

51. National Academy, *Surveys of the Territories*, 3-5.

52. Schuchert and Le Vene, *Marsh*, 252-254.

53. *Congressional Record*, 45 Cong., 3 Sess., February 11, 1879, 1209-1210.

54. *Ibid.*, 1203-1207.

55. Smith, *Mississippi Valley Historical Review*, 34:52 (1947).

56. 45 Cong., 3 Sess., 1564.

57. Stegner, *Powell*, 239.

58. For the law, see *Geological Survey*, 8 and 113.

59. Darrah, *Powell*, 246.
60. Quoted in Stegner, *Powell*, 240.
61. Henry Adams, quoted *ibid.*, 241.
62. Quoted in *Geological Survey*, 10.
63. *Ibid.*, 9–11; Stegner, *Powell*, 244–245.
64. Stegner, *Powell*, 247.
65. Darrah, *Powell*, 276.
66. *Ibid.*, 247, 274.
67. *Geological Survey*, 124.
68. Darrah, *Powell*, 268–269.
69. Stegner, *Powell*, 272.
70. "The Sphere of the United States Geological Survey," *Science*, 1:185–186 (1883).
71. Stegner, *Powell*, 272–273.
72. Darrah, *Powell*, 280.
73. Allison Commission, *Testimony*, 196, 688; Schuchert and Le Vene, *Marsh*, 317.
74. Darrah, *Powell*, 278.
75. Schuchert and Le Vene, *Marsh*, 270, 280–284.
76. Darrah, *Powell*, 275.
77. *Geological Survey*, 12.
78. Stegner, *Powell*, 279–280.

Chapter XI. The Allison Commission and the Department of Science

1. Joint Commission to Consider the Present Organization of the Signal Service, Geological Survey, Coast and Geodetic Survey, and the Hydrographic Office of the Navy Department, with a View to Secure Greater Efficiency and Economy of Administration of the Public Service in said Bureau . . . , *Testimony*, March 16, 1886, 49 Cong., 1 Sess., Sen. Misc. Doc. 82 (Ser. 2345), hereafter cited as Allison Commission, *Testimony*.
2. See p. 190.
3. Wallace Stegner goes the length of comparing science as an issue in the 1880's with welfare in the 1930's, *Beyond the Hundredth Meridian: John Wesley Powell and the Second Opening of the West* (Boston, 1954), 283.
4. H. H. Varrell, "Theodore Lyman," *Dictionary of American Biography*, XI, 519.
5. Theodore Lyman, Washington, to O. C. Marsh, July 9, 1884, in Allison Commission, *Testimony*, 2*.
6. *Ibid.*, 1*.
7. *Ibid.*, 6*–7*.
8. *Ibid.*, 7*.
9. *Ibid.*, 8*.
10. *Ibid.*, 8*–9*.
11. *Ibid.*, 23–26.
12. *Ibid.*, 26–27.
13. *Ibid.*, 27–29.
14. *Ibid.*, 30.
15. *Ibid.*, 999–1000.
16. *Ibid.*, 66–70.

17. *Science*, 7:155 (1886).
18. "Reformation of Scientific Legislation," *Science*, 5:325 (1885).
19. *Science*, 7:156 (1886).
20. "A National University," *Science*, 6:509 (1885).
21. *Ibid.*, 509–510.
22. [Alexander Agassiz], "The National Government and Science," *Nation*, 41:526 (1885).
23. See W. H. Dall, *Spencer Fullerton Baird: A Biography* (Philadelphia, 1915), 403–404.
24. W. C. Darrah, *Powell of the Colorado* (Princeton, 1951), 293.
25. For a résumé of charge and answer, see *Science*, 6:301 (1885); "Answers to Charges Affecting the Geological Survey," *Science*, 6:424–425 (1885).
26. F. M. Thorn, quoted in *Science*, 8:359 (1886).
27. *Ibid.*, 360.
28. *Science*, 6:204 (1885).
29. Alexander Agassiz, "The Coast-Survey and 'Political Scientists,'" *Science*, 6:253–255 (1885).
30. "The President and Professor Agassiz," *Science*, 6:302–303 (1885).
31. *Science*, 8:359–360 (1886); *National Cyclopedia of American Biography*, XXII, 198–199.
32. H. A. Herbert, Washington, to Alexander Agassiz, November 27, 1885, in Allison Commission, *Testimony*, 1013–1014.
33. Alexander Agassiz, Cambridge, Mass., to H. A. Herbert, December 2, 1885, *ibid.*, 1014.
34. Stegner, *Powell*, 291.
35. Agassiz to Herbert, December 2, 1885, in Allison Commission, *Testimony*, 1014–1015.
36. J. W. Powell, Washington, to W. B. Allison, February 26, 1886, in Allison Commission, *Testimony*, 1078.
37. *Ibid.*, 1708; the reference is evidently to H.M.S. *Challenger* reports.
38. Powell to Allison, February 26, 1886, in Allison Commission, *Testimony*, 1079.
39. *Ibid.*, 1074.
40. *Ibid.*, 1079–1081.
41. *Ibid.*, 1082.
42. Joint Commission, *Report*, June 8, 1886, 49 Cong., 1 Sess., Sen. Report 1285 (Ser. 2361), 125, hereafter cited as Allison Commission, *Report*.
43. H. A. Herbert, *Restricting the Work and Publications of the Geological Survey . . .*, May 5, 1886, 49 Cong., 1 Sess., H. R. Report 2214 (Ser. 2441), 16.
44. Allison Commission, *Report*, 68, 88.
45. *Science*, 7:383 (1886).
46. See C. V. Woodward, *Origins of the New South, 1877–1913*, in W. H. Stephenson and E. M. Coulter, eds., *A History of the South* (Baton Rouge, La., 1951), IX, Herbert being specifically mentioned on p. 275.
47. Allison Commission, *Report*, 53.
48. *Ibid.*, 13.
49. *Ibid.*, 54.
50. J. S. Billings, "Scientific Men and their Duties," *Science*, 8:545–547 (1886).
51. *Ibid.*, 549.

Chapter XII. Conservation

1. In preparing this chapter I have benefited greatly from Samuel P. Hays, "The First Conservation Movement," an unpublished doctoral dissertation in the Harvard University Archives.

2. See p. 91.

3. Quoted in E. W. Sterling, "The Powell Irrigation Survey, 1888–1893," *Mississippi Valley Historical Review*, 27:422 (1940); the story of the irrigation survey is also told in W. C. Darrah, *Powell of the Colorado* (Princeton, 1951), 299–314; Wallace Stegner, *Beyond the Hundredth Meridian: John Wesley Powell and the Second Opening of the West* (Boston, 1954), 294–350.

4. Stegner, *Powell*, 303.

5. Darrah, *Powell*, 300–301.

6. Quoted in Sterling, *Mississippi Valley Historical Review*, 27:426 (1940).

7. Quoted, *ibid.*, 428–429.

8. *Ibid.*, 433; Darrah, *Powell*, 310.

9. Stegner, *Powell*, 337.

10. Darrah, *Powell*, 345.

11. *Ibid.*

12. Sterling, *Mississippi Valley Historical Review*, 27:433–434 (1940).

13. Hays, "Conservation Movement," 416–417.

14. Quoted in W. H. Dall, *Spencer Fullerton Baird: A Biography* (Philadelphia, 1915), 420–422.

15. Quoted, *ibid.*, 423.

16. *Ibid.*, 429.

17. R. H. Connery, *Government Problems in Wildlife Conservation* (New York, 1935), 118; D. S. Jordan, "Spencer Fullerton Baird and the United States Fish Commission," *Scientific Monthly*, 17:105 (1923).

18. See *The United States Bureau of Fisheries: Its Establishment, Functions, Organization, Resources, Operations, and Achievements* (Washington, 1908), 33.

19. F. R. Lillie, *The Wood's Hole Marine Biological Laboratory* (Chicago, 1944), 24–26.

20. Sir Lyon Playfair, quoted in "Science and the State," *Science*, 6:325 (1885).

21. Connery, *Wild Life Conservation*, 117–120.

22. S. F. Baird, quoted in Dall, *Baird*, 431.

23. *Ibid.*, 432; Connery, *Wild Life Conservation*, 123.

24. Jenks Cameron, *The Bureau of the Biological Survey: Its History, Activities and Organization* (Institute for Government Research, *Service Monograph No. 54*, Baltimore, 1929), is an outstanding volume in its series.

25. See pp. 161–162.

26. Cameron, *Biological Survey*, 20–21.

27. *Ibid.*, 23.

28. G. S. Miller, Jr., "Mammalogy and the Smithsonian Institution," Smithsonian Institution, *Annual Report for 1928* (Washington, 1929), 405–406.

29. Quoted in Cameron, *Biological Survey*, 29.

30. D. H. Smith, *The Forest Service: Its History, Activities and Organization* (Institute for Government Research, *Service Monograph No. 58*, Washington, 1930), 6.

31. *Ibid.*, 7–10.

32. *Ibid.*, 11–12; Gifford Pinchot, *Breaking New Ground* (New York, 1947), 133–135.

33. Smith, *Forest Service*, 9.

34. *Ibid.*, 13.

35. For a contemporary statement, see E. J. James, "The Government in its Relations to the Forests," *Report on the Forest Conditions of the Rocky Mountains* (Forestry Division, *Bulletin No. 2*, Washington, 1888), 23–39.

36. Smith, *Forest Service*, 15–16.

37. *Ibid.*, 18–19.

38. Pinchot, *Breaking New Ground*, 20–35.

39. *Ibid.*, 89.

40. *Ibid.*, 93.

41. *Ibid.*, 107–113.

42. *Ibid.*, 117.

43. *Ibid.*, 119–122, for Pinchot's analysis.

44. "Report of the Commission . . . upon a Forest Policy for the Forested Lands of the United States," National Academy of Sciences, *Report for the Year 1897* (Washington, 1898), 42–43.

45. *Ibid.*, 49.

46. *Ibid.*, 51.

47. Pinchot, *Breaking New Ground*, 130–131.

48. *Ibid.*, 123; *Science* [N. S.], 6:694 (1897).

49. Pinchot, *Breaking New Ground*, 136.

50. *Ibid.*, 134.

51. *Ibid.*, 143–144.

52. Smith, *Forest Service*, 27.

53. Pinchot, *Breaking New Ground*, 255.

54. Smith, *Forest Service*, 205.

55. Pinchot, *Breaking New Ground*, 152–153.

56. *Ibid.*, 161–172.

57. *Ibid.*, 172.

58. *Ibid.*, 175.

59. *Ibid.*, 181.

60. E. E. Morison, ed., *The Letters of Theodore Roosevelt* (Cambridge, Mass., 1954), III, 463.

61. Pinchot, *Breaking New Ground*, 188–191.

62. *The U. S. Reclamation Service: Its History, Activities and Organization* (Institute for Government Research, *Service Monograph No. 2*, New York, 1919), 17.

63. Theodore Roosevelt, Washington, to E. A. Hitchcock, June 17, 1902, in Morison, ed., *Letters of Roosevelt*, III, 277.

64. *Reclamation Service*, 23–24.

65. *Ibid.*, 70.

66. Pinchot, *Breaking New Ground*, 257–260.

67. James Wilson, Secretary of Agriculture, quoted, *ibid.*, 261.

68. Smith, *Forest Service*, 205.

69. *The United States Geological Survey: Its History, Activities, and Organization* (Institute of Government Research, *Service Monograph No. 1*, New York, 1918), 26.

70. Pinchot, *Breaking New Ground*, 322.

71. Walter Hough, "William John McGee," *DAB*, XII, 47.

72. Pinchot, *Breaking New Ground*, 359.

73. *Ibid.*, 331.

74. *Ibid.*, 326–333.

75. Inland Waterways Commission, *Preliminary Report*, February 26, 1908, 60 Cong., 1 Sess., Sen. Doc. 325.

76. *Ibid.*, 451–490.

77. *Ibid.*, 31.

78. Pinchot, *Breaking New Ground*, 355–368.

79. Quoted, *ibid.*, 380–381.

80. Theodore Roosevelt, New York, to Mrs. Emma Baker Kennedy, April 28, 1911, in Morison, ed., *Letters of Roosevelt*, VII, 247.

81. L. M. Wolfe, *Son of the Wilderness: The Life of John Muir* (New York, 1945), 310–316.

82. Robert Shankland, *Steve Mather of the National Parks* (New York, 1951), 47–53.

83. *Ibid.*, 60, 245, 257–258.

84. Cameron, *Biological Survey*, 37.

85. Theodore Roosevelt, Washington, to Florence Lockwood La Farge, February 2, 1907, in Morison, ed., *Letters of Roosevelt*, V, 578.

86. Quoted in Cameron, *Biological Survey*, 38.

87. W. H. Osgood, "Clinton Hart Merriam, 1855–1942," National Academy of Sciences, *Biographical Memoirs*, 24: 16 (1947).

88. Quoted in Cameron, *Biological Survey*, 39.

89. *Ibid.*, 45.

90. Osgood, "Merriam," National Academy of Sciences, *Memoirs*, 24: 22 (1947).

91. *Ibid.*, 94–95.

92. *Ibid.*, 100–101. The Supreme Court ruled on this power in Missouri vs. Holland, 252 U.S., 416.

93. For an examination of the relation of forestry to science see I. W. Bailey and H. A. Spoehr, *The Role of Research in the Development of Forestry in America* (New York, 1929).

94. Smith, *Forest Service*, 39.

95. *Ibid.*, 38.

96. See Bailey and Spoehr, *Development of Forestry*, 16–19.

97. Pinchot, *Breaking New Ground*, 308.

Chapter XIII. Medicine and Public Health

1. E. E. Hume, *Victories of Army Medicine: Scientific Accomplishments of the Medical Department of the United States Army* (Philadelphia, 1943), 22–26.

2. See p. 129.

3. See Hume, *Army Medicine*, 45–51; E. E. Hume, "The Army Medical Library of Washington, the Largest Medical Library That Has Ever Existed," *Isis*, 26:423–447 (1937); M. C. Leikind, "Army Medical Museum and Armed Forces Institute of Pathology in Historical Perspective," *Scientific Monthly*, 79:71–78 (1954).

4. W. F. Willcox, "John Shaw Billings," *Dictionary of American Biography*, II, 266–269; F. H. Garrison, *John Shaw Billings: A Memoir* (New York, 1915), 337.

5. Leikind, *Scientific Monthly*, 79:73–76 (1954); Hume, *Army Medicine*, 141–142.

6. R. D. Leigh, *Federal Health Administration in the United States* (New York, 1927), 92; R. C. Williams, *The United States Public Health Service, 1798–1950* (Washington, 1951), 46–47.

7. Leigh, *Federal Health Administration*, 93–94.

8. *Ibid.*, 94–96.

9. *Ibid.*, 287–289.

10. *Ibid.*, 463–468.

11. *Ibid.*, 469–472.

12. *Ibid.*, 467–484; for briefer, more recent accounts see R. H. Shryock, *American Medical Research, Past and Present* (New York, 1947), 43–44, and Williams, *Public Health Service*, 76–79.

13. Leigh, *Federal Health Administration*, 473–474.

14. *Ibid.*, 476–478.

15. Garrison, *Billings*, 163.

16. Quoted in Leigh, *Federal Health Administration*, 369.

17. *Ibid.*, 474–476.

18. *Ibid.*, 293–294, 478–481. Quote is from Dr. H. I. Bowditch.

19. V. Y. Bowditch, *Life and Correspondence of Henry Ingersoll Bowditch* (Boston, 1902), II, 247, 251.

20. Leigh, *Federal Health Administration*, 478.

21. Quoted, *ibid.*, 479.

22. *Ibid.*, 480–481.

23. "Excerpts from the Hoover Report on Federal Medical Services," *New York Times*, February 28, 1955, p. 12, points out the need for a Federal Advisory Council of Health.

24. J. M. Phalen, "George Miller Sternberg," *DAB*, XVII, 590–592.

25. J. A. Phalen, "Walter Reed," *DAB*, XV, 459–461.

26. Hume, *Army Medicine*, 27.

27. *Ibid.*, 28–29; J. A. Tobey, *The Medical Department of the Army: Its History, Activities and Organization* (Institute for Government Research, Service Monograph No. 45, Baltimore, 1927), 25–26.

28. Hume, *Army Medicine*, 99–104.

29. For a general account of the yellow fever story, with bibliography, see Hume, *Army Medicine*, 92–99; a more highly colored version is L. N. Wood, *Walter Reed: Doctor in Uniform* (New York, 1943).

30. Gerstle Mack, *The Land Divided: A History of the Panama Canal and Other Isthmian Canal Projects* (New York, 1944), 520–536.

31. Hume, *Army Medicine*, 158–160; J. M. Phalen, "Bailey Kelly Ashford," *DAB*, XXI, 32–33; B. K. Ashford, *A Soldier in Science* (New York, 1934).

32. Phalen, "Ashford," *DAB*, XXI, 333.

33. R. B. Fosdick, *The Story of the Rockefeller Foundation* (New York, 1952), 30–43, 58–70.

34. R. E. Dyer, "The Research Program of the United States Public Health Service," J. S. Simmons, ed., *Public Health in the World Today* (Cambridge, Mass., 1949), 94; Williams, *Public Health Service*, 177.

35. L. F. Schmeckebier, *The Public Health Service: Its History, Activities and Organization* (Institute for Government Research, Service Monograph No. 10, Baltimore, 1923), 16–24.

36. For public discussion of proposals, see *Science* [N. S.], 6:913, 918 (1897).

37. Schmeckebier, *Public Health Service*, 24.

38. Williams, *Public Health Service*, 181.

39. Schmeckebier, *Public Health Service*, 26.

40. *Ibid.*, 222.

41. Shryock, *Medical Research*, 268.

42. Schmeckebier, *Public Health Service*, 27–28.

43. Leigh, *Federal Health Administration*, 485.

44. Washington, 1909.

45. *Ibid.*, 126.

46. Leigh, *Federal Health Administration*, 486.

47. Theodore Roosevelt, Washington, to Irving Fisher, May 8, 1907, in E. E. Morison, ed., *The Letters of Theodore Roosevelt* (Cambridge, Mass., 1951–1954), V, 664.

48. Schmeckebier, *Public Health Service*, 36–37.

49. Leigh, *Federal Health Administration*, 486.

50. *Ibid.*, 489.

51. Schmeckebier, *Public Health Service*, 37.

52. Williams, *Public Health Service*, 270–279.

53. Neither the figures in Schmeckebier, *Public Health Service*, or in Surgeon General of the Public Health Service, *Annual Report for 1915* (Washington, 1915), 355, give a clear breakdown of research expenditures. An unpublished study of the National Science Foundation arrives at a figure of $416,000 for research expenditures for social security, welfare, and health functions, in 1915. This is only 7.5 per cent of agricultural research and 9.8 per cent of natural resources' research. Information courtesy of Mrs. Mildred C. Allen.

Chapter XIV. The Completion of the Federal Scientific Establishment

1. Joint Commission to Consider the Present Organization of the Signal Service (etc.) . . . , *Testimony*, March 16, 1886, 49 Cong., 1 Sess., Sen. Misc., Doc. 82 (Ser. 2345), 370.

2. G. A. Weber, *The Bureau of Standards: Its History, Activities and Organization* (Institute for Government Research, *Service Monograph No. 35*, Baltimore, 1925), 28–30.

3. *Ibid.*, 31–32.

4. "Henry S. Pritchett," *National Cyclopedia of American Biography*, XXIX, 124–125; S. C. Prescott, "Samuel Wesley Stratton," *Dictionary of American Biography*, XVIII, 127–128; H. S. Pritchett, "The Story of the Establishment of the National Bureau of Standards," *Science* [N.S.], 15:282 (1902).

5. Weber, *Bureau of Standards*, 37–38.

6. Pritchett, *Science* [N.S.], 15:281 (1902).

7. *Ibid.*, 282.

8. Weber, *Bureau of Standards*, 248.

9. *Ibid.*, 248–249.

10. Director of the Bureau of Standards, *Annual Report for 1906* (Washington, 1906), 18.

11. *Ibid.*, 3–15.

12. Director of the Bureau of Standards, *Annual Report for 1915* (Washington, 1915), 98–139, 143.

13. E. B. Rosa, "National Bureau of Standards and Its Relation to Scientific and Technical Laboratories," *Science* [N.S.], 21:173 (1905).

14. H. W. Wiley, *The History of a Crime Against the Food Law* (Washington, 1929), 344. Although Wiley's most vituperative remarks relate to patent matters in the 1920's, he included the whole history of the Bureau of Standards in his condemnation.

15. Weber, *Bureau of Standards*, 48–49.

16. For a recent discussion of this relation, see Ad Hoc Committee for the Evaluation of the Present Functions and Operations of the National Bureau of Standards, *A Report to the Secretary of Commerce* (Washington, 1953), 59.

17. See, for example, Standards, *Annual Report for 1905* (Washington, 1905), 7–8.

18. Weber, *Bureau of Standards*, 45.

19. *Ibid.*, 30.

20. Standards, *Annual Report for 1915*, 145–146.

21. Director of the Bureau of Standards, *Annual Report for 1916* (Washington, 1916), 58.

22. E. B. Rosa, "The Function of Research in the Regulation of Natural Monopolies," *Science* [N.S.], 27:579–593 (1913).

23. W. S. Holt, *The Bureau of the Census: Its History, Activities and Organization* (Institute for Government Research, *Service Monograph No. 53*, Washington, 1929), 1–20; W. F. Willcox, "Development of the American Census and Its Methods," *Studies in American Demography* (Ithaca, 1940), 61–94; C. D. Wright, *The History and Growth of the United States Census* (Washington, 1900).

24. J. P. Nichols, "Francis Amasa Walker," *DAB*, XIX, 342–344; J. P. Munroe, *A Life of Francis Amasa Walker* (New York, 1923), 101–210.

25. Tenth Census, *Reports*, XII (Washington, 1886).

26. W. C. Darrah, *Powell of the Colorado* (Princeton, 1951), 268–269.

27. Tenth Census, *Reports*, IX (Washington, 1884).

28. *Ibid.*, XIII (Washington, 1885).

29. *Ibid.*, XV (Washington, 1886).

30. *Ibid.*, X (Washington, 1884).

31. *Ibid.*, III (Washington, 1883).

32. See, for example, "The Census Report of 1880," *Science*, 4:119 (1884); Holt, *Bureau of the Census*, 34.

33. Holt, *Bureau of the Census*, 28.

34. See S. N. D. North, *Coöperation and Unification in Federal and State Statistical Work* (Washington, 1903), 3–5.

35. For the movement leading up to this act see W. F. Willcox, "The Development of the American Census Office since 1890," *Political Science Quarterly*, 29: 438–459 (1914).

36. Holt, *Bureau of the Census*, 37.

37. See S. A. Stouffer, "Problems of the Bureau of the Census in Their Relation to Social Science," in National Resources Committee, *Research — A National Resource* (Washington, 1938), 201.

38. For the early evolution of the pattern of handling cotton production figures, see Holt, *Bureau of the Census*, 39–40.

39. In preparing this section I have benefited from an unpublished study by

Nathan Reingold prepared for the National Science Foundation and from an informative interview with Dr. Arno C. Fieldner of the Bureau of Mines.

40. F. W. Powell, *The Bureau of Mines: Its History, Activities and Organization* (Institute for Government Research, *Service Monograph No. 3*, New York, 1922), 1–2; *Report on the Operations of the Coal-Testing Plant* (U. S. Geological Survey, *Professional Paper No. 48*, Washington, 1906), 23–24.

41. Gifford Pinchot, *Breaking New Ground* (New York, 1947), 56.

42. *Operations of Coal-Testing Plant*, 28.

43. A. C. Fieldner, A. W. Gauger, and G. R. Yohe, "Gas and Fuel Chemistry," *Industrial and Engineering Chemistry*, 43:1040 (1951).

44. Committee on Mines and Mining, *Establishing a Bureau of Mines in the Interior Department*, 60 Cong., 1 Sess., Sen. Report 692, 16.

45. Powell, *Bureau of Mines*, 3.

46. See A. C. Fieldner, "Achievements in Mine Safety Research and Problems Yet to Be Solved," Bureau of Mines, *Information Circular 7573.*

47. *Ibid.*, 4.

48. See p. 275.

49. See Director of the Bureau of Mines, *Annual Report for 1911* (Washington, 1912), 6.

50. Director of the Bureau of Mines, *Annual Report for 1913* (Washington, 1914), 5.

51. Ernestine Adams, "History and Analysis of Bureau of Mines' Petroleum and Natural Gas Division," *Petroleum Engineer* (January, 1949), 10–11.

52. Director of the Bureau of Mines, *Annual Report for 1915* (Washington, 1915), 19–20.

53. Director of the Bureau of Mines, *Annual Report for 1914* (Washington, 1914), 28–29.

54. *Ibid.*, 31–32.

55. See Frank Cameron, *Cottrell: Samaritan of Science* (Garden City, New York, 1952), 156–158, 178–180.

56. P. H. Oehser, *Sons of Science: The Story of the Smithsonian Institution and Its Leaders* (New York, 1949), 97–98.

57. *Ibid.*, 155; Remington Kellogg, "A Century of Progress in Smithsonian Biology," *Science*, 104:132–141 (1946).

58. Smithsonian Institution, *Annual Report for 1901* (Washington, 1902), 6–7.

59. C. G. Abbot, "Samuel Pierpont Langley," *DAB*, X, 594.

60. Jeremiah Milbank, Jr., *The First Century of Flight in America: An Introductory Survey* (Princeton, 1943), 191.

61. Oehser, *Sons of Science*, 133.

62. *Ibid.*, 134–136; I. B. Holley, Jr., *Ideas and Weapons: Exploitation of the Aerial Weapon by the United States during World War I: A Study in the Relationship of Technological Advance, Military Doctrine, and the Development of Weapons* (New Haven, 1953), 26.

63. *Ibid.*, 141–151, 157–158.

64. In preparing this section I have benefited from an unpublished study of the National Advisory Committee for Aeronautics by Nathan Reingold, prepared for the National Science Foundation.

65. Smithsonian Institution, *Annual Report for 1913* (Washington, 1914), 7–8.

66. *Ibid.*, 117–119.

67. Board of Regents of the Smithsonian Institution, *Memorial on the Need of a National Advisory Committee for Aeronautics in the United States*, February 1, 1915, 63 Cong., 3 Sess., H. R. Doc. 1549, 2.

68. *Ibid.*, 1.

69. G. W. Gray, *Frontiers of Flight: The Story of NACA Research* (New York, 1948), 11.

70. Franklin D. Roosevelt, Washington, to L. P. Padgett, February 12, 1915, Committee on Naval Affairs, *National Advisory Committee for Aeronautics*, February 19, 1915, 63 Cong., 3 Sess., H. R. Report 1423, 2–3.

71. The law is quoted in Smithsonian Institution, *Annual Report for 1915*, 14.

Chapter XV. Patterns of Government Research in Modern America

1. This episode is summarized in A. W. MacMahon and J. D. Millett, *Federal Administrators: A Biographical Approach to the Problem of Departmental Management* (New York, 1939), 380.

2. See Bureau of Science, Philippine Islands, *Annual Reports*, 1903–1916 (Manila, 1903–1917); for an informal account, see E. D. Merrill, "Autobiographical: Early Years, The Philippines, California," *Asa Gray Bulletin* [N.S.], 2:346–360 (1953).

3. The national university movement was reported periodically in *Science*. See especially C. D. Walcott, "Relations of the National Government to Higher Education and Research," *Science* [N.S.], 13:1004–1011 (1901).

4. See, for example, editorials in *Science*, 4:1, 109 (1884).

5. Simon Newcomb, "Conditons Which Discourage Scientific Work in America," *North American Review*, 174:153–154 (1902).

6. A typewritten report of the committee is in Box 1937, Pinchot Papers, Library of Congress.

7. L. M. Short, *The Development of National Administrative Organization in the United States* (Baltimore, 1923), 400–402.

8. Gifford Pinchot, *Breaking New Ground* (New York, 1947), 240–243.

9. *Ibid.*, 243.

10. *Conduct of Scientific Work under United States Government*, January 18, 1909, 60 Cong., 2 Sess. H.R. Doc. 1337, 3.

11. *Ibid.*, 4.

12. *Ibid.*, 2.

13. C. D. Walcott and others, Washington, to Theodore Roosevelt, July 20, 1903, Box 1937, Pinchot Papers, Library of Congress; Pinchot, *Breaking New Ground*, 242.

14. *Ibid.*, 242.

15. See Richard Hofstadter and C. D. Hardy, *The Development and Scope of Higher Education in the United States* (New York, 1952).

16. R. H. Shryock points out this split in *American Medical Research, Past and Present* (New York, 1947), 32, 249.

17. "Opportunities for Young Men in Science," *Science* [N.S.], 26:875–880 (1908).

18. Walcott, *Science* [N.S.], 13:1006 (1901).

19. Carnegie Institution of Washington, *Year Book 1902* (Washington, 1903), xv.

20. *Ibid.*, xiii.

21. Carnegie Institution of Washington, *Year Book 1915* (Washington, 1916), i.

22. *Ibid.*, 4.

23. "The Carnegie Institution," *Science* [N.S.], 16:481, 483 (1902).

24. See Abraham Flexner, *Funds and Foundations: Their Policies Past and Present* (New York, 1952), 24–76.

25. See Merle Curti and Kendall Birr, *Prelude to Point Four: American Technical Missions Overseas, 1838–1938* (Madison, Wisc., 1954), *passim*.

26. Carl Snyder, "America's Inferior Position in the Scientific World," *North American Review*, 174:59 (1902).

27. Newcomb, *North American Review*, 174:145 (1902).

28. *Ibid.*, 155.

29. C. S. Slichter, "Recent Criticism of American Scholarship," *Transactions of the Wisconsin Academy of Sciences, Arts and Letters*, 14:9–12 (1902).

30. W J McGee, "Fifty Years of American Science," *Atlantic Monthly*, 82:320 (1898).

Chapter XVI. The Impact of World War I

1. I. B. Holley, Jr., *Ideas and Weapons: Exploitation of the Aerial Weapon by the United States during World War I: A Study in the Relationship of Technological Advance, Military Doctrine, and the Development of Weapons* (New Haven, 1953), 27–29.

2. F. L. Paxson, *American Democracy and the World War*, I, *Pre-War Years, 1913–1917* (Boston, 1936), 114.

3. Throughout this chapter I have benefited from the use of an unpublished study of science and the Army, prepared for the National Science Foundation by Nathan Reingold.

4. M. M. Johnson, Jr. and C. T. Haven, *For Permanent Victory: The Case for an American Arsenal of Peace* (New York, 1942), 81–84.

5. Allan Westcott, ed., *American Sea Power Since 1775* (Philadelphia, 1947), 300–304.

6. W. R. MacLaurin, *Invention and Innovation in the Radio Industry* (New York, 1949), 82.

7. E. E. Morison, *Admiral Sims and the Modern American Navy* (Boston, 1942), 319.

8. Paxson, *Pre-War Years*, 201.

9. *Ibid.*, 287.

10. See L. M. Short, *The Development of National Administrative Organization in the United States* (Baltimore, 1923), 441–450.

11. Council of National Defense, *First Annual Report, 1917* (Washington, 1917), 5–6.

12. Council of National Defense, *Second Annual Report, 1918* (Washington, 1918), 5.

13. G. B. Clarkson, *Industrial America in the World War: The Strategy Behind the Line* (Boston, 1923), 418–419.

14. *Ibid.*, 379–381.

15. L. N. Scott, *Naval Consulting Board of the United States* (Washington, 1920), 286.

16. *Ibid.*, 13–15.

17. *Ibid.*, 14–15.

18. *Ibid.*, 15.

19. *Ibid.*, 111–112.

20. *Ibid.*, 113.

21. Irvin Stewart, *Organizing Scientific Research for War: The Administrative History of the Office of Scientific Research and Development* (Boston, 1948), 33.

22. Scott, *Naval Consulting Board*, 122–147.

23. R. A. Millikan, *Autobiography* (New York, 1950), 129.

24. G. E. Hale, *National Academies and the Progress of Research* ([1915], reprinted from various issues of *Science*, 1913–1915).

25. Simon Flexner and J. T. Flexner, *William Henry Welch and the Heroic Age of American Medicine* (New York, 1941), 366.

26. National Academy of Sciences, *Annual Report for 1916* (Washington, 1917), 12.

27. Millikan, *Autobiography*, 124–125.

28. *Proceedings of the National Academy of Sciences*, 2:508 (1916).

29. *Ibid.*, 509.

30. Flexner and Flexner, *Welch*, 367.

31. *Proceedings of the National Academy of Sciences*, 2:510 (1916).

32. *Ibid.*, 507–508.

33. Flexner and Flexner, *Welch*, 367–368.

34. *Proceedings of the National Academy of Sciences*, 2:603 (1916).

35. *Ibid.*, 602–603.

36. *Ibid.*, 738–740.

37. Millikan, *Autobiography*, 139.

38. *Ibid.*, 147.

39. *Ibid.*; National Academy of Sciences, *Annual Report for 1918* (Washington, 1919), 50.

40. Millikan, *Autobiography*, 147.

41. R. A. Millikan, quoted in I. B. Cohen, "American Physicists at War from the First World War to 1942," *American Journal of Physics*, 13:337 (1945).

42. National Academy of Sciences, *Annual Report for 1918*, 61.

43. *Ibid.*, 61–62, 99–101; *Annual Report for 1917* (Washington, 1918), 69–70.

44. Council of National Defense, *Report, 1917*, 48.

45. Scott, *Naval Consulting Board*, 120.

46. Flexner and Flexner, *Welch*, 369.

47. National Academy of Sciences, *Annual Report for 1918*, 61–62.

48. Millikan, *Autobiography*, 144, 152.

49. National Academy of Sciences, *Annual Report for 1918*, 41–50.

50. *Ibid.*, 41.

51. Millikan, *Autobiography*, 148–149.

52. I. J. Cox, "George Owen Squier," *Dictionary of American Biography*, XVII, 489–490.

53. *Proceedings of the National Academy of Sciences*, 2:604 (1916).

54. G. O. Squier, Washington, to Dr. George E. Hale, July 2, 1917, in National Academy of Sciences, *Annual Report for 1917*, 47.

55. Millikan, *Autobiography*, 165.

56. *Ibid.*

57. *Ibid.*, 156.

58. National Academy of Sciences, *Annual Report for 1918*, 73.

59. Quoted in I. B. Cohen, *American Journal of Physics*, 13:337 (1945).

60. Millikan, *Autobiography*, 151.

61. R. M. Yerkes, *The New World of Science: Its Development During the War* (New York, 1920), is a summary treatment by NRC leaders in various fields, but Holley, *Ideas and Weapons*, clearly indicates the complexity of the interrelationships of science to the war even in a single area of activity. The multitude of individual projects is faintly indicated by the closely packed 300-page *War Work of the Bureau of Standards* (Bureau of Standards, *Miscellaneous Publication No. 46* [Washington, 1921]).

62. F. H. Martin, *Fifty Years of Medicine and Surgery: An Autobiographical Sketch* (Chicago, 1934), 368–393.

63. National Academy of Sciences, *Annual Report for 1917*, 66–67.

64. Flexner and Flexner, *Welch*, 370.

65. National Academy of Sciences, *Annual Report for 1918*, 88–89.

66. See R. H. Shryock, *American Medical Research, Past and Present* (New York, 1947), 273–275; F. H. Garrison, *Notes on the History of Military Medicine* (Washington, 1922), 196–205; J. R. Darnall, "Contributions of the World War to the Advancement of Medicine," *Journal of the American Medical Association*, 115:1443–1451 (1940), emphasizes the positive aspects of the record.

67. R. M. Yerkes, "How Psychology Happened into the War," in Yerkes, ed., *New World of Science*, 352.

68. R. M. Yerkes, "What Psychology Contributed," *ibid.*, 373.

69. Holley, *Ideas and Weapons*, makes a uniquely penetrating analysis of the research and development process in discussing the aerial weapon in World War I as a study in the relation of technological advance, military doctrine, and the development of weapons.

70. Holley, *Ideas and Weapons*, 112–116.

71. Millikan, *Autobiography*, 178.

72. National Advisory Committee for Aeronautics, *Annual Report for 1918* (Washington, 1920), 24.

73. Holley, *Ideas and Weapons*, 116.

74. *Ibid.*, 106.

75. R. A. Millikan, "Contributions of Physical Science," in R. M. Yerkes, ed., *New World of Science*, 39.

76. Scott, *Naval Consulting Board*, 68–69. This account differs in detail from that in Millikan, *Autobiography*, which was written much later.

77. Millikan, *Autobiography*, 161–164; Scott, *Naval Consulting Board*, 67–83.

78. Millikan, *Autobiography*, 164.

79. *Ibid.*, 172–176.

80. National Academy of Sciences, *Annual Report for 1917*, 52–53.

81. A. A. Fries and C. J. West, *Chemical Warfare* (New York, 1921), 32.

82. *Ibid.*, 34, 35.

83. A. W. Lane and L. H. Wall, eds., *The Letters of Franklin K. Lane* (Boston, 1922), 264.

84. Fries and West, *Chemical Warfare*, 38–39.

85. *Ibid.*, 53–61.

86. C. J. West, "The Chemical Warfare Service," in Yerkes, ed., *New World of Science*, 148.

87. Frank Cameron, *Cottrell: Samaritan of Science* (Garden City, New York, 1952), 187–194; V. H. Manning, *Petroleum Investigations and Production of Helium* (Bureau of Mines, *Bulletin 178C*, Washington, 1919), 75–87.

88. See Millikan, *Autobiography*, 179; National Academy of Sciences, *Annual Report for 1918*, 74.

89. See A. A. Noyes, "The Supply of Nitrogen Products for the Manufacture of Explosives," in Yerkes, ed., *New World of Science*, 123–147; W. Haynes, *American Chemical Industry* (New York, 1945), II, chaps. 6–11.

90. See H. E. Howe, "Optical Glass for War Needs," in Yerkes, ed., *New World of Science*, 103–120.

91. *Ibid.*, 108; *War Work of Bureau of Standards*, 183–193.

92. Benedict Crowell and R. F. Wilson, *How America Went to War: An Account From Official Sources of The Nation's War Activities, 1917–1920, The Armies of Industry*; I (New Haven, 1921), 147–148.

93. Haynes, *American Chemical Industry*, II, 229; National Academy of Sciences, *Annual Report for 1918*, 75–76.

94. "Work of the Department of Agriculture," *Science* [N.S.], 46:607–610 (1917).

95. William Hard, "America Prepares," *New Republic*, 10:253 (1917).

96. See "Some Suggestions for National Service on the Part of Zoologists and Zoological Laboratories," *Science* [N.S.], 45:627–630 (1917); W. P. Taylor, "The Vertebrate Zoologist and National Efficiency," *Science* [N.S.], 46:123–126 (1917); Neil E. Stevens, "American Botany and the Great War," *Science* [N.S.], 48:177 (1918).

97. F. B. Jewett, *Industrial Research* (Washington, 1918), 2–4.

98. Millikan, *Autobiography*, 210.

Chapter XVII. Transition to a Business Era

1. G. E. Vincent to R. A. Millikan, February 5, 1918, in Millikan, *Autobiography* (New York, 1950), 180–181.

2. *Ibid.*, 182.

3. *Ibid.*, 183–184.

4. *Ibid.*, 185.

5. National Academy of Sciences, *Annual Report for 1918* (Washington, 1919), 40–41.

6. National Academy of Sciences, *Annual Report for 1919* (Washington, 1920), 65.

7. *Ibid.*, 75.

8. National Academy of Sciences, *Annual Reports, 1920* (Washington, 1921), 50; *1921* (Washington, 1922), 54; *1922* (Washington, 1923), 37; *1923* (Washington, 1924), 35.

9. *A History of the National Research Council, 1919–1933* (NRC, *Reprint and Circular Series, No. 106*, Washington, 1933), 8.

10. National Academy of Sciences, *Annual Report for 1919*, 70.

11. National Academy of Sciences, *Annual Report for 1924* (Washington, 1925), 53.

12. National Academy of Sciences, *Annual Report for 1919*, 76–77.

13. E. C. Brunauer, *International Council of Scientific Unions, Brussels and Cambridge* (U. S. Department of State, *Publication 2413*, Washington, 1945), 10.

14. Nathan Reingold, "Science and the United States Army," Manuscript, pp. 108–113.

15. M. S. Watson, *Chief of Staff: Prewar Plans and Preparations* (United

States Army in World War II. The War Department, Washington, 1950), 15–17.

16. C. A. Herty, *The Reserves of the Chemical Warfare Service* (NRC, *Reprint and Circular Series, No. 16*, Washington, 1921), 2.

17. *Ibid.*, 5.

18. *Ibid.*, 3–5.

19. *Ibid.*, 11.

20. See A. H. Taylor, *The First Twenty-five Years of the Naval Research Laboratory* (Washington [1948]); Joint Board on Scientific Information Policy, *Radar: A Report on Science at War* (Washington, 1946), 4; Henry Guerlac, *Radar in World War II* (Microfilm Publication, Washington, 1947), part 1, chap. 3.

21. See Vannevar Bush, *Modern Arms and Free Men* (New York, 1949), 17–26, for a gloomy picture of weapons research between the wars.

22. Vernon Kellogg, in National Academy of Sciences, *Annual Report for 1924*, 49–50.

23. Frank Cameron, *Cottrell: Samaritan of Science* (Garden City, New York, 1952), 243–248.

24. National Advisory Committee for Aeronautics, *Annual Report for 1920* (Washington, 1921), 20.

25. National Advisory Committee for Aeronautics, *Annual Report for 1923* (Washington, 1924), 52; *Annual Report for 1927* (Washington, 1928), 11.

26. See G. W. Gray, *Frontiers of Flight: The Story of NACA Research* (New York, 1948), 19–33.

27. For a general discussion, see C. H. Wooddy, "The Growth of Governmental Functions," *Recent Social Trends in the United States* (New York, 1933), II, 1274–1330; for an agency-by-agency analysis see C. H. Wooddy, *The Growth of the Federal Government, 1915–1932* (New York, 1934).

28. R. H. Shryock, *American Medical Research, Past and Present* (New York, 1947), 271; C. D. Leake, "Cooperative Research: A Case Report," *Science* [N.S.], 62:251–256 (1925).

29. R. D. Leigh, *Federal Health Administration in the United States* (New York, 1927), 491–548; J. A. Tobey, *The National Government and Public Health* (Baltimore, 1926).

30. R. C. Williams, *The United States Public Health Service, 1798–1950* (Washington, 1951), 150.

31. *Ibid.*, 193–200.

32. Smithsonian Institution, *Annual Report for 1927* (Washington, 1928), 16.

33. See J. M. Gaus and L. O. Wolcott, *Public Administration and the United States Department of Agriculture* (Chicago, 1940), 47–57.

34. See G. C. Fite, *George N. Peek and the Fight for Farm Parity* (Norman, Okla., 1954).

35. *Recent Social Trends*, II, chap. 25.

36. *Ibid.*, I, xi.

37. H. R. Bartlett, "The Development of Industrial Research in the United States," *Research — A National Resource*, II (Washington, 1941), 37.

38. *History of NRC, 1919–1933*, 18.

39. Arthur D. Little, quoted in H. U. Faulkner, *The Decline of Laissez Faire, 1897–1917* (*Economic History of the United States*, VII, New York, 1951), 133.

40. F. B. Jewett, *Industrial Research* (Washington, 1918), 13.

41. See Herbert Hoover, *Memoirs: Years of Adventure, 1874–1920* (New York, 1951).

42. Quoted in Fite, *George N. Peek*, 127; see also Herbert Hoover, *Memoirs: The Cabinet and the Presidency, 1920–1933* (New York, 1952), 109.

43. Secretary of Commerce, *Annual Report for 1926* (Washington, 1926), 2–28.

44. *Ibid.*, 3.

45. Bureau of Standards, *Annual Report for 1923* (Washington, 1923), 299–302.

46. Bureau of Standards, *Annual Report for 1924* (Washington, 1924), 35–36.

47. Secretary of Commerce, *Annual Report for 1926*, 9.

48. *Ibid.*, 10.

49. Herbert Hoover, *Memoirs, 1920–1933*, 227–229.

50. Secretary of Commerce, *Annual Report for 1927* (Washington, 1927), 33–35.

51. *Ibid.*, 45–51; W. R. MacLaurin, *Invention and Innovation in the Radio Industry* (New York, 1949), 225–227.

52. Hoover, *Memoirs, 1920–1933*, 44.

53. Herbert Hoover, "The Nation and Science," *Science* [N.S.], 65:26–29 (1927).

54. "The National Research Endowment," *Science* [N.S.], 63:158 (1926).

55. *New York Times*, April 21, 1926, 1.

56. J. M. Cattell, "Scientific Research in the United States," *Science* [N.S.], 63:188 (1926).

57. Hoover, *Science*, 65:28 (1927).

58. Secretary of Commerce, *Annual Report for 1926*, 22.

59. National Academy of Sciences, *Annual Report for 1931* (Washington, 1932), 87.

60. Hoover, *Memoirs, 1920–1933*, 73.

Chapter XVIII. The Depression and the New Deal

1. All figures for total government research expenditures for past years must be arbitrary. The National Science Foundation figures, which should bring some order out of this chaos, minimize the impact of the depression because they show only the years 1930 and 1935. Other sources are summarized in E. R. Gray, "Federal Expenditures for Research, 1937 and 1938," National Resources Committee, *Research — A National Resource* (Washington, 1938) I, 91.

2. "Mayor-Elect La Guardia on Research," *Science* [N.S], 78:509–510 (1933).

3. Director of Finance, Department of Agriculture, *Report*, 1941 (Washington, 1941), 14.

4. "Scientific Notes and News," *Science* [N.S.], 79:76 (1934).

5. "Government Research," *Science* [N.S.], 79:205 (1934).

6. Secretary of Commerce, *Annual Reports, 1929–1940* (Washington, 1929–1941).

7. Vannevar Bush, *Science, the Endless Frontier* (Washington, 1945), 80, 82.

8. W. S. Myers, ed., *The State Papers and Other Public Writings of Herbert Hoover* (Garden City, New York, 1934), II, 423–424.

9. F. D. Roosevelt, Washington, to Vannevar Bush, November 17, 1944, in Bush, *Science, the Endless Frontier*, viii.

10. L. M. Passano, "Ploughing under the Science Crop," *Science* [N. S.], 81:46 (1935).

11. Harvey Cushing to F. D. Roosevelt, August 21, 1933, in J. F. Fulton, *Harvey Cushing: A Biography* (Springfield, Ill., 1946), 664.

12. F. D. Roosevelt to Harvey Cushing, May 9, 1934, *ibid.*

13. F. D. Roosevelt to Harvey Cushing, August 25, 1934, *ibid.*

14. F. D. Roosevelt to Harvey Cushing, August 25, 1936, *ibid.*, 666.

15. Commission on Organization of the Executive Branch of the Government, *Report on Federal Medical Services* (Washington, 1955), 13–15.

16. H. A. Wallace, *Research and Adjustment March Together* (Washington, 1934), 2.

17. H. A. Wallace, *New Frontiers* (New York, 1934), 279.

18. Secretary of Agriculture, *Report for 1935* (Washington, 1935), 84.

19. H. A. Wallace, "The Social Advantages and Disadvantages of the Engineering-Scientific Approach to Civilization," *Science* [N.S.], 79:2 (1934).

20. *Ibid.*, 5.

21. *Ibid.*, 3.

22. K. T. Compton, "The Government's Responsibilities in Science," *Vital Speeches*, 1:425–426 (1935).

23. In the preparation of this section I have benefited from the use of a manuscript on the forerunners of the National Science Foundation by Frank Freidel.

24. Isaiah Bowman, Creation of Science Advisory Board, memorandum attached to letter H. A. Wallace to F. D. Roosevelt, July 22, 1933, Franklin D. Roosevelt Library, Hyde Park, New York.

25. Science Advisory Board, *Report, 1933–1934* (Washington, 1934), 7.

26. *Ibid.*, 15.

27. *Ibid.*, 12.

28. *Ibid.*, 13.

29. *Ibid.*, 15.

30. *Ibid.*, 13.

31. *Ibid.*, 17–43.

32. Science Advisory Board, *Report, 1934–1935* (Washington, 1935), 313.

33. *Ibid.*, 33–34.

34. *Ibid.*, *1933–1934*, 53.

35. *Ibid.*, 63.

36. *Ibid.*, 113.

37. *Ibid.*, *1934–1935*, 43.

38. *Ibid.*, *1933–1934*, 15.

39. Karl T. Compton and Alfred D. Flinn to Harold L. Ickes, September 15, 1933, *ibid.*, 267.

40. *Ibid.*, 269–271.

41. *Ibid.*, 275–282.

42. K. T. Compton, "Report of the Science Advisory Board," *Science* [N.S.], 81:15 (1935).

43. Science Advisory Board, *Report, 1933–1934*, 8.

44. See Fulton, *Cushing*, 659.

45. National Academy of Sciences, *Report for 1934-1935* (Washington, 1936), 27; J. C. Merriam, "The Role of Science in National Planning," National Planning Board, *Final Report, 1933-1934* (Washington, 1934), 40-53.

46. V. E. Baugh, comp., *Central Office Records of the National Resources Planning Board* (National Archives, *Preliminary Inventory No. 50*, Washington, 1953), 1.

47. H. L. Ickes, *Secret Diary* (New York, 1953), I, 171-172.

48. K. T. Compton, "Put Science to Work!" *Technology Review*, 37:133 (1935).

49. *Ibid.*, 134.

50. *Ibid.*, 135.

51. *Ibid.*, 156.

52. Memorandum, Federal Science Program, attached to letter K. T. Compton to F. D. Roosevelt, December 15, 1934, Franklin D. Roosevelt Library, Hyde Park, New York.

53. H. B. Ward, ed., "The Pittsburgh Meetings," *Science* [N.S.], 81:110–111 (1935).

54. Quoted in Science Advisory Board, *Report, 1934-1935*, 37.

55. F. A. Delano to Karl T. Compton, January 17, 1935, National Resources Planning Board Records, National Archives.

56. Memorandum, F. A. Delano to Harold L. Ickes, January 25, 1935, *ibid.*

57. Harold L. Ickes to F. D. Roosevelt, January 31, 1935, *ibid.*

58. *Ibid.*

59. Memorandum, F. D. Roosevelt to the Secretary of the Interior, February 12, 1935, *ibid.*

60. Science Advisory Board, *Report, 1934-1935*, 84.

61. *Ibid.*, 81.

62. National Academy of Sciences, *Report for 1935-1936* (Washington, 1937), 24.

63. *Ibid.*, 25.

64. National Academy of Sciences, *Report for 1939-1940* (Washington, 1941), 7.

65. K. T. Compton, "Program for Social Progress," *Scientific Monthly*, 44:6 (1937).

66. Harold L. Ickes, letters to Robert T. Crane, W. W. Campbell, and George F. Zook, February 23, 1935, National Resources Planning Board Records, National Archives.

67. Science Committee, Minutes, March 25, 1935, National Resources Planning Board Records, National Archives.

68. Baugh, *Central Office Records of the NRPB*, 2.

69. Science Committee, Minutes, January 12, 1936, National Resources Planning Board Records, National Archives.

70. Memorandum, "Joint Dinner of the Science Committee, National Resources Committee, and the Executive Committee of the Government Relations and Science Advisory Committee of the National Academy of Sciences," January 18, 1936, National Resources Planning Board Records, National Archives.

71. National Resources Committee, *Problems of a Changing Population* (Washington, 1936); National Resources Committee, *Technological Trends and National Policy* (Washington, 1937).

72. Science Committee, Action Minute, March 6, 1937, National Resources Planning Board Records, National Archives.

73. F. D. Roosevelt to F. A. Delano, July 19, 1937, in *Research — A National Resource*, I, 2.

74. *Ibid.*, 167-193; National Resources Planning Board, *Research — A National Resource* (Washington, 1940), II.

75. W. F. Ogburn, Chicago, to F. A. Delano, November 22, 1938, National Resources Planning Board Records, National Archives.

76. Science Committee, Minutes, September 30, 1939, National Resources Planning Board Records, National Archives.

77. D. S. Howard, *The WPA and Federal Relief Policy* (New York, 1943); for administration of WPA, see A. W. MacMahon, J. D. Millett, and G. Ogden, *The Administration of Federal Work Relief* (Chicago, 1941).

78. Corrington Gill, *Wasted Manpower* (New York, 1939), 179.

79. Works Projects Administration, *Final Report, 1935-1943* (Washington [1946]), 66.

80. Federal Works Agency, WPA, *Bibliography of Research Projects Reports* (WPA Technical Series, *Research and Records Projects Bibliography No. 1*, Washington, 1940), 4.

81. H. L. Hopkins, *Activities of the Works Progress Administration: Summary of Testimony* . . . April 9, 1936, 74 Cong., 2 Sess., Sen. Doc. 226, 4.

82. MacMahon *et al*, *Administration of Relief*, 11-12.

83. Howard, *WPA*, 107.

84. *Ibid.*, 356.

85. A statement of research policy contemporary to the period of this chapter appears in Tennessee Valley Authority, *Annual Report for 1937* (Washington, 1937), 43-50; the most specific analysis of research as a separate function of TVA appears in the President's Scientific Research Board, *Science and Public Policy* (Washington, 1947), II, 281-290, hereafter cited as Steelman, *Report*.

86. National Science Foundation figure.

87. *Ibid.*

88. Ickes, *Diary*, I, 328, 344, 347, 350.

89. Secretary of Agriculture, *Report for 1936* (Washington, 1936), 12.

90. Wellington Brink, *Big Hugh: The Father of Soil Conservation* (New York, 1951); H. H. Bennett, *Soil Conservation* (New York, 1939).

91. Brink, *Big Hugh*, 102.

92. L. C. Gray, quoted in J. M. Gaus and L. O. Wolcott, *Public Administration and the United States Department of Agriculture* (Chicago, 1940), 135.

93. Quoted, *ibid.*, 115.

94. *Research and Related Sciences in the United States Department of Agriculture* (Washington, 1951), 18-19.

95. Secretary of Agriculture, *Report for 1935*, 86.

96. Secretary of Agriculture, *Report for 1936*, 109-110; B. T. Shaw, "The Agricultural Research Administration — Its Philosophy and Organization" (Mimeographed, 1950), 4.

97. Secretary of Agriculture, *Report for 1938* (Washington, 1938); Steelman, *Report*, II, 122-125.

98. *United States Government Organization Manual, 1953-1954* (Washington, 1955), 235.

99. See H. S. Mustard, *Government in Public Health* (New York, 1945), 63.

100. R. H. Shryock, *American Medical Research, Past and Present* (New York, 1947), 272-273.
101. R. H. Heindel, *The Integration of Federal and Non-Federal Research as a War Problem* (Washington, 1942), 18-20.
102. Henry Guerlac, *Radar in World War II* (Microfilm Publication, Washington, 1947), part 1, chaps. 3 and 4; Joint Board on Scientific Information Policy, *Radar: A Report on Science at War* (Washington, 1945), 3-8.
103. M. S. Watson, *Chief of Staff: Prewar Plans and Preparations (United States Army in World War II. The War Department*, Washington, 1950), 40.
104. Quoted, *ibid.*, 42.

Chapter XIX. Prospect and Retrospect at the Beginning of a New Era

1. For science and the federal government in World War II see J. P. Baxter, 3rd, *Scientists Against Time* (Boston, 1946); Irvin Stewart, *Organizing Scientific Research for War* (Office of Scientific Research and Development, *Science in World War II*, Boston, 1948).
2. Quoted, *ibid.*, 36.
3. See H. D. Smyth, *Atomic Energy for Military Purposes* (Princeton, 1945).
4. National Science Foundation, *Federal Funds for Science*, III, *The Federal Research and Development Budget, Fiscal Years 1953, 1954, and 1955* (Washington [1954]), 14-15.
5. On the postwar period, see D. K. Price, *Government and Science* (New York, 1954); Vannevar Bush, *Science, the Endless Frontier* (Washington, 1945); The President's Scientific Research Board, *Science and Public Policy* (Washington, 1947).
6. See R. H. Heindel, *The Discussion of Federal Research Problems in Congress and the 1943 Appropriation* (Washington, 1942).

INDEX

Abbe, Cleveland, 187, 189, 191, 192

Abbott, Henry L., 242, 243

Abert, J. J., 64, 70, 72

Abert, J. W., 93

Academic research, 157. *See also* Research, Universities, *and individual universities by name*

Adams, John, 7, 379

Adams, John Quincy, 67, 73, 112, 247, 296; scientific ability and program of internal improvement, 39–43; urged reform of patent system, 46; urged exploring expedition to northwest coast, 56; program approached realization, 64; on Smithson bequest, 68, 69, 77

Adams Act (1906), 172

Aeronautical Division (Bureau of Standards), 339

Aeronautics, research in, 127–128, 283–287

Aeronautics Branch (Department of Commerce), 339

Agassiz, Alexander, 205, 242, 290; on government science and the theory of *laissez faire*, 220–224; Powell's rebuttal and refutation, 224–227; interest in Wood's Hole laboratory, 237

Agassiz, Louis, 61, 87, 104, 115, 215; *Contributions to the Natural History of the United States*, 88; Scientific Lazzaroni, 118, 136; Smithsonian regent, 136; National Academy officer, 142

Age of Enlightenment, 7

Agricultural Adjustment Act of 1938, 365

Agricultural Adjustment Administration, 363

Agricultural Colleges, Association of American, 170

Agricultural colleges, research, 337

Agricultural Economics, Bureau of, 335–336, 349

Agricultural missions to foreign lands, 299

Agricultural Research Administration, 365

Agricultural Society, United States, 113

Agriculture, period of improvement, 8; early dream of an American Society for, 16; demand for scientific aid to, 110–112

Agriculture, Department of, 158, 295, 326, 329, 352, 353, 362, 363, 371, 374, 375; genesis, 149–150; early years, 152–157; development of bureaus and problem approach, 157–161; coöperation with the states, 169–172; artesian-well survey, 234; interest in irrigation, 235–236; nominal control of wildlife research, 239; forestry, 244–246; work on standards, 272; objections to Bureau of Standards, 275; growth, 278; comparison with Bureau of Mines, 282–283; its creation started era of bureau-building, 289–290; research for war, 304; war conservation efforts, 323; increasing research, 337–340; impact of depression, 345; aid to science and research during depression, 348–349; conservation and research during later New Deal, 364–365. *See also individual Bureaus and Divisions by name*

Agrostology, Division of, 167

Air Corps, creation of, 317; appropriations for research and development, 367

Aircraft Production, Board of, 317

Albatross, Fish Commission vessel, 237

Allegheny Observatory (Pittsburgh), 284

Allen, Mrs. Mildred C., 331, 332, 333

Allison, W. B., 215

Allison Commission, 190, 192, 290, 293; proposals and activities of, 215–231 *passim*

American Academy of Arts and Sciences (Boston), 7, 8, 39, 43, 75, 97, 115, 144, 349, 356, 377; connection with government, 115, 116; Bache's address as retiring president, 116–118; defends Coast Survey, 222; committee on forest preservation, 239; Committee of One Hundred on National Health, 269; work on standards, 273

American Chemical Society, work on standards, 273

American Council on Education, 359

American Ethnology, Bureau of, 235, 283

American Expeditionary Force, aviation research, 317; poison-gas research, 320

American Fish Cultural Association, artificial propagation efforts, 237

American Forest Congress, 240, 249

American Institute of Electrical Engineers, work on standards, 273

American Journal of Science, 107

American Medical Association, on national department of health, 258–259; on yellow fever, 266

American Ornithologists' Union, English-sparrow problem and others, 238–239

American Philosophical Society, 7, 8, 17, 57, 71, 74, 75, 88, 97, 99, 116, 377; scientific schemes of, 19; sponsorship of explorations, 25, 27; idea of a coast survey, 29

American Physical Society, work on standards, 273

American Public Health Association, 258–261

American Telephone and Telegraph Company, 310, 342

American University (Washington), 320

Ames, Joseph S., 313

Anaconda Smelter Smoke Commission, 283

Animal-disease problems, 164

Animal Industry, Bureau of, 164–166

Antisell, Thomas, 153

Antisubmarine research, 318–319

Anti-Vivisection League, 270

Appropriation bills, creation and control of scientific bureaus primarily through, 216, 217, 233, 234, 239, 242, 247, 286, 291, 380

Armistice (World War I), effect on science and research, 302, 324–325

Army, United States, part played in explorations, 26, 28; interest revived in trans-Mississippi exploration, 35; carried on tradition of Lewis and Clark expedition, 63; in trans-Mississippi West, 92–95; application of science to technological problems, 126–127; use of medicine, 128–130; war activities of Topographical Engineers, 134; meteorology under, 187; permanent Signal Corps, 192; surveying in West, 195; federal medical care, 256; work on standards, 272; research for war, 302–303; World War I commissions to scientists, 314–315; appropriation cut—end of World War I, 325; postwar research cuts, 331–332; state of research, beginning of World War II, 367–368; World War II research, 371; postwar research, 371, 374

Army Chemical Warfare Service, creation and activity, 320–321; postwar cuts, 332

Army Engineers, Corps of, 193–194, 289; geological survey, 196; and conservationists, 250–251; civil works and war research, 303; poison-gas research, 320

Army Medical Corps, 128–129, 289; nominal decline after Civil War, 256; research, 256–257; period of great accomplishment, 263–267; growth, 290; poison-gas research, 320

Army Medical Library, 129, 130, 256–257, 264, 265, 289, 290, 291, 347–348

Army Medical Museum, 129, 130, 256–257, 264, 265, 289, 290

Army Medical School, 264–265, 290

Army Signal Corps, 192, 329; research on aviation and radio, 303; commissioning of scientists, 314; lost aviation control, 317; poison-gas research, 320

Army Signal Service, 192, 193; Joint Commission to study, 215. See also Signal Service

Arthur, Chester A., 262

Ashford, Bailey K., 266–267

Association of American Geologists and Naturalists, 74, 115

Association of Economic Entomologists, 162

Astrophysical Laboratory (Smithsonian), 284

Atomic Energy Commission, 374

Atomic energy program, postwar, 374, 381

Atwater, W. O., 171

Aviation, World War I research, 317–318; industrial research, 339, 340. See also Aeronautics

Bache, Alexander Dallas, 46, 50, 79–80, 116, 222, 297, 298, 300, 377, 381; Coast Survey under, 100–105; quest for a central scientific organization, 116–118, 149; chief of the Scientific Lazzaroni, 118; on Sanitary Commission, 129; Civil War activities, 132–133; the professional scientist in government service, 135, 136; on Navy Department Permanent Commission, 137; activities as first president of National Academy, 139, 142, 144, 146, 147–148, 216

Bacteriology, early research work and development of, 263–267

Baekeland, L. H., 306, 307

Bailey, Jacob Whitman, 46, 87, 104

Baird, Spencer Fullerton, 93, 116, 139, 213, 218, 220, 284; Director of National Museum, and Assistant Secretary of Smithsonian, 83, 85, 236; work on Fish Commission, 236–238

Baker, Newton D., 367

Ballinger, Richard, 251

Balloons, use during Civil War, 127–128
Bankhead-Jones Act (1935), 364–365
Banks, Sir Joseph, 10, 11
Barlow, Joel, 23, 294; plan for a national university, 23, 67
Barnard, J. G., Coast Survey secret commission, 133; on Navy's Permanent Commission, 137;
Barnes, J. K., 256
Bartram, William, 8
Baruch, Bernard, 305
Beagle, surveys, 57
Becher, G. F., 278
Bell, Alexander Graham, 285, 286
Beltsville Research center (Maryland), 171, 365
Bennett, Hugh H., 363, 364
Bentham, Jeremy, 259
Benton, Thomas Hart, 49; and Smithson bequest controversy, 69, 79
Bigelow, John M., 94
Billings, John Shaw, 154, 263, 267, 297, 298, 300, 347; work on Army Medical Library, 129, 256–257; against a department of science, 230–231; on National Board of Health, 259, 260, 262; Report on the Mortality and Vital Statistics of the United States, 278
Biltmore, N. C., Pinchot's forestry work, 241
Biological Survey, 169, 238–239; basic research policy questioned and redirected, 253; regulatory power, 253–254; effect of depression on, 345–346
Blunt's Coast Pilot, 108
Bomford, George, 123
Bond, George P., 139
Bond, William Cranch, 61
Boondoggling, 362
Boston (Massachusetts), cultural influence in early republic, 8
Botany, Division of, 167, 177
Boudinot, Elias, 18
Bougainville, L. A. de, 56
Bowditch, Henry Ingersoll, report on national department of health, 259; on National Board of Health, 260, 262
Bowditch, Nathaniel, Practical Navigator, 108
Bowers, G. M., 292
Bowman, Isaiah, 350, 351, 354
Brewer, W. H., 242, 278
Briggs, Lyman J., 372
Brookings Institution, 336
Browne, Daniel J., 112
Buchanan, James, 113

Buckle, H. T., 228
Budget, Bureau of the, 352
Building trades, aid from Bureau of Standards, 276
Bureau chief, effective control of science by, 380–381
Bureaus, development of, 157–158, 289–293; proposed reorganization 217–231, impact of depression, 344–350. See also individual bureaus by name
Burrell, G. A., 320
Bush, Vannevar, 369, 370, 371, 372, 373, 374, 375
Business, setting the tone of the country, 338; Hoover's Conference, 338–339. See also Industrial research
Bussey Institution (Harvard), 158
Butler, Nicholas Murray, 294

Cabell, James L., chairman of National Board of Health, 260, 262
Calhoun, John C., 229; on Smithson bequest, 67, 69, 70
California Institute of Technology, 309
Cambridge (Massachusetts), activities of Lazzaroni, 136
Campbell, W. W., 351, 354
Canada, boundary line ascertained, 33
Capron, Horace, 154, 155
Carnegie, Andrew, gifts for research, 297–298, 308, 310, 312, 329
Carnegie Institution (Washington), 297–298, 322, 377
Carter, Henry R., 265, 266
Carty, John J., 310, 327, 341–343
Cass, Lewis, 49
Cattell, J. M., 342
Census, Bureau of the, 290, 295, 338, 362; postwar, 335; influence of industrial growth on, 271; development of, 277–279; vital-statistics program, 260
Centennial Exposition (Philadelphia), 283
Central Pacific Railroad, 195
Central research organization, evolution of, 293–296, 305–315
Central scientific organ, later New Deal need for, 366–367; retrospect of 150 years of effort, 375–379
Chamberlain-Kahn Act, 334
Chandler, William E., 219
Chemistry, Bureau of, 169, 176–181; work on standards, 275
Chemistry and Soils, Bureau of, 352
Chesapeake, 24, 31
Chicago, University of, 297, 319
Chief of Engineers, Office of, 36

Chief of Naval Operations, Office of, 303
Chile, nitrates for explosives from, 321, 322
Chipman, John, 78
Choate, Rufus, 76–77, 84
Christian Science Church, 270
Churchman, John, sought subsidy from Congress, 9–11
Churchman, *Magnetic Atlas*, 11
Civil engineering, 37
Civil service merit system, 173, 174, 376
Civil War, technological problems and research, 120–148, 289, 376, 378, 380
Clark, Alvan, and Sons, 186
Clark, C. D., 242
Clark, George Rogers, 25
Clayton, H. Helm, 192
Cleveland, Grover, 222, 223, 238, 242, 292
Coast and Geodetic Survey, 203, 209, 289, 292, 295, 338, 375, 377; Allison proposed reorganization and Commission investigation, 215–229; relations of Agassiz family with, 220; appropriations cut, 235; office of weights and measures in, 271–273
Coast and Interior Survey, unsuccessful proposal, 216–217
Coast Survey, 43, 86, 100; plans for, 29–33; responded to urgent needs, 52, 53; set off chains of scientific demands, 64–65; under Bache, 100–105; appropriations for, 104; services during war, 132; under Peirce, 202–203; name changed to Coast and Geodetic Survey, 203
Collège de France, 300
Colleges, shelter for research, 44
Colorado Rockies, 199–200
Columbia University, 284, 297; Institute of Tropical Medicine and Hygiene, 267
Columbia River, United States claim to, 25
Columbian Exposition, 272
Columbian Institute, 34, 71
Commerce, Department of, 192, 352; encouragment of industrial research, 336–340
Commerce and Labor, Department of, 295; National Bureau of Standards under, 274; Census Bureau under, 279
Committee of Organization of Government Scientific Work, 294–295, 296
Compton, Arthur H., 372
Compton, Karl T., 370; work on Science Advisory Board, 250–262

Comstock, Cyrus, 216
Comstock Lode, 224, 226
Conant, James B., 370
Congress, United States, patronage during first decade, 9–11; reluctance to aid science, 14, 38–39, 41–42; sets up Joint (Allison) Commission to study scientific bureaus, 215; control of science through appropriation bills, 216, 217; new attitude toward bureaus, 291–292; current willingness to support research, 380
Conservation, 377, 381; activities 1865–1916, 232–255; World War I impact on concepts, 322–323; Hoover's activities as Secretary of Commerce, 339; during the later New Deal, 363–364
Constitution, United States, position of science in, 5–6
Constitutional Convention, 380; position of science considered, 3–5
Construction and Repair, Bureau of (Navy), 329
Consumption, heavy drain on natural resources, 232
Contract research, growth of, 371, 374
Contributions to Knowledge (Smithsonian Institution), 86
Cook, Captain James, 25, 56
Coolidge, Calvin, 330, 338
Cooper, Thomas, 68, 70
Cope, Edward Drinker, 208, 222, 235
Cornell University, 297, 319; forestry school established, 244
Cottrell, Frederick G., 283, 321
Coues, Elliott, 131
Council of National Defense, 305, 312–313, 316, 327, 328, 370
Coville, F. V., 246
Cow Colleges, 160
Craven, T. A., 96
Crawford, William H., 32
Cuba, yellow fever commission sent to, 260; disease during Spanish-American War, 264; yellow fever research, 265–266
Curtice, Cooper, 166
Cushing, Harvey, 347, 348
Cutbush, Edward, 34, 71

Dabney, Charles W., 173, 174, 294
Dahlgren, John A., ordnance experiments, 123–124; Chief, Bureau of Ordnance, 124–125; and National Academy, 139, 143
Dallas, Alexander J., 80

Dallas, George Mifflin, 80

Dampier, William, 27

Dana, James Dwight, 142, 205; Reynolds expedition, 58; editor *American Journal of Science*, 107; Scientific Lazzaroni, 118

Daniels, Josephus, 306, 307

Darwin, Charles, *Origin of Species*, 165

Davis, Arthur Powell, 212, 235

Davis, Charles Henry, 381

Davis, Charles Henry, head of *Nautical Almanac*, 107–108; American Academy activity, 116; Warship Study Board, 126; Coast Survey Secret Commission, 133; head of Bureau of Navigation, 133; Office of Detail, 134; the professional scientist in government service, 135, 136; on Navy Department Permanent Commission, 137; plan for National Academy, 138–139; Isthmian Canal problem, 186; bureau chief, 381

Davis, E. H., and E. G. Squire, *Ancient Monuments of the Mississippi Valley*, 87

Davis, Jefferson, 49, 113, 131; railroad surveys, 94–95

Davis, John, on accepting Smithson's bequest, 67–68

Davis, John W., 342

Davis, W. M., 212

Defense, Department of, 374

DeForest, Lee, 304

Delano, Frederick A., 355, 356, 357

DeLong, G. W., 193

Democracy, science in, 300–301, 379–381

Democrats, 347

Depot of Charts and Instruments, 61, 62, 65, 105

Depression and New Deal, effect on government science, 344–368, 381

Detail, Office of, 134

Dickerson, Mahlon, 58, 59

Dictatorships, use of science, 379

Diller, Isaac R., 125

Douglas, Stephen A., 84

Draper, John W., 139, 141

Dreadnought type of battleship, 304

Dudley Observatory (Albany), 119

Dunbar, William, 28

Dunn, Gano, 310, 327, 341, 351

Du Ponceau, Peter S., 57

Du Pont, Samuel F., 133

D'Urville, Dumont, 57

Dutton, Clarence E., 201, 212, 233

Economic Ornithology and Mammalogy, Division of, 238

Edison, Thomas A., 306, 307

Edgewood Arsenal, 320, 332

Education, an internal improvement, 22, 23; included in planning for research, 359

Electric telegraph, discovery, 48

Electrical standards, work on, 272

Eliot, Charles William, 185

Eliot, Charles W., 2nd, 355, 356, 357

Ellesmere Island, 193

Ellsworth, Henry L., commissioner of patents, 47, 73; custodian of government collection, 74; agricultural activities, 110–111

Ellsworth, Chief Justice Oliver, 47

Emmons, S. F., 278

Emory, W. H., 93, 94

Engineering and Research, Division of (National Research Council), 337

Engineering Foundation, 310, 312

Engineering societies, 315, 329

Entomology, problem approach, 161

Entomology, Division of (Department of Agriculture), 161, 162, 163, 238

Entomology and Plant Quarantine, Bureau of (Department of Agriculture), 362, 375

Era of Good Feelings, scientific activities during, 33–34

Ericsson, John, 122; designed *Monitor*, 126

Erie Canal, 41

Erni, Henri, 152–153

Espy, James F., 69, 70, 109

Ethnology, Bureau of, 211

Ethnology, research in, 207

Europe, Age of Enlightenment, 6, 7; advance of science, 40–41; question of inferiority to, 342–343

Evans, John, 96

Evans, Oliver, 13

Experiment-station system, in Bureau of Mines, 282

Experiment stations, aided by federal land grants, 169, 170, 171

Experiment Stations, Office of (Department of Agriculture), 171, 337

Experimental Evolution, Department of (Carnegie Institution), 298

Exploration, recognized part of activity of government, 24–29; trans-Mississippi by Army, 35; expeditions, 56–61; scientific demands caused by, 64–65; overseas, 95–100, 184–187; polar, 192–193

Exploring Expedition, 56–61
Extension Service (Department of Agriculture), 181–183

Faraday, Michael, 80
Featherstonhaugh, G. W., 63
Federal funds, Jefferson's plan for use, 22–23
Federal Quarantine Act, 163
Federal Radio Commission, 340
Federal regulation, law of 1838 introduces, 50
Federalists, attitude toward science, 21–22
Felch, Cheever, 32
Fellowships in science, 327, 330
Felton, C. C., 118
Fernow, Bernhard E., 240–241, 244
Ferrel, William, 190; *Recent Advances in Meteorology*, 189
Financial aid for research, early petitions rejected, 14
Finlay, Carlos, 265
Fish Commission, 217, 236–238
Fisher, Irving, *National Vitality*, 269, 336
Fisheries, Bureau of (Department of Commerce and Labor), 238, 292, 295, 338
Fiske, Bradley, 304
Fixed Nitrogen Research Laboratory, 334
Fleischmann, Charles Lewis, 68
Flexner, Simon, 268
Food Administration, 305
Food and Drug Administration, 352
Food production, World War I, 323
Food Research Laboratory, 181
Ford, Henry, 319
Ford's Theatre (Washington), 129
Forest Products Laboratory (Madison, Wisconsin), 254
Forest protection, early interest in, 239–244
Forest Service (Department of Agriculture), 169, 290, 334, 339, 362, 363, 365, 377; established, 249; applied research, 252–255; decentralization and experiment stations, 254; *esprit de corps*, 292; research for war, 304
Forestry, Bureau of, name changed to Forest Service, 249
Forestry, Division of, 177; early ridicule of, 240; work of Pinchot as head, 244–246
Foundations, an estate of science, 297–299; aid to peacetime National Research Council, 329

France, government science in, 23, 221; airplane development, 286
Franklin, Benjamin, 3, 4, 7, 8
Franklin Institute (Philadelphia), 50, 88
Franklin, Sir John, 98, 193
Frazer, John F., 119
Frémont, John Charles, 64, 92–93
French Academy, 221, 308
Fulton, Robert, *Demologus*, 122

Gage, Lyman J., 273
Gallatin, Albert, 22, 30, 31, 68
Galloway, Beverly T., 167
Gamgee, John, 154
Gannett, Henry, 212, 244
Gardens and Grounds, Division of, 168
Garfield, James A., 200, 210
Garfield, James R., 294, 295
Gas mask, development of, 320
Gedney, T. R., 54
General Education Board, 298
General Electric Company, 310, 319, 342
General Land Office, 197, 198, 199, 232, 246, 292; Powell's attack on, 206; shut down for irrigation survey, 233–234; bad reputation, 244; lost control of forest reserves, 249
General Survey Act, repealed, 64
Genêt, Edmund (Citizen), 26
Geographic Society of New York, 99
Geographical and Geological Survey of the Rocky Mountain Region, 201
Geographical and Topographical Survey of the Colorado River, 200–201
Geological and Geographical Survey of the Territories, 198
Geological Survey, 177, 326, 339, 352, 362, 377; creation, 208–211; policies of Powell as director, 211–212; Allison Committee to study, 215; proposed bureau, 217; attacks on, 222–224, 226, 228–229; Powell's defense, 226–227; irrigation survey, 233–234; restricted in scope, 235; survey of forest reserves, 243; attempts to keep alive scientific planning, 244; to control Newlands Act reclamation, 248; function of land classification revived, 249–250; comparison with National Board of Health, 262; friction with Bureau of Standards, 275; early mining research, 280–281, 282; growth of, 290–291; foreign mission to China, 299; research for war, 304; Geological Survey of the Fortieth Parallel, 195
Geologists, early an organized group, 46

Geology, opposition to government activity in, 224; Powell's defense of, 226–227

Geophysical Laboratory (Carnegie Institution), 322

Germany, government science in, 221; airplane development, 286; university system, 300; our dependence on technology, 322

Gibbs, Wolcott, 119, 129, 136, 142, 241, 242

Gilbert, G. K., 201, 212

Gilliss, James M., 61, 62, 65, 97, 105, 133, 139

Gilman, Daniel Coit, 293–294

Girard College, 79, 80

Glover, Townend, 112, 153, 161

Goldberger, Joseph, 270

Goldsborough, L. M., 62

Goode, George Brown, 237, 283

Gorgas, W. C., work on yellow fever control, 265, 266

Gould, Benjamin Apthorp, 103, 108, 119, 129, 136, 142

Government Relations, Division of (National Research Council), 329

Government Relations and Science, Advisory Committee on (National Academy), 358

Government research and science, early activities, 9, 33, 48–49, 51; aid to private research, 50, 224–227; laissez-faire policy, 220–224; and industry, 287–288; patterns of, 289–301; one of the estates of science, 297; retrospect of 150 years, 375–379; and democracy, 379–381

Grand Canyon, 200

Grange movement, 169

Grant, Ulysses S., 186, 199; on scandal of surveys, 203

Graves, Henry S., 245

Gray, Asa, 58, 61, 87, 88, 139, 141, 144, 198

Gray, Robert, 25

Grazing, research in, 246

Great Britain, precedent for patronage of science, 10, 11; trading interest in Canada, 25; airplane production, 286; government organization of industrial research, 337

Great depression, impact of, 344–350

Great Lakes Survey, 134

Greely, A. W., 192, 193

Greenwich Observatory (Great Britain), 69

Gregory, F. H., 126

Guyer, Frederick, 14

Hague, Arnold, 242, 243, 244

Hale, George Ellery, 341, 350, 373; National Research Council activities, 308–314, 326, 327, 328, 330

Hall, Charles F., 99, 193

Halley, Edmund, 10

Hammond, William A., 129

Harding, Warren, 338

Hare, Robert, 80

Harper, William Rainey, 294

Harriman, Mrs. E. H., 253

Harrington, Mark W., 192

Harrison, William Henry, 192

Harvard College, 108, 284, 297, 319

Harvard Museum of Comparative Zoology, 215, 220

Harvard Observatory, 62, 103

Harvey, W. H., 87

Hassler, Ferdinand Rudolph, 300; Coast Survey plan, 29–33; activity in field of weights and measures, 52; activities in Coast Survey, 52–55, 100; on role of Smithsonian, 70

Hatch, William H., 170

Hatch Act, 170

Haupt, Herman, 120–121

Hayden, Ferdinand V., 94, 201, 252; Bulletin, 198; survey of Nebraska, 198–199; and Wheeler investigation, 204; struggle with Powell, 221, 222, 225

Hayes, I. I., 99, 193

Hayes, Rutherford B., 210, 260

Hazen, William B., 189–191, 192

Health, Department of, unsuccessful proposals, 258–259, 269

Health laws, early action to enforce, 16–17

Helium, research and production (World War I), 321

Henderson, L. J., 46

Henry, Joseph, 155, 187, 218, 220, 221, 237, 284, 296, 297, 379, 381; research into electricity, 45; discoveries, 48, 49; first Secretary of Smithsonian, 79–82, 88, 109–110; American Academy activity, 116; Scientific Lazzaroni, 118; approved of balloons, 128; Civil War efforts, 131, 132; ideal government scientist, 135; objections to a national academy, 136; on Navy Department Permanent Commission, 137; activities as President of National Academy, 139, 140, 147–148, 204–205; assistance to

Powell, 200, 201; report on forest trees, 239; electrical unit named for, 272
Henry Mountains (Utah), 201
Henshaw, Henry W., 253
Herbert, Hilary, 191; attacks on Powell and the Geological Survey, 222, 223, 228–229, 235; laissez-faire policy for science, 227–228
Herndon, W. L., 96
Hetch Hetchy Valley, 252
Hewitt, Abram S., 205, 259, 298
Hilgard, E. W., 171
Hilgard, J. E., consultant for Permanent Commission, 137; superintendent of Coast Survey, 222; forced to resign, 222–223
Hill, G. W., 185
Hiroshima, 369
Holden, E. S., 184
Holmes, Joseph A., Coal Testing Commission, 280; Director of Bureau of Mines, 281, 282; analysis of functions, 282–283
Homestead Act, 150, 158, 197, 202
Hooker, Sir Joseph, 198
Hookworm, research on control of, 266, 267
Hoover, Herbert, 336, 379; as Food Administrator, 305; as Secretary of Commerce, 338–340; efforts to stimulate basic research, 340–343; as President, 344, 346–347, 349
Hoover Dam, 339
Hopkins, Harry, 362
Hopkins, John, bequest for hospital, 257
Hough, Franklin B., 239, 240
Hough, W. J., 78
House, E. M., 327
Houston, D. F., 182
Howard, Leland O., 162, 173–174, 265
Howard University (Washington), 294
Hubbell, Horatio, 68
Hudson, William L., 59
Hughes, Charles Evans, 292, 341, 342
Humboldt, Alexander, 95–96
Humphreys, A. A., 128, 208
Hydrographic Office of the Navy Department, 184, 187; Joint Commission to study, 215; aided by Powell, 223
Hygienic Laboratory, 297; established, 267–268; function; 268–269, 270; moved to Bethesda (Maryland), 365

Ickes, Harold L., 353–357, 363
Illinois Natural History Society, 199

Index Catalogue (of Medical Library), 257, 347
Index Medicus, 257, 347
Indians, of trans-Mississippi West, 200; Powell's interest, 201–202; need for study of, 207
Industrial League of Illinois, 113
Industrial research, beginnings, 271, 288; as an estate of science, 309–310, 337; and Department of Commerce, 336–340; efforts to direct into basic science, 340–343; effect of depression, 346
Inland Waterways Commission, 250–251
Institute for Government Research, Service Monographs, 336
Institute of Tropical Medicine and Hygiene, 267
Interior, Department of the, 338, 339, 352, 363; shift of Patent Office to, 112; forest reserve administration and other conservation activities, 244, 246; Census Bureau under, 279; Bureau of Mines established, 281; growth, 290
Internal improvements, 36; Jefferson's state-administered program, 22–23; Adams's program, 40–43; uncertain constitutional position of, 49
International Council of Scientific Unions, 330
International Polar Conference (Hamburg), 193
International Research Council, 311, 330
Interoceanic Commission, Isthmian Canal program, 186
Irrigation Investigations, Office of, 236, 248
Irrigation Survey, Powell's, 232–236
Isherwood, B. F., Experimental Researches in Steam Engineering, 125
Isthmian Canal problem, 186

Jackson, Andrew, 44
Jackson, Charles T., 92
Jacobin societies, 21
James, Edwin, 35
Jefferson, Thomas, 7, 379, 380; administrator of patent law of 1790, 12; President American Philosophical Society, 19, 21; a great exponent of science, 20–24; position on Constitution's relation to science, 22; plan for a Philosophical Society, 24; organizer of Lewis and Clark expedition, 25–27
Jewett, Charles C., 83
Jewett, Frank B., 324, 337, 370
Johnson, Andrew, 78, 83, 89

Johnson, Walter R., 50, 69
Johns Hopkins University, 157, 237, 296
Joint Commission, to study work of scientific bureaus, 215. *See also* Allison Commission
Jones, G. W., 78
Jones, Thomas ap Catesby, 59
Jordan, David Starr, 294
Judd, Charles H., 360

Kane, Elisha Kent, 99
Kendall, Amos, 53
Kennedy, John P., 97–98
Kennicott, Robert, 131
Kentucky Resolutions, 22
Kew Observatory (Great Britain), 273
Kilborne, F. L., 166
King, Clarence, 205, 224; head of Geological Survey, 196, 210; work on *Statistics and Technology of the Precious Metals*, 278; mining conservation research, 280
King, Henry, 72, 73
Kinyoun, J. J., 267–268
Knapp, Seaman A., 181, 298
Knowledge, Powell's theory of, 227
Koch, Robert, 165, 263

Labor Statistics, Bureau of (Department of Labor), 335
Lacey Act, wildlife control, 253
La Guardia, Fiorello H., 344
Laissez-faire theory in government science, 220–224, 227
Lamar, Lucius Quintus Cincinnatus, 220
Lambert, William, 34–35
Lamson-Scribner, Frank, 167
Land classification, 91–92; aim of government science in West, 206
Land-grant colleges, 149, 150–151, 160, 169, 172, 297, 362
Langley, Samuel Pierpont, 298; Secretary of Smithsonian, 284; *Experiments in Aerodynamics*, 284; work on aeronautical research, 284–285
Langley Aerodynamical Laboratory, established, 285–286; merged with NACA activities at Langley Field, 287
Langley Field (Virginia), NACA activities at, 287, 318, 334
Langley's Folly, 285
Langmuir, Irving, 318
Lapham, Increase A., 187
Laveran, C. L. A., 165
Law, Thomas, 71

Lawrence Scientific School (Harvard), 157
Lazear, J. W., 265
Lazzaroni. *See* Scientific Lazzaroni
League of Nations, 325, 328, 330
Leavenworth, M. K., 63
Ledyard, John, 25
Lee, Daniel, 112
Leopard, 24
Lesley, J. P., 143
Lewis, Captain Meriwether, 27
Lewis and Clark Expedition, 375; appropriation grant, 26–27; precedents arising from, 27–28, 43, 57
Lewisite gas, 320
Library of Congress, 84, 85
Lick Observatory (University of California), 184
Liebig, Justus, *Chemistry in Its Applications to Agriculture and Physiology*, 111–112
Lighthouse Commission, 235
"Light-houses of the skies," 69–70
Lincoln, Abraham, 124, 125
Lincoln, Robert Todd, 191, 193
Livingston, Robert R., 22
Lodge, Henry Cabot, 284
Logan, John A., 190
Long, Stephen H., 35, 37
Loomis, Elias, 187
Lord, N. W., 280
Loring, George, 164
Louisiana Purchase, 25
Louisiana Purchase Exposition (Saint Louis), 280
Louisiana State Board of Health, 261, 262
Lovell, Joseph, 37
Lowe, Thaddeus S. C., 128
Lyman, Theodore, 190, 191, 314; work on Allison Commission, 215; defeated, 222
Lynch, W. F., 96

McCallum, D. C., 120–121
McClellan, George B., 134
Macedonian, Reynolds expedition, 58
McGee, W J, 235; work for conservation, 250, 251; on America as a nation of science, 301
Machine gun, Allied development, 302
Mackenzie, Alexander, 251
McKinley, William, 175, 238, 242, 246, 274, 292
McNeill, William G., 37

Madison, James, 31; plan to encourage the sciences, 4, 5, 10; interest in a national university, 15, 33

Manning, Van H., 320

Mapping, duplication in, 215

Marine Biological Laboratory (Wood's Hole), 237

Marine Hospital Service, 257–258; expansion into a public health agency, 259, 267–268; and quarantine, 261–263

Markoe, Francis, 70, 72

Marsh, G. P., 77

Marsh, Othniel C., 212–213, 245; head of National Academy, 205; inquiry on policies by Joint Commission, 215–216; dropped from Geological Survey, 235

Mason, Max, 319

Massachusetts, first state board of health, 258

Massachusetts Institute of Technology, 278, 297, 309, 334

Mather, Stephen T., 252

Maury, Matthew Fontaine, 96, 136, 184; at Naval Observatory, 105–107; *Wind and Current Charts*, 105; *Sailing Directions*, 106

Mead, Elwood, 236

Meade, George Gordon, 106, 134

Medical and Surgical History of the War of the Rebellion, 129, 257

Medical research, Civil War, 128–130; 1865–1916, 256–270; World War I, 315–316

Mellon, Andrew, 341

Mellon Institute (Pittsburgh), 306

Merchant seamen, federal medical care, 16–17, 256, 257–258

Merriam, Charles E., 355, 356, 357

Merriam, C. Hart, 237; head Division of Economic Ornithology and Mammalogy, 238–239; close relations with Theodore Roosevelt, 247; basic wildlife research, 253; research endowed by Mrs. Harriman, 253

Merriam, J. C., 327

Merrill, E. D., 175

Meteorology, part of Smithsonian program, 88; activities in, 109; under the army, 187–192; issue of place in the Signal Service, 215, 314; proposed bureau, 217

Mexican Boundary Survey, 94

Michaux, André, 7, 25–26

Michelson, A. A., 184–185, 315

Michigan, University of, 297

Michler, Nathaniel, 96

Microscopy, Division of (Department of Agriculture), 167, 168

Military Academy (West Point), established, 29; Calhoun's reorganization, 36; source of competent engineers, 37; employment of specialists, 46

Military research, status before World War I, 302–305; postwar setback and cuts, 331–334; lack of readiness before World War II, 367–368

Military services, explorations and surveys, 32, 43; science introduced into its education, 43; aeronautics during Civil War, 127–128; decline of science in, 184–194. See also Army and Navy

Mill, John Stuart, 160; principle of centralization of knowledge, 227

Millikan, Robert A., 341, 351; World War I, 309, 323, 324; head of National Research Council, 311, 313–315; Army commission, 314, 317, 318–319; peacetime scientific activities, 326, 327

Mines, Bureau of (Department of Commerce), 275, 290, 291, 334, 337, 352, 365; influence of industrial growth upon, 271; development of, 280–283; research for war, 304; poison-gas research, 320; development of helium, 321; transfer from Department of Interior to Department of Commerce, 338

Mint, setting up, 17–18

Missouri River, exploration, 26

Mitchell, S. Weir, *Gunshot Wounds and Other Injuries of Nerves*, 130

Mitchell, Wesley C., 336, 355–357

Mitchill, Samuel Latham, 28, 35

Monitor, 126

Monroe, James, 32

Morgan, John T., 229

Mormons, 202

Morrill, Justin S., backed land-grant college act, 113, 149–150

Morrill Act, 169, 170

Morris, Gouverneur, 5

Morrow, Dwight L., 335

Morse, Samuel F. B., 48, 89

Mount Wilson Observatory, 186, 308, 311

Muir, John, conservation activities, 240, 241, 252

Muscle Shoals (Alabama), 362; synthetic nitrates production, 322

Museum of Comparative Zoology (Harvard), 215, 220

Myer, Albert J., 128, 188–189

Nagasaki, 369

Nahant (Massachusetts), submarine detection experiments, 318

Nation, 161

National Academy of Sciences, 263, 289, 341, 350–359, 368, 369, 370–372, 377–379; genesis of, 135–141; scientific adviser to the government, 141–146; fight for survival, 146–148; active role in survey controversy, 205, 207–208; Committee's proposals for reorganization of bureaus, 216–217; proposals for a department of science, 217, 230; forest conservation efforts, 241–244; for a national public health organization, 259–260; work on standards, 272, 273; advisory nature of, 294; committee for central scientific organization, 295–296; failure of government to aid, 300; no representation on Naval Consulting Board, 306; part of World War I research, 308–310, 312; and peacetime National Research Council, 327–330

National Advisory Committee for Aeronautics, 291, 297, 315, 337, 339, 370, 371, 378; influence of industrial growth upon, 271; development of, 283–287; research for World War I, 304, 318; postwar activities, 334; New Deal activities, 366

National Board of Health, 366, 376; complete failure of, 258–263, 269, 291

National Bureau of Standards, 281, 352; influence of industrial growth upon, 271; development of, 271–277. *See also* Standards, Bureau of

National Cancer Institute, 365–366

National Conservation Commission, 251

National Defense Research Committee, 369–371, 372; becomes branch of OSRD, 371

National Forests, 249

National Health, Committee of One Hundred on, 269

National Herbarium, 155, 156, 168

National Industrial Recovery Program, 353–354

National Institute, 216, 371, 372, 377, 378; drive to obtain collections, 72; care of Wilkes collections, 72–73; question of legal right to collections, 73–74; failure of, 74–76

National Institute for the Promotion of Science, 60; established, 70–71; name changed, 72

National Institutes of Health, 365

National Inventors Council, 308

National Laboratories (AEC), 374

National League for Medical Freedom, 270

National Museum, 289, 297; inclusion in Smithsonian, 85–86; proposed bureau for, 217; continued growth and value, 283–284

National Observatory, proposed bureau, 217

National Park Service (Department of Interior), 252, 334, 362

National Physical Laboratory (Great Britain), 273

National Planning Board, 354–355

National Research Committee, 377

National Research Council, 350–359, 368–373, 377–379; World War I activities, 309–315, 316, 317–324; shift to peacetime activities, 326–330; separation from government, 343

National Research Endowment, 341

National Research Foundation, 375

National Research Fund, 340–343, 377

National Resources Board, activities, 355–357; science committee, 358–359

National Resources Committee, 294, 359, 362, 368

National Resources Planning Board, 361

National Science Foundation, 375, 379; graphs by, 331, 332, 333

National university, idea of, 14–15, 33, 216, 377; Joel Barlow's plan, 23–24; Adams' approval of, 40; unpopularity, 42; question of constitutionality, 67–68; reappearance of idea, 220; Alexander Agassiz on, 221; failure to materialize, 293

Natural history, early dabbling in, 7–8

Natural philosophy, early dabbling in, 7–8

Natural resources, emphasis on conservation of, 232–255

Natural sciences, 2 *and passim*

Nautical Almanac, 87, 107–108, 114, 133, 185, 289

Naval Academy (Annapolis), Adams's recommendation for, 39, 41

Naval Astronomical Expedition to Chile, 97

Naval Consulting Board, 306–308, 309, 312, 313, 315, 318, 333

Naval Observatory, 62, 114, 133, 289, 302; surreptitious creation, 62–63; history of, 105–106; program of fundamental research in astronomy, 184; proposed transfer to Coast Survey, 218; cut in appropriations, 235; retained functions after establishment of Bureau of Standards, 276

Naval Research Laboratory, 307, 333

Navigation, Bureau of (Navy Department), 133; Isthmian Canal problem, 186

Navy, United States, role in overseas explorations, 95, 186–187; technological changes during Civil War, 122–126; reorganization of, 125; application of science to technological problems, 133–134; federal medical care, 256; work on standards, 272; research for war, 303–304, 306; helium gas process, 321; postwar cut in appropriations, 325, 333; World War II research, 371; postwar, 374

Navy Department, 352; Coast Survey switched to, 53; attempt to mount an expedition, 56, 57; reorganization, 124; Permanent Commission, 137; proposed change in scientific bureaus, 218

Nevada, 304

New Deal, Hoover blamed failure of National Research Fund on, 343; effects on government science, 344–368; shaping of scientific policy, 347

New London (Connecticut), submarine research, 319

New York Herald, 235

New York Lyceum of Natural History, 44

New York Times, 342

Newberry, J. S., 129, 205, 212

Newcomb, Simon, 205, 216, 292, 297; head of *Nautical Almanac*, 185; views on reorganization of scientific bureaus, 218–219; hedged on a Department of Science, 230; on the importance of the National Academy, 294; defense of American government science, 300

Newell, Frederick H., 235; director of Reclamation Service, 248; on Inland Water Ways Commission, 250

Newlands Act (1902), 248, 291

Newlands Bill (1908), 251

Newton, Sir Isaac, 6

Newton, Isaac, first commissioner of agriculture in Patent Office, 152

Nicholas, John, 15

Nicollet, Joseph N., 64

Nitrates, development of, 321–322, 334

Normad-Aitchungs-Commission (Germany), 273

North American Review, 299

North Pacific Exploring Expedition, 98

Northern Regional Research Laboratory (Peoria, Illinois), 365

Northwest, Adams's plan for naval expedition, 42

Northwest Boundary Survey, 94

Northwest Passage, 25

Norton, J. P., 269

Noyes, A. A., 327

Nuclear fission, 372

Oath question, Civil War, 372

Observatory, National, Adams's recommendation of, 39, 41, 69; hostility toward construction of, 52–53

Office of Naval Research, 374

Office of Scientific Research and Development, 315, 377–379; differences from National Research Council, 324; activities, 371–375; demobilization problem, 374

Oklahoma, 304

Ordnance, technological problems of, 123, 127

Ordnance, Bureau of (Navy Department), 124

Ordnance Department (Army), 320

Owen, David Dale, survey of mineral lands, 63; geological reconnaissance, 72, 91–92; and Smithson bequest controversy, 77–78

Owen, Robert Dale, 45, 77; and Smithson bequest controversy, 77; plan of organization for Smithsonian, 81; criticism of Smithsonian, 83

Pacific Ocean, voyage of exploration, 56–60

Page, Charles G., 49

Page, John, 10–11, 13

Page, Thomas J., 96

Paine, E. H., 187, 188

Paleontology, opposition to government activity, 224; defended by Powell, 226

Panama, yellow fever research, 266

Parran, Thomas, 354

Parrott, Robert P., 127

Parry, C. C., 94, 155

Pasteur, Louis, 154, 165, 263

Pasteur Institute, 300

Patent clause, significance of, 9–11

Patent laws, administrative machinery set up by, 12–14

Patent office, 43; Adams recommends, 40; reorganization and expansion of, 46–47;

agricultural program, 111, 113; transferred to Commerce Department, 338
Pathology, Division of (Department of Agriculture), 168
Patronage, decline in scientific bureaus, 292
Patterson, Robert, 29, 31
Paulding, Hiram, 126
Peabody, George, 99
Peace, benefits to science from, 301
Peale, Charles Willson, 9, 21–22
Peale, Titian, 35
Peirce, Benjamin, 49, 115, 222; Coast Survey's consultant, 103, 107, 108; consultant on *Nautical Almanac*, 116; Scientific Lazzaroni, 118, 136; consultant for Permanent Commission, 137; officer of National Academy, 141; director of Coast Survey, 202–203
Peirce, Charles Sanders, 271–272
Pelican Island (Florida), bird reservation, 253
Pendleton Act, 173
Pennsylvania, University of, 12, 50, 284
Pennsylvania Assembly, early assistance to science, 9
Permanent Commission (Navy), 306, 307
Perrine, Henry, 110, 168
Perry, Matthew C., 97, 122
Petroleum Division (Geological Survey), 281
Philadelphia (Pennsylvania), cultural center of new republic, 8; capital shifted from, 20
Philippine Bureau of Science, 293
Physical observatory, a proposed bureau, 217
Physical Sciences, Division of (National Research Council), secrecy problem, 372
Physikalische-Technische Reichsanstalt (Germany), 273
Pickering, Charles, 58, 60, 73, 74
Pike, Zebulon M., 28
Pinchot, Gifford, 290, 292, 294, 300, 360, 363, 377, 381; early forestry training, 241–244; confidential forest agent, 244; head of Division of Forestry, 244–247; conservation efforts with Theodore Roosevelt, 248, 249–251; dismissed by Taft, 251; managed use of natural resources, 252; decentralization of Forest Service, 254; compromise between research and practice, 255; efforts toward central organization, 295, 296
Pinckney, Charles, 4, 5, 10

Planning, relation to research, 354–355, 357, 359
Plant Industry, Bureau of (Department of Agriculture), 167–169, 181
Pleuropneumonia, 164, 165
Poinsett, Joel Roberts, 65, 72, 74; Reynolds expedition, 59; on use of Smithson bequest, 69–71
Poinsettia, 70
Point Barrow (Alaska), Signal Service station, 193
Poison gas, German development of, 302; World War I research, 319–321
Polar explorations, 192–193
Politics as factor in government science, 379–380
Polytechnic School (France), 36–37, 69
Pomology, Division of (Department of Agriculture), 168
Populist Party, 175
Portland Cement Association, 339
Postwar readjustment, in science and research, 369–375
Potomac, 57
Potomac River, 40
Powell, John Wesley, 290, 296, 300, 344, 377, 381; survey by, 199–202; *Report on the Lands of the Arid Region*, 202, 204, 232; sided with Hayden in Investigation, 204; Survey controversy, 204–206; ideas on nature of government science, 206–207; work on ethnology, 210; formative policies, 211–214; inquiry on policies by Joint Commission, 215; on proposed reorganization of bureaus, 217–218; for moderate centralization, 221; attacked by Agassiz and Herbert, 223–224, 227–228; government research stimulates private, 224–227, 229; irrigation survey, 232–236; analysis of his success, 262–263; classification of Indian languages, 278; research in mining conservation, 280
President, final authority over bureaus, 378
Preston, W. C., 67
Priestley, Joseph, 9; belief in congeniality of free institutions to science, 300
Princeton, 122
Princeton University, 237, 284
Pritchett, Henry S., 273, 298
Private research, stimulated by government research, 224–227
Problem approach, used by scientific bureaus, 158–163, 206–207, 230, 262, 296, 323, 352–353

Progressive Era, 380; conservation during, 246–249; legacy of conservationists, 251–255; government science and large-scale industry during, 287–288

Promote the general welfare clause, 12

Psychology, success in World War I, 316–317

Public Administration Clearing House, 351

Public health, early attempts, 38–39; from 1865 to 1916, 256–270

Public Health and Marine Hospital Service, established, 268, 269–270; name changed to Public Health Service, 270

Public Health Service, 291, 375; evolution of, 267–270; vital statistics under, 279; postwar activities, 334–335; New Deal activities, 365

Public Works Administration, 354, 355

Puerto Rican Anemia Commission, 267

Puerto Rico, medical research in, 266–267

Pumpelly, Raphael, *Report on the Mining Industries of the United States*, 278

Pupin, Michael, 310

Pure food and drug laws, enforcement of, 275

Pure food and drug program, 177–181

Quarantine system, question of, 258, 259, 261, 262

Quételet, L. A. J., 106

Radar, development of, 333, 367

Radio, work on standards for, 276; industrial research, 339–340

Radio Division (Department of Commerce), 340

Railroad surveys, 94–95

Reclamation, Bureau of (Department of Interior), 281

Reclamation Service, 290, 339; separate bureau in Interior Department, 248–249; applied research, 252–253

Recovery Program of Science Progress, 353

Red Cross, 129

Reed, Walter, 290; medical research, 256, 264, 267; death, 265

Referee Board, 180

Relief funds, question of use for research and science, 348, 359, 362

Remsen, Ira, 180, 260

Republican government, favorable to science, 9

Republican Party, conservation as issue, 251

Research, early petitions for aid rejected, 14; sharp increase in expenditures, 363. *See also* Science

Research — A National Resource, 358–361, 363, 366, 368

Research and Development, Assistant Secretary of Defense for, 374

Research and Development Board, 374

Research Information Service, 312–313

Research organization, evolution of, 305–315

Reynolds, Jeremiah N., 42, 56–57

Rice University, 319

Riley, C. V., 173; Department of Agriculture entomologist, 161, 162; head of Division of Entomology, 238

Rittenhouse, David, observatory of, 8; President of American Philosophical Society, 17; Master of the Mint, 17–18; leader in Jacobin societies, 21

Robbins, Asher, 68–69, 70

Rockefeller Foundation, 181, 310, 312; aid to medical research, 267; established, 298; fellowships, 327; aid to National Research Council, 329; aid to Science Advisory Board, 351

Rockefeller Institute for Medical Research, 326

Rodman, Thomas J., 123, 127

Rogers, Fairman, 142

Rogers, Henry Darwin, 115

Rogers, William Barton, 205; National Academy incorporator, 139; M. I. T. founder, 143

Roosevelt, Franklin Delano, 379; as acting Secretary of Navy, 286; attitude on science and research as president, 344–375 *passim*

Roosevelt, James, 347

Roosevelt, Théodore, 179, 180, 266, 279, 308, 381; personal background for science, 247; champion of conservation, 248; relations with Pinchot, 249–251; encouragement of basic research, 253; for grouping of health services, 269; attempt at central scientific organization. 294; for preparedness, 305

Root, Elihu, 305, 308, 312, 327, 341, 342

Ross, Sir James Clark, 57

Royal College of Physicians (Great Britain), 260

Royal Institution of London, 69, 300; model for Carnegie Institution, 297

Royal Society (Great Britain), 308

Rush, Benjamin, 17; ardent patriot, 7; leader in Jacobin societies, 21

Rush, Richard, 41; claimed Smithson legacy, 68; views on role of Smithsonian, 68–69
Russia, airplane production, 286
Rutherford, Sir Ernest, 318

St. Elizabeth's Hospital, 294
Saint Louis, Exposition, 250
Salmon, D. E., 164, 165
Sanitary Commission (Army), 129
Sargent, Charles Sprague, *Report on the Forest of North America*, 240, 278; conservation efforts, 240–244, 252
Saunders, William, 153, 167
Say, Thomas, 35–36
School of Mines, 69
Schoolcraft, H. R., 63
Science, 6–7 *and passim*; diffusion and amateurishness in early Republic, 7–8, 9; utilitarian aspect, 8; environment favorable to, 9; reluctance of government to become active in, 13–14; ambiguous position of institutions in the government, 33, 151; contributions of first forty years, 43; 18th century norm and the new trends, 44–46, 64; problem of getting and training personnel, 54; need for coördinated effort, 115–118; application to technological problems, 120–127, 148; quest for central organization, 116–118, 215–221, 229–230, 293–296; the estates of, 296–299, 376; center of the stage by end of World War II, 369; retrospect of 150 years, 375–379; in a democracy, 379–381
Science, Department of, proposals, 215–220, 377; Agassiz not opposed to idea, 221; Allison Committee found unnecessary, 229–230; failure to materialize, 293
Science, journal in Department of Agriculture, 156, 171, 190; views on reorganization of scientific bureaus, 220; on attacks on Coast Survey, 222; defense of Powell's position, 229; on Woodrow Wilson's appointees, 293; on depression, 345
Science, The Endless Frontier, 375
Science Advisory Board, 350–358, 368, 371, 377, 378
Scientific Lazzaroni, 118; activities of, 135–136, 138; in National Academy, 142
Scientific Research, Division of (Marine Hospital Service), 268
Scientists, trend from amateurish status to specialization, 44, 45, 46

Schumacher, 115
Schurz, Carl, 204, 210
Scott, Percy, 304
Secrecy and security problem, 372–373
Sedgwick, W. T., 268
Seed-distribution scandal, 168
Service academies, 289
Shaeffer, George C., 141
Shaler, N. S., 212
Sheffield Scientific School (Yale), 157, 277
Sheridan, Philip, 191
Sherman, John, 164
Sherman, Roger, 4, 5
Sherman, William Tecumseh, 208
Shufeldt, R. W., 186
Shufeldt, R. W. (the younger), 219–220
Sibley, John, 28
Sierra Club, 252
Signal Service (Department of Agriculture), 188, 189; question of civilian versus military control, 190. *See also* Army Signal Service
Silliman, Benjamin, 49, 80; *American Journal of Science and the Arts*, 46, 87; supporter for Reynolds expedition, 57; officer of National Academy, 142
Sims, William S., 304
Sinclair, Upton, *The Jungle*, 167
Smith, Caleb, 149
Smith, Erwin Frink, 167
Smith, F. O. J., 48
Smith, Hoke, 241
Smith, James, 38
Smith, Joseph, 126
Smith, Theobald, 166, 175, 297, 298
Smithson, James, bequest and controversy over, 66–70
Smithsonian Institution (Washington), 221, 289, 291, 297, 341, 372, 377–379; establishment, 66, 76–79; *Contributions to Knowledge*, 82, 86; early policies and activities, 83–90; services during Civil War and effect of war on, 130–131; proposed control over scientific bureaus, 216, 218, 220; cut in appropriations, 235; Baird director of, 237; report on forest trees, 239; Theodore Roosevelt's tribute to, 251; godmother of aeronautical research, 271; continuing growth and value, 283–284; and National Advisory Committee for Aeronautics, 283–287; a universal institution, 330; postwar doldrums, 335
Snyder, Carl, 299–300

Social Science Research Council, 357, 359

Social sciences, 2; growing research activities, 335–336; Wallace's aid to, 349–350; growing impact, 351–360, 368

Social Sciences, Division of (Rockefeller Foundation), 351

Social Security Act of 1935, 365

Soil Conservation Service (Department of Agriculture), 363–364

Soil Erosion Service (Department of Interior), 363

Soils, Bureau of (Department of Agriculture), 177

Southard, Samuel L., 42, 57, 67

Spanish-American War, effect on Army medical corps, 264–267; on medicine, 268; Army-Navy aeronautical research, 284

Specialization, 45

Sperry, Elmer A., 306

Sprague, Frank J., 306

Springfield rifle, 303

Squier, E. G., and E. H. Davis, *Ancient Monuments of the Mississippi Valley*, 87

Squier, G. O., 313–314, 329

Standards, development of, 17–18, 271–277

Standards, Bureau of, 290–291, 295, 297, 315, 326, 329, 337, 338; research for war, 304, 322; business research associates at, 339; effect of depression, 346. *See also* National Bureau of Standards

Statistics, early development of, 278–279

Steam engine, science's debt to, 46

Steam Engineering, Bureau of (Navy), 125

Steelman Report, 294

Sternberg, George M., 290, 298; medical research, 256–257, 260, 263–267; Army Medical Corps work on yellow fever, 263, 266; Army Surgeon General, 264

Stevens, Robert L., 123

Stewart, W. D., 258

Stewart, W. M., 233–235

Stiles, Charles W., 267

Stockton, Robert F., 122

Stratton, S. W., 310, 313, 329; director of Office of Weights and Measures, 273–274; director of National Bureau of Standards, 274; on establishing a radio laboratory, 276

Submarine, German development of, 302; World War I, research, 318–319

Submarine Signal Company, 318

Subsidies for scientific research, precedents for granting, 10; granted by Congress under patent clause, 10, 11

Sumner, William Graham, 157

Surgeon General's Library, 256–257, 347–348. *See also* Army Medical Library

Survey, revived in 1832, 52–55. *See also* Coast and Geodetic Survey *and* Geological Survey

Survey Act (1824), 36

Survey controversy, issue of military versus civilian control, 203–205

Symmes, John Cleves, 41, 56

Taft, William Howard, 251; for independent health department, 269; interest in aeronautical research, 285

Tanks, Allied development of, 302

Tappan Bill, 76–77, 78

Taylor, A. Hoyt, 333

Taylor, David W., 304, 329

Technological Branch (Geological Survey), mining research, 280–281

Technological change, government unprepared with basic policy, 46–51

Technology of weapons and industry, application of science to, 302

Teller, Henry M., 233

Tennessee Valley Authority, anticipated by Inland Waterways Commission, 250; research activity, 362–363

Terrestrial Magnetism, Department of (Carnegie Institution), 298

Texas fever, 164, 165–166

Thayer, Sylvanus, 36

Thompson, Almon, 201, 233

Thorn, F. M., 292; attacked Coast Survey, 222; made Superintendent, 223

Thornton, William, 13

Throop College of Technology, 309

Tick Eradication, Division of (Department of Agriculture), 166

Timber Culture Act, 240

Topographical Bureau, 64

Topographical Engineers, Corps of (Army), 36, 65; scientific corps for Army, 63, 64; for Western expeditions, 92–93; war activities and abolition of, 134

Topography, opposition to government activity in, 224; Powell's defense, 226

Torrey, John, 36, 46, 139

Totten, Joseph G., member of National Institute, 70; Smithson bequest controversy, 79

Trans-Mississippi West, 195; government efforts in, 63; Army in, 92–95
Treadwell, Daniel, 127
Treasury Department, 209; inclusion of Coast Survey in, 30, 52, 53–55; work on standards, 272; National Bureau of Standards in, 274
Trowbridge, Augustus, 314
Trowbridge, W. P., 205
Trumbull, Lyman, 209
Turner, Frederick Jackson, 232
Turner, Jonathan B., 113
Turner's Land Hospital, 130
Typhoid, Army efforts to conquer, 264

United States Agricultural Society, 149, 152
United States Botanic Garden, 61
United States Entomological Commission, 161
United States Military Philosophical Society, 29
Universities, an estate of science, 296–297
Uranium, Advisory Committee on, 372

Vaccine institution, national episode of, 38
Van Buren, Martin, 58, 59, 68, 70
Vancouver, George, 25
Vasey, George, 156, 167
Vaughan, Victor C., 316
Vegetable Physiology and Pathology, Division of (Department of Agriculture), 167
Venereal Diseases, Division of (Public Health Service), 334
Veterinary Division (Department of Agriculture), 164
Vincent, George, 326, 327
Virginia, University of, 41
Virginia, 126
Virginia Resolutions, 22

Wadsworth, James W., 159
Walcott, C. D., 248, 309, 310, 327, 329; director of Geological Survey, 235; aid to forestry conservation, 242–243, 244; mining research, 280; Secretary of Smithsonian, 285; on Taft Committee for Aeronautical research laboratory, 285, 286; efforts to establish NACA, 286–287; efforts at central organization, 294–296
Walker, Francis A., 297; superintendent of several censuses, 277–278

Walker, Robert J., 67–68, 80
Walker, Sears C., 87
Wallace, Henry A., aid to science and research as Secretary of Agriculture, 348–350, 352, 364
War Department, 352; reorganization, 35–37; aid to science, 36–38; allotment for aeronautical research, 284–285. See also Army, United States
War Department Technical Committee, 331–332
War Industries Board, 305–306
War of 1812, 31
Ward, Lester, Dynamic Sociology, 212
Warren, G. K., 93
Wartime research, World War I, 315–323; World War II, 370–374
Washington, George, encouraged progress in science, 11; advocated national university, 14–15
Waterhouse, Benjamin, 22
Watertown Arsenal (Massachusetts), 272, 276
Watts, Frederick, 155
Wayland, Francis, 68
Weapons research, 377; new emphasis on, 302–304; World War I, 317; World War II, 370–374
Weather Bureau, 169, 177, 314, 350, 352; taken by Agriculture Department from Army, 192, 290, 303
Weights and measures, standards for, 17–18; Adams's report on, 39; government policy for, 52
Weights and Measures, Office of (Coast Survey), 271, 272–273
Welch, William H., 264–268, 308, 309, 312, 314, 316
Weld, Isaac, 67
Welles, Gideon, 306; reorganization of Navy Department, 124–125; set up Permanent Commission, 137
West Point (New York), United States Military Academy, 29
Western Electric Company, 318
Wetherill, C. M., 152
Weyerhaeuser Lumber Company, 245
Wheeler, George M., 196–197, 204
Whistler, G. W., 37
White, J. D., Engineering Corporation, 310
Whitney, Eli, 13, 126
Whitney, J. D., 196, 197
Whitney, Milton, 177
Whitney, Willis R., 306, 318
Wildlife research, 238–239

Wiley, Harvey W., 298, 364; work in Bureau of Chemistry, 176–180; work on pure food and drug laws, 247; work on standards, 275

Wilkes, Charles, 62; preparation for naval expedition, 22; commander of Reynolds Expedition, 58–61; collections, 70, 72, 74, 86, 380

Wilkinson, James, 28

Williams College, 237

Willis, Bailey, 212, 299

Willits, Edwin, 173

Wilson, Edwin B., 359

Wilson, Henry, 138–141

Wilson, Tama Jim, 348; Secretary of Agriculture, 175, 199; gave free hand to Pinchot, 244

Wilson, Woodrow, 308, 309, 312, 325; feared effect of aeronautical experiments on neutrality, 286; scientist appointees, 292–293; neutrality policy changed to preparedness, 305; established permanent National Research Council, 327

Wisconsin, University of, 319

Wood, Leonard, 266, 290

Woodbury, Levi, 76

Wood's Hole (Massachusetts), Marine Biological Laboratory, 237

Woodward, J. J., 256–257, 259

Woodworth, John M., 258–259

Works Projects Administration, use of scientists and skill, 361–362

World War I, 376, 377, 380; impact on science, 302–325

World War II, 377, 378, 380; effect on scientific research, 367, 368, 369–375

Wright, Charles, 94

Wright, Frances, 45

Wright, Orville, squabble with Smithsonian, 287

Wright brothers, 285

Wyman, Jeffries, 139

Yale College, 284, 319

Yale School of Forestry, 245, 297

Yellow fever, efforts to conquer, 258–261; Army Medical Corps success in, 263–266

Yellow Fever Board, work in Cuba, 264–265

Yellowstone Park, 252

Yerkes Observatory, 186

Yosemite Park, 252

Young, Owen D., 341, 342

THREE CENTURIES
OF
SCIENCE IN AMERICA

An Arno Press Collection

Adams, John Quincy. **Report of the Secretary of State upon Weights and Measures.** 1821.

Archibald, Raymond Clare. **A Semicentennial History of the American Mathematical Society: 1888-1938** *and* **Semicentennial Addresses of the American Mathematical Society.** 2 vols. 1938.

Bond, William Cranch. **History and Description of the Astronomical Observatory of Harvard College** *and* **Results of Astronomical Observations Made at the Observatory of Harvard College.** 1856.

Bowditch, Henry Pickering. **The Life and Writings of Henry Pickering Bowditch.** 2 vols. 1980.

Bridgman, Percy Williams. **The Logic of Modern Physics.** 1927.

Bridgman, Percy Williams. **Philosophical Writings of Percy Williams Bridgman.** 1980.

Bridgman, Percy Williams. **Reflections of a Physicist.** 1955.

Bush, Vannevar. **Science the Endless Frontier.** 1955.

Cajori, Florian. **The Chequered Career of Ferdinand Rudolph Hassler.** 1929.

Cohen, I. Bernard, editor. **The Career of William Beaumont and the Reception of His Discovery.** 1980.

Cohen, I. Bernard, editor. **Benjamin Peirce: "Father of Pure Mathematics" in America.** 1980.

Cohen, I. Bernard, editor. **Aspects of Astronomy in America in the Nineteenth Century.** 1980.

Cohen, I. Bernard, editor. **Cotton Mather and American Science and Medicine: With Studies and Documents Concerning the Introduction of Inoculation or Variolation.** 2 vols. 1980.

Cohen, I. Bernard, editor. **The Life and Scientific Work of Othniel Charles Marsh.** 1980.

Cohen, I. Bernard, editor. **The Life and the Scientific and Medical Career of Benjamin Waterhouse: With Some Account of the Introduction of Vaccination in America.** 2 vols. 1980.

Cohen, I. Bernard, editor. **Research and Technology.** 1980.

Cohen, I. Bernard, editor. **Thomas Jefferson and the Sciences.** 1980.

Cooper, Thomas. **Introductory Lecture** *and* **A Discourse on the Connexion Between Chemistry and Medicine.** 2 vols. in one. 1812/1818.

Dalton, John Call. **John Call Dalton on Experimental Method.** 1980.

Darton, Nelson Horatio. **Catalogue and Index of Contributions to North American Geology: 1732-1891.** 1896.

Donnan, F[rederick] G[eorge] and Arthur Haas, editors. **A Commentary on the Scientific Writings of J. Willard Gibbs** *and* Duhem, Pierre. **Josiah-Willard Gibbs: A Propos de la Publication de ses Mémoires Scientifiques.** 3 vols. in two. 1936/1908.

Dupree, A[nderson] Hunter. **Science in the Federal Government: A History of Policies and Activities to 1940.** 1957.

Ellicott, Andrew. **The Journal of Andrew Ellicott.** 1803.

Fulton, John F. **Harvey Cushing: A Biography.** 1946.

Getman, Frederick H. **The Life of Ira Remsen.** 1940.

Goode, George Brown. **The Smithsonian Institution 1846-1896: The History of its First Half Century.** 1897.

Hale, George Ellery. **National Academies and the Progress of Research.** 1915.

Harding, T. Swann. **Two Blades of Grass: A History of Scientific Development in the U.S. Department of Agriculture.** 1947.

Hindle, Brooke. **David Rittenhouse.** 1964.

Hindle, Brooke, editor. **The Scientific Writings of David Rittenhouse.** 1980.

Holden, Edward S[ingleton]. **Memorials of William Cranch Bond, Director of the Harvard College Observatory, 1840-1859, and of his Son, George Phillips Bond, Director of the Harvard College Observatory, 1859-1865.** 1897.

Howard, L[eland] O[sslan]. **Fighting the Insects: The Story of an Entomologist, Telling the Life and Experiences of the Writer.** 1933.

Jaffe, Bernard. **Men of Science in America.** 1958.

Karpinski, Louis C. **Bibliography of Mathematical Works Printed in America through 1850.** Reprinted with **Supplement** and **Second Supplement.** 1940/1945.

Loomis, Elias. **The Recent Progress of Astronomy: Especially in the United States.** 1851.

Merrill, Elmer D. **Index Rafinesquianus: The Plant Names Published by C.S. Rafinesque with Reductions, and a Consideration of his Methods, Objectives, and Attainments.** 1949.

Millikan, Robert A[ndrews]. **The Autobiography of Robert A. Millikan.** 1950.

Mitchel, O[rmsby] M[acKnight]. **The Planetary and Stellar Worlds: A Popular Exposition of the Great Discoveries and Theories of Modern Astronomy.** 1848.

Organisation for Economic Co-operation and Development. **Reviews of National Science Policy: United States.** 1968.

Packard, Alpheus S. **Lamarck: The Founder of Evolution; His Life and Work.** 1901.

Pupin, Michael. **From Immigrant to Inventor.** 1930.

Rhees, William J. **An Account of the Smithsonian Institution.** 1859.

Rhees, William J. **The Smithsonian Institution: Documents Relative to its History.** 2 vols. 1901.

Rhees, William J. **William J. Rhees on James Smithson.** 2 vols. in one. 1980.

Scott, William Berryman. **Some Memories of a Palaeontologist.** 1939.

Shryock, Richard H. **American Medical Research Past and Present.** 1947.

Shute, Michael, editor. **The Scientific Work of John Winthrop.** 1980.

Silliman, Benjamin. **A Journal of Travels in England, Holland, and Scotland, and of Two Passages over the Atlantic in the Years 1805 and 1806.** 2 vols. 1812.

Silliman, Benjamin. **A Visit to Europe in 1851.** 2 vols. 1856

Silliman, Benjamin, Jr. **First Principles of Chemistry.** 1864.

Smith, David Eugene and Jekuthiel Ginsburg. **A History of Mathematics in America before 1900.** 1934.

Smith, Edgar Fahs. **James Cutbush: An American Chemist.** 1919.

Smith, Edgar Fahs. **James Woodhouse: A Pioneer in Chemistry, 1770-1809.** 1918.

Smith, Edgar Fahs. **The Life of Robert Hare: An American Chemist (1781-1858).** 1917.

Smith, Edgar Fahs. **Priestley in America: 1794-1804.** 1920.

Sopka, Katherine. **Quantum Physics in America: 1920-1935** (Doctoral Dissertation, Harvard University, 1976). 1980.

Steelman, John R[ay]. **Science and Public Policy: A Report to the President.** 1947.

Stewart, Irvin. **Organizing Scientific Research for War: The Administrative History of the Office of Scientifc Research and Development.** 1948.

Stigler, Stephen M., editor. **American Contributions to Mathematical Statistics in the Nineteenth Century.** 2 vols. 1980.

Trowbridge, John. **What is Electricity?** 1899.

True. Alfred. **Alfred True on Agricultural Experimentation and Research.** 1980.

True, F[rederick] W., editor. **The Semi-Centennial Anniversary of the National Academy of Sciences: 1863-1913** *and* **A History of the First Half-Century of the National Academy of Sciences: 1863-1913.** 2 vols. 1913.

Tyndall, John. **Lectures on Light: Delivered in the United States in 1872-73.** 1873.

U.S. House of Representatives. **Annual Report of the Board of Regents of the Smithsonian Institution...A Memorial of George Brown Goode together with a selection of his Papers on Museums and on the History of Science in America.** 1901.

U.S. National Resources Committee. **Research: A National Resource.** 3 vols. in one. 1938-1941.

U.S. Senate. **Testimony Before the Joint Commission to Consider the Present Organizations of the Signal Service, Geological Survey, Coast and Geodetic Survey, and the Hydrographic Office of the Navy Department.** 2 vols. 1866.